Thermal Explosion

Wiley-ASME Press Series

Thermal Spreading and Contact Resistance: Fundamentals and Applications
Yuri S. Muzychka and M. Michael Yovanovich

Analysis of ASME Boiler, Pressure Vessel, and Nuclear Components in the Creep Range
Maan H. Jawad and Robert I. Jetter

Voltage-Enhanced Processing of Biomass and Biochar
Gerardo Diaz

Pressure Oscillation in Biomedical Diagnostics and Therapy
Ahmed Al-Jumaily and Lulu Wang

Robust Control: Youla Parameterization Method
Farhad Assadian and Kevin R. Mellon

Metrology and Instrumentation: Practical Applications for Engineering and Manufacturing
Samir Mekid

Fabrication of Process Equipment
Owen Greulich and Maan H. Jawad

Engineering Practice with Oilfield and Drilling Applications
Donald W. Dareing

Flow-Induced Vibration Handbook for Nuclear and Process Equipment
Michel J. Pettigrew, Colette E. Taylor, and Nigel J. Fisher

Vibrations of Linear Piezostructures
Andrew J. Kurdila and Pablo A. Tarazaga

Bearing Dynamic Coefficients in Rotordynamics: Computation Methods and Practical Applications
Lukasz Brenkacz

Advanced Multifunctional Lightweight Aerostructures: Design, Development, and Implementation
Kamran Behdinan and Rasool Moradi-Dastjerdi

Vibration Assisted Machining: Theory, Modelling and Applications
Li-Rong Zheng, Dr. Wanqun Chen, and Dehong Huo

Two-Phase Heat Transfer
Mirza Mohammed Shah

Computer Vision for Structural Dynamics and Health Monitoring
Dongming Feng and Maria Q. Feng

Theory of Solid-Propellant Nonsteady Combustion
Vasily B. Novozhilov and Boris V. Novozhilov

Introduction to Plastics Engineering
Vijay K. Stokes

Fundamentals of Heat Engines: Reciprocating and Gas Turbine Internal Combustion Engines
Jamil Ghojel

Offshore Compliant Platforms: Analysis, Design, and Experimental Studies
Srinivasan Chandrasekara and R. Nagavinothini

Computer Aided Design and Manufacturing
Zhuming Bi and Xiaoqin Wang

Pumps and Compressors
Marc Borremans

Corrosion and Materials in Hydrocarbon Production: A Compendium of Operational and Engineering Aspects
Bijan Kermani and Don Harrop

Design and Analysis of Centrifugal Compressors
Rene Van den Braembussche

Case Studies in Fluid Mechanics with Sensitivities to Governing Variables
M. Kemal Atesmen

The Monte Carlo Ray-Trace Method in Radiation Heat Transfer and Applied Optics
J. Robert Mahan

Dynamics of Particles and Rigid Bodies: A Self-Learning Approach
Mohammed F. Daqaq

Primer on Engineering Standards, Expanded Textbook Edition
Maan H. Jawad and Owen R. Greulich

Engineering Optimization: Applications, Methods and Analysis
R. Russell Rhinehart

Compact Heat Exchangers: Analysis, Design and Optimization using FEM and CFD Approach
C. Ranganayakulu and Kankanhalli N. Seetharamu

Robust Adaptive Control for Fractional-Order Systems with Disturbance and Saturation
Mou Chen, Shuyi Shao, and Peng Shi

Robot Manipulator Redundancy Resolution
Yunong Zhang and Long Jin

Stress in ASME Pressure Vessels, Boilers, and Nuclear Components
Maan H. Jawad

Combined Cooling, Heating, and Power Systems: Modeling, Optimization, and Operation
Yang Shi, Mingxi Liu, and Fang

Applications of Mathematical Heat Transfer and Fluid Flow Models in Engineering and Medicine
Abram S. Dorfman

Bioprocessing Piping and Equipment Design: A Companion Guide for the ASME BPE Standard
William M. (Bill) Huitt

Nonlinear Regression Modeling for Engineering Applications: Modeling, Model Validation, and Enabling Design of Experiments
R. Russell Rhinehart

Geothermal Heat Pump and Heat Engine Systems: Theory and Practice
Andrew D. Chiasson

Fundamentals of Mechanical Vibrations
Liang-Wu Cai

Introduction to Dynamics and Control in Mechanical Engineering Systems
Cho W.S. To

Thermal Spreading and Contact Resistance: Fundamentals and Applications
Yuri S. Muzychka and M. Michael Yovanovich

Mixed-flow Pumps: Modeling, Simulation, and Measurements
Wei Li, Leilei Ji, Ramesh Agarwal, Weidong Shi, and Ling Zhou

Thermal Explosion: Theory and Application
Vasily B. Novozhilov

This Work is a co-publication between John Wiley & Sons Ltd. and ASME Press

Thermal Explosion

Theory and Application

Vasily B. Novozhilov
Honorary Research Professor in Mathematics
Victoria University
Melbourne, Victoria 8001
Australia

and

N.N. Semenov Federal Research Centre for Chemical Physics
Russian Academy of Sciences
119 991 Moscow, Russia

ASME PRESS
WILEY

This edition first published 2026

This Work is a co-publication between John Wiley & Sons Ltd. and ASME Press.

Registered Offices
John Wiley & Sons, Inc., 111 River Street, Hoboken, NJ 07030, USA
John Wiley & Sons Ltd, New Era House, 8 Oldlands Way, Bognor Regis, West Sussex, PO22 9NQ, UK

For details of our global editorial offices, customer services, and more information about Wiley products visit us at www.wiley.com.

The manufacturer's authorized representative according to the EU General Product Safety Regulation is Wiley-VCH GmbH, Boschstr. 12, 69469 Weinheim, Germany, e-mail: Product_Safety@wiley.com.

Wiley also publishes its books in a variety of electronic formats and by print-on-demand. Some content that appears in standard print versions of this book may not be available in other formats.

Library of Congress Cataloging-in-Publication Data applied for

Hardback ISBN: 9781119830528

Cover Design: Wiley
Cover Images: Designed by Inga Novozhilov

Set in 9.5/12.5 pt STIXTwoText by Straive, Pondicherry, India

TO MY TEACHERS

Igor Fedorovich Sharygin, in mathematics
and
Boris Vasilievich Novozhilov, in physics

Contents

About the Author

Professor Vasily B. Novozhilov

Professor Vasily B. Novozhilov was born in 1963 in Moscow. He graduated with an M.Sc. in Applied Mathematics from the Russian State University of Oil and Gas in 1986. He later received a PhD in Physical and Mathematical Sciences (with a specialization in Mechanics of Fluid, Gas and Plasma) from the Moscow Aviation Institute in 1993.

Professor V.B. Novozhilov has held research positions at the Russian Academy of Sciences (the Institute of Physics of the Earth and the Institute for Problems in Mechanics) and the University of Sydney (Australia). Furthermore, he has held academic appointments at Nanyang Technological University (Singapore), as a Professor in Fire Dynamics at the University of Ulster (UK), and as a Professor of Mathematics at Victoria University (Australia). From 2014 to 2017, he was the Director of the Centre for Environmental Safety and Risk Engineering at Victoria University, Australia.

Major research interests of Professor V.B. Novozhilov are mathematical methods in mechanics, heat and mass transfer and chemical physics.

He is a leading expert in theoretical and computational methods in the areas of combustion and fire research, in particular, Computational Fluid Dynamics modelling of compartment fires. He has also made important contributions to the application of dynamical system methods in fire dynamics. Professor V.B. Novozhilov has also been greatly involved with analytical methods in heat transfer theory, including applications to ultra-fast heat transfer processes.

He is the author of 69 journal papers and 4 book chapters. He co-authored (together with Professor B.V. Novozhilov) the monograph 'Theory of Solid-Propellant Nonsteady Combustion', published by Wiley in 2020.

He was Colloquium Co-Chair at the Thirty Sixth International Symposium on Combustion (2016) and a Keynote Speaker at the Sixty-Eighth International Astronautical Congress (2017).

Currently, Professor Vasily B. Novozhilov is an Honorary Research Professor in Mathematics at Victoria University in Melbourne, Australia. He also has a research affiliation with the N.N. Semenov Federal Research Centre for Chemical Physics of the Russian Academy of Sciences in Moscow, Russia.

Preface

In 2028, one hundred years will have passed since the publication of the N.N. Semenov's paper (Semenov 1928a), which laid the foundation for the thermal explosion theory. It has been fascinating to see how the approach of this study, essentially amounting to a single first-order ordinary differential equation, has been influential in establishing new research directions on the subject and attracting the attention of generations of scientists. The reasons for this, nearly one hundred years after the publication, are clear.

First, Semenov's theory captures the very physical essence of the problem, explaining it and making predictions in a succinct mathematical form. Owing to these virtues, Semenov's theory became classical. Second, since the proposed approach was very simple and involved a number of assumptions, the theory allowed for generalizations.

The present book reviews, along with the Semenov's theory itself, its major logical extensions. There is still no monograph written on the subject, and making such a monograph available has been the primary aim of the development of the present book.

Originally, the theory was practically applied to identify safe conditions for storage or chemical processing of various chemically reactive substances.

Consequently, the major generalizations have been concerned with traditional reactive systems (in particular, gas mixtures and solid energetic materials) and developed primarily in two main directions: (1) consideration of complicated heat transfer regimes between the system and its surroundings, and (2) effects of complicated chemical kinetics.

Several examples of the first stream of developments may be mentioned here.

One practical problem related to civil explosives is their application for underground works. In these circumstances, an explosive substance is being subjected to increased ambient temperature and, while being inert at the ground-level temperature, may unintentionally explode in the process of transportation into deep wells. It is well known that temperatures in the wells of up to 8 km depth may reach 350 °C, or up to 400 °C in more deeper wells.

A similar problem arises upon studying thermophysical and kinetic properties of solid substances in specifically designed laboratory experiments, applying the technique of Thermogravimetric Analysis (TGA). In these experiments, samples are heated in such a way that their temperature increases linearly (or less often, according to some other law, e.g. parabolic, exponential, etc.).

It is essential that the sample maintains integrity during heating. The onset of a thermal explosion may result in the destruction of the sample. Besides, knowledge of the conditions causing thermal explosion allows the accuracy of the experiment under different heating rates to be estimated. Therefore, the possibility of thermal explosion needs to be taken into account when designing such laboratory experiments.

The next example is the concept of conjugate thermal explosion, proposed and further developed by the author. In this problem, conjugate heat transfer between different chemically reacting materials needs to be investigated. Analysis of the problem leads to the prediction of conditions under which such type of thermal explosion occurs. These conditions are significantly different from the analogous conditions of the classical thermal explosion theory.

Examples of the second major line of developments, identified above, are modern Computational Fluid Dynamics (CFD) simulations of thermal explosion, which take into account very detailed chemical kinetic mechanisms, and are often performed for complicated regimes (e.g. turbulent flows of reactive media).

As time has passed, it has gradually become apparent that manifestations of thermal explosion are much more diverse than were assumed traditionally. Thermal explosion may occur in different media and at a very wide range of circumstances. It has become especially evident with industrial developments which occurred later in the twentieth and at the beginning of the twenty-first century. Indeed, whilst chemical processing and storage of reactive materials is still highly relevant, a number of modern technologies have emerged which, whether in traditional or new forms, inevitably use chemical energy sources. In various circumstances, these technologies are prone to thermal explosion.

These technological advancements brought up new and challenging applications of the thermal explosion theory. Their overview provided the second motivation for writing this book.

There are at least three areas of such developments.

The first one, that has emerged over the last few decades, concerns the handling of new types of chemical and biological substances. The increasing use of environmentally friendly fuels, combined with their low (compared to traditional energy sources) calorific values, results in the need to store, transport and burn them in very large quantities. This circumstance presents a significant challenge from a safety point of view. Storages of such fuels (various biosolids, wood chips, Refuse Derived Fuels (RDF), Refuse Paper and Plastic Fuels (RPF) and similar substances) have been known to be a source of fires, accompanied by casualties and very difficult to extinguish. The ignition of such fires is caused by a thermal explosion in the system. Development of a thermal explosion in such substances has peculiar features (e.g. the presence of microbiological activity that initiates chemical reactions with more substantial heat release) that are still not completely understood quantitatively.

As the application area of these fuels widens, there will be a greater need for a more reliable prediction of thermal explosion conditions, and for the development of safe fuel handling procedures.

The theory of thermal explosion is capable of predicting to what extent storage sizes and ranges of storage conditions must be limited in order to avoid catastrophic accidents.

The second area of development is related to the application of new generations of electric batteries and other innovative devices utilizing chemical energy. The most remarkable example of these recent technologies is lithium-ion electric batteries of various capacity. The accidents involving Boeing 787 and Samsung Galaxy Note 7 lithium-ion batteries are most commonly known. The cause of these accidents is still the same, that is thermal explosion developing in electric batteries in a quite specific way. The same safety concern applies to other devices such as fuel cells and microcombustors.

The development of the above technologies is expected to continue, demanding an understanding of their thermal explosion scenarios and efficient means to prevent them.

A very different and important third area where considerations of thermal explosion have recently proved to be very helpful is fire protection design. It has become apparent over the last two decades that the most severe, dangerous and most often resulting in fatalities type of compartment fire (the so-called *Fire Flashover*) is a very specific form of thermal explosion. The problem has become more severe in recent years with a fundamental shift of combustible load in compartments towards plastics and similar highly flammable materials. The available data suggests that the time to flashover (which is effectively the time available to occupants to leave the property) has decreased in modern compartments by several times, or perhaps even up to an order of magnitude.

Understanding the flashover mechanism from the point of view of thermal explosion theory, and the application of respective quantitative models, can play a vital role in identifying scientifically justified ways towards safer building designs.

There are also other areas where the application of methods of the thermal explosion theory will be important, for example in handling and storage of highly energetic propellants of the new generation (metalized propellants).

There are some, probably inevitable, limitations to the scope of the book. Thus, multi-phase systems are not considered (except for explosion prevention application in Chapter 9). Extension of the analysis to multi-phase case in any context, not just related to thermal explosion, results in a dramatic increase in complexity. Correspondingly, it would require a considerable increase of the volume of the book.

The author's own scientific preferences are also reflected in the scope, predominantly via noticeable attention given to applications for fire safety.

It should be noted, finally, that while the book is essentially confined to the analysis of chemical sources of energy generation, the phenomenon of thermal explosion may also be considered in a more broad sense. Some very interesting and important manifestations of identical behaviour have been found in systems where energy generation progresses due to physical mechanisms, such as in some processes related to the mechanics of polymers.

Thus, the theory of thermal explosion provides a unified and powerful framework for understanding, prediction and practical handling of phenomena in very different fields of human activity. This fact emphasizes the strength and vitality of the theory created almost one hundred years ago.

This book is organized into nine Chapters.

The first chapter is introductory. It seeks to provide very general exposition of the subject. It introduces the concept of Thermal Explosion, provides an informal description of this phenomenon in relation to other similar processes and discusses its definition adopted in the present book. It presents historical remarks and also provides a necessary discussion on terminology. Prerequisite facts concerning chemical kinetics are also briefly reviewed in this chapter.

The second chapter presents classical mathematical results of the thermal explosion theory, obtained in the two major formulations: (i) Semenov formulation (uniform temperature of the mixture), and (ii) Frank-Kamenetskii formulation (distributed system where the temperature distribution in the vessel, for subcritical conditions, forms part of the problem solution). Critical conditions for thermal explosion are derived in the three basic cases possessing particular types of symmetry. Both steady-state and nonsteady versions of the theory are considered. The chapter also discusses other important properties of thermal explosion, such as an induction period. Comparison between the Semenov and the Frank-Kamenetskii approaches is presented at the end of the chapter.

The third chapter provides an extensive discussion of the major results obtained in the mathematical theory of thermal explosion. In particular, it considers several specific topics, such as generalized boundary conditions, dynamical regimes, stability of solutions of the thermal explosion theory and critical conditions of thermal explosion in regions of arbitrary shape. The chapter also investigates and discusses mathematical connections of the theory with the theory of catastrophes and control theory. Some results are discussed in a very detailed way, while the rest are reviewed so that the reader can obtain the full knowledge of the facts established in the theory of thermal explosion up to the present moment.

The fourth chapter concerns the further extensions of the theory. Importantly, the problems considered here are confined to a quiescent medium, with the absence of any dynamical effects, for example related to internal flow. Manifestations of thermal explosion in such situations are versatile. Most important results obtained for various media and in various formulations are presented in this chapter. A very important topic considered in this chapter is effects of chemical kinetics. Several other considered problems describe conjugate thermal explosion (controlled by heat exchange between reactive systems with different properties), diffusion thermal explosion (where reactants are separated initially) and the so-called spotted thermal explosion (originating from a hot spot of limited size in the mixture). Finally, experimental validation of the theory of thermal explosion is discussed.

In contrast to the preceding chapter, the fifth chapter presents an extensive treatment of the dynamical regimes of thermal explosion. The distinctive feature of such regimes is that chemical transformations are accompanied by the flow of the reacting mixture. At the beginning of the chapter thermal explosion development in a flow reactor is considered. In subsequent sections, major emphasis is put on the analysis of the two major types of flow conditions, that is (i) natural convection and (ii) forced convection. The influence of such conditions on the development of thermal explosion is discussed in detail using specific examples of both types of flow regimes.

The following chapter is devoted to discussion of dynamical models of compartment fire flashover, developed within the framework of the thermal explosion theory. The chapter starts with a description of the phenomenon of fire flashover, and a discussion of its mechanism. Consideration of fire flashover as a specific form of thermal explosion is justified. The models considered here are constructed using the widely known zone approach to fire modelling and differ essentially by the number of variables describing the state of the system (i.e. by the dimension of their phase space). Major attention is paid to one- and two-variable models. Attempts to develop models with a higher dimension of phase space, as well as some other recent flashover model developments are also discussed.

The seventh chapter discusses several very specific practical applications of the concept of thermal explosion. The major aim of this chapter is to emphasize the possibility of thermal explosion development in some modern industrial processes and devices, and to review quantitative models that allow such dangerous developments to be predicted. The starting section of this chapter provides an exposition of available experimental data. Then, the three major classes of reactive media are considered, namely granular media, biosolid fuels and electric batteries. The major focus of the discussion in the latter case is lithium-ion batteries.

The last two chapters deal with the theoretical and practical considerations related to the prevention of thermal explosion.

The eighth chapter discusses potential approaches theoretically. Here, several control problems in the theory of thermal explosion are formulated. The major question considered in this chapter is how external conditions for a particular reactive system need to be changed so that thermal explosion in the system is prevented. Investigation of this question leads to the proposition of several control strategies. The first of them is an instantaneous change of parameters, in the simplest case, the ambient temperature. The second strategy is a smooth control, that is a continuous change of control parameters in the course of thermal explosion development. Smooth control may be exerted in two forms: autonomous and non-autonomous (the mixed type is not considered) reflecting the type of the relevant differential equation. For both these types of control analytical solutions, providing rigorous estimations of the system evolution under control conditions are obtained.

The final, ninth chapter discusses some practical approaches to the prevention of thermal explosion. The active prevention strategies follow the theoretical developments of Chapter 8. It is shown that the problem of thermal explosion control is closely related to the concept of thermal management. Control methods are classified into passive and active, and definitions of both types are presented. Further sections provide examples of passive and active methods. As far as active methods are concerned, the strategies of reacting system inertization and dilution with cooled media are considered, along with their basic mathematical models. The third active method, prevention of fire flashover, is considered in greater detail. In particular, analytical solutions describing heat exchange between water spray and the fire-generated hot smoke layer are obtained. These solutions are then used in the flashover development model in order to predict critical water flow rates required to suppress flashover.

It is necessary to make remarks regarding some format conventions.

Predominantly, non-dimensional variables are used throughout the book. To indicate dimensional variables, these are typed with the wave as an ascent.

Formulae are numbered consecutively within each chapter. This results in double-digit notation, with the first digit being the chapter number.

Definitions and Theorems are numbered consecutively within each chapter, that is the first digit in their numbers being the respective chapter number.

In conclusion, it is my pleasure to acknowledge the significant contribution of the people who invaluably helped in the preparation of this book.

I am greatly indebted to the Wiley Editorial staff for their patience in the course of the book preparation, and for their encouragement and advice.

Special thanks are given to Inga Novozhilov. It is certain that without her careful and dedicated work, the manuscript could not have been adequately prepared. She kindly provided a number of figures used in the book, including the cover image, and also assisted in the preparation of the text.

Finally, I would like to thank my wife, Natalia Golubnichaya, for her love, patience and continuous support throughout the project.

Vasily B. Novozhilov
Melbourne
2020–2025

type of the relevant differential equation. For both these types of control analytical solutions, providing rigorous estimations of the system evolution under control conditions are obtained.

The final, ninth chapter discusses some practical approaches to the prevention of thermal explosion: the active prevention strategies follow the theoretical developments of Chapter 8. It is shown that the problem of thermal explosion control is closely related to the concept of thermal management. Control methods are classified into passive and active, and definitions of both types are presented. Further sections provide examples of passive and active methods. As far as active methods are concerned, the strategies of reacting system inertization and dilution with cooled media are considered, along with their basic mathematical models. The third active method, prevention of the flashover, is considered in greater detail. In particular, analytical solutions describing heat exchange between water spray and the fire-generated hot smoke layer are obtained. These solutions are then used in the flashover development model in order to predict critical water flow rates required to suppress flashover.

It is necessary to make remarks regarding some format conventions.

Predominantly, non-dimensional variables are used throughout the book. To indicate dimensional variables, these are typed with the wave as an ascent.

Formulae are numbered consecutively within each chapter. This results in double digit notation, with the first digit being the chapter number.

Definitions and Theorems are numbered consecutively within each chapter, that is the first digit in their numbers being the respective chapter number.

In conclusion, it is my pleasure to acknowledge the significant contribution of the people who [illegible] helped in the preparation of this book. [illegible]

I am greatly indebted to the Wiley Editorial staff for their patience in the course of the book preparation, and for their encouragement and advice.

Special thanks are given to Inna Novozhilov. It is certain that without her careful and dedicated [illegible] quickly prepared. She also provided a number of [illegible] of the text. [illegible] and constant support throughout the project.

[illegible]

[illegible]

[illegible]

Important Notation and Abbreviations

Abbreviations

BMS	battery management system
BVP	boundary value problem
CDP	Cresyl diphenyl phosphate
CFD	computational fluid dynamics
DDB	2,5-ditertbutyl-1,4,-dimethoxybenzene
DEC	diethyl carbonate
def	by definition
DMC	dimethyl carbonate
DMMP	dimethyl methylphosphonate
DOP	diphenyl octyl phosphate
DPA	diphenylamine
EC	ethylene carbonate
EGD	energy generation-dissipation
EMC	ethyl methyl carbonate
EV	electric vehicle
FR	flame retardant
HEV	hybrid electric vehicle
HLC	heat loss capacity
HMPA	hexamethylphosphoramide
HRR	heat release rate
HTE	heat transfer equation
ISC	internal short circuit
LB	Lattice Boltzmann
LiB	lithium-ion battery
LFL	lower flammability limit
LHS	left-hand side
MFE	methyl nonafluorobutyl ether
ODE	ordinary differential equation
PC	propylene carbonate
PDE	partial differential equation
PE	polyethylene
PMMA	polymethylmethacrylate

PP	polypropylene
PSR	perfectly stirred reactor
PTS	positive temperature coefficients
RDF	refuse-derived fuel
RHS	right-hand side
RPF	refuse paper and plastic fuels
SEI	solid–electrolyte interface
SIT	spontaneous ignition test
TBP	tributyl phosphate
TDB	4-tertbutyl-1,2-dimethoxybenzene
TEMPO	2,2,6,6-tetramethylpiperidine-1-oxyl
TGA	thermogravimetric analysis
TG-DTA	thermogravimetry differential thermal analysis
TMP	trimethyl phosphate
TMS	thermal management system
TPP	triphenyl phosphate
UFL	upper flammability limit
ZN	Zeldovich-Novozhilov
3CA	3-chloroanisole

Mathematical Functions

Arctg	Arctangent function
Arsh	Inverse hyperbolic sine function
arcsin	Inverse sine function
ch	Hyperbolic cosine function
cos	Cosine function
erfc	Complimentary error function
exp	Exponential function
g	Legendre transformation
J_n	Bessel function of the first kind of the order $n \in \mathbb{N} \cup \{0\}$
ln	Natural logarithmic function
log	Decimal logarithmic function
sec*h*	Hyperbolic secant function
sh	Hyperbolic sine function
sin	Sine function
th	Hyperbolic tangent
Y_n	Bessel function of the second kind of the order $n \in \mathbb{N} \cup \{0\}$
δ	δ–function

Mathematical Symbols

$BV^2(G)$	BV^2 Space of functions, vanishing outside the region $G \subset \mathbb{R}^n$
D	Discriminant
d	Derivative; differential

$\tilde{M}$	mass, $[M]$
M_i, $i \in \mathbb{N}$	molecular species
$\tilde{m}$	mass, $[M]$
m	symmetry index; chemical reaction order (Chapters 1 and 4); stirring intensity (Chapter 5); mass of smoke layer (Chapter 6)
$\dot{\tilde{m}}_f$	fire mass burning rate, $[M][T]^{-1}$
$\dot{\tilde{m}}_f''$	fire mass burning rate per unit area, $[M][L]^{-2}[T]^{-1}$
$\dot{\tilde{m}}_{out}$	mass flow rate from fire compartment, $[M][T]^{-1}$
$\dot{\tilde{m}}_p$	mass flow rate in fire plume, $[M][T]^{-1}$
m_r	geometrical real parameter (Chapter 5)
m_s	symmetry index (Chapter 4)
N	natural parameter
n	natural parameter
$\vec{n}$	outward pointing unit normal to the surface $\tilde{S}$ surrounding volume $\tilde{V}$
Nu	Nusselt number
Nv	Novozhilov number
$\tilde{P}$	part of compartment cross-sectional perimeter corresponding to locations of contact between walls and hot smoke, $[L]$
P	perimeter or surface area of the region G
P_0	perimeter (surface area) of the circle (ball) of the same area (volume) as region G
P_1	perimeter (surface area) of the circle (ball) with the same value of first eigenvalue as region G
Pe	Peclet number
Pr	Prandtl number
$\tilde{p}$	pressure, $[M][L]^{-1}[T]^{-2}$; parameter (Chapter 5)
$\tilde{Q}$	heat of reaction, $[M][L]^2[T]^{-2}[N]^{-1}$
$\dot{\tilde{Q}}_{spr}$	energy transfer rate from spray to flow, $[M][L]^2[T]^{-3}$
$\dot{\tilde{Q}}_0$	HRR of self-sustained fire, $[M][L]^2[T]^{-3}$
$\dot{\tilde{q}}''$	heat flux from steadily burning initial fire, $[M][T]^{-3}$
$\dot{\tilde{q}}''_{rad}$	radiative heat flux, $[M][T]^{-3}$
$\tilde{R}$	universal gas constant, $[M][L]^2[T]^{-2}[\theta]^{-1}[N]^{-1}$; fire radius (Chapter 6)
$\tilde{r}$	characteristic spatial scale of reacting system, $[L]$
r	radial polar coordinate
Ra	Rayleigh number
Re	Reynolds number
Re_d	particle Reynolds number
r_i, $i \in \mathbb{N}$	radical species
$\tilde{r}_m$	stirrer rotational speed, $[T]^{-1}$
$\tilde{S}$	surface area, $[L]^2$
s	parameter (Chapter 5); soot conversion factor (Chapter 7)
S_{cr}	degree of subcriticality
Se	Semenov number

$\tilde{T}$	temperature, $[\theta]$
$\tilde{t}$	time, $[T]$
Td	Todes number
$\tilde{t}_{ad}$	adiabatic time scale, $[T]$
$\tilde{U}$	horizontal droplet velocity component, $[L][T]^{-1}$
u	real function defined on the region $G \subset \mathbb{R}^n$, $n = 2, 3$
$\vec{\tilde{u}}$	flow velocity, $[L][T]^{-1}$
u_r	radial velocity component
u_φ	angular velocity component
$\tilde{V}$	volume, $[L]^3$; vertical droplet velocity component, $[L][T]^{-1}$ (Chapter 9)
V	view factor
$\vec{\tilde{V}}$	droplet velocity, $[L][T]^{-1}$
$\tilde{W}$	compartment width, $[L]$
$\tilde{w}$	chemical reaction rate, $[N][L]^{-3}[T]^{-1}$
$\tilde{X}_n$, $n \in \mathbb{N}$	eigenfunctions
x	spatial coordinate; auxiliary variable (Chapters 2 and 3)
$\vec{\tilde{x}}$	spatial coordinate, $[L]$
$\tilde{y}$	spatial coordinate, $[L]$
y	spatial coordinate; auxiliary variable (Chapters 2 and 3)
$\tilde{Z}$	fire upper (smoke) layer thickness, $[L]$
$\tilde{z}$	vertical spatial coordinate, $[L]$
Zn	Zeldovich number

Greek

α	real numerical parameter; real numerical parameter on the closed curve (Chapter 5)
α_i, $i \in \mathbb{N}$	real numerical parameters
α_m	real numerical parameter
$\tilde{\beta}$	temperature coefficient of expansion, $[\theta]^{-1}$
β	parameter
β_i; $i \in \mathbb{N}$	real numerical parameters
β_m	real numerical parameter
γ	real numerical parameter
γ_r	fraction of fire HRR dissipated by radiation
$\tilde{\Delta}$	smoke layer thickness, $[L]$
δ	Frank-Kamenetskii parameter
δ_1	real parameter
ε	small real positive parameter; integer parameter (Chapter 3); emissivity (Chapters 6 and 7)
$\tilde{\zeta}$	spatial coordinate of the boundary of the outer region (Chapter 4)

ζ	spatial coordinate
$\tilde{\eta}$	spray evaporation parameter, $[T][M]^{-1}$
η	reacted fraction of substance; progress variable (Chapters 3 and 4); real parameter (Chapter 3)
η_0	criterion of autocatalysis
$\tilde{\eta}_n$, $n \in \mathbb{N}$	eigenvalues
Θ	temperature
θ	temperature
$\tilde{\theta}_p$	excess temperature at the centreline of fire plume, $[\theta]$
ϑ	excess temperature; parameter (Chapter 4)
$\tilde{\kappa}$	thermal diffusivity, $[L]^2[T]^{-1}$
$\tilde{\kappa}_c$	convective thermal diffusivity, $[L]^2[T]^{-1}$
$\tilde{\lambda}$	thermal conductivity, $[M][L][T]^{-3}[\theta]^{-1}$
λ	real parameter; wavelength (Chapter 5)
λ_{sg}	ratio of thermal conductivities
$\tilde{\mu}$	dynamic viscosity, $[M][L]^{-1}[T]^{-1}$; molecular weight, $[M][N]^{-1}$ (Chapter 7)
μ	eigenvalue (Chapter 5)
$\tilde{\mu}_k$, $k \in \mathbb{N}$	square root of positive eigenvalue
μ_i, $i \in \mathbb{N}$	products stoichiometric coefficients
Ξ	energy transfer rate due to full spray evaporation
$\vec{\xi}$; ξ	spatial coordinate; spatial coordinate, one-dimensional case
$\tilde{\rho}$	density, $[M][L]^{-3}$
$\tilde{\sigma}$	Stefan–Boltzmann constant, $[M][T]^{-3}[\theta]^{-4}$
σ	real parameter
τ	time
τ_{sg}	square root of the ratio of time scales
υ_i, $i \in \mathbb{N}$	reactants stoichiometric coefficients
φ	universal approximating function; kinetic function (Chapters 3 and 4); angular polar coordinate (Chapter 5); real numerical parameter (Chapter 8); flashover critical boundary (Chapter 9)
χ	combustion efficiency
$\tilde{\Psi}$	mass flow rate of the spray, $[M][T]^{-1}$
Ψ	auxiliary function
$\tilde{\psi}$	constant rate of ambient temperature increase, $[\theta][T]^{-1}$
ψ	constant rate of ambient temperature increase; stream function (Chapter 5); auxiliary function (Chapter 9)
Ω	region in space $\mathbb{R}^n$; $n = 2, 3$
$\partial\Omega$	boundary of the region Ω
$\tilde{\omega}$	frequency of ambient temperature oscillation, $[T]^{-1}$ (Chapter 3)
ω	wave number
ω_k	ratio of thermal diffusivities
ω_λ	ratio of thermal conductivities
$\tilde{\mho}$	volumetric flow rate, $[L]^3[T]^{-1}$

Gothic

$\aleph$ ratio of the critical Frank-Kamenetskii parameter and the first eigenvalue of auxiliary BVP problem (Chapter 3)
$\wp$ auxiliary function
$\wp_0$, $\wp_1$ real numerical parameters

Superscripts

a	analytical
c	chemical
cr	critical
lim	limit
maj	majorizing function
max	maximum
min	minimum
p	physical
0	initial
(2)	fraction of compartment wall area contained in fire upper (smoke) layer
(i), $i \in \mathbb{N}$; $i \in \mathbb{N} \cup \{0\}$	equation root number; symmetry index (Chapters 3 and 4)

Subscripts

a	ambient; approximate
abs	absolute zero
c	convection
cr	critical
$ctrl$	control
d	droplet
eff	effective
ex	exit
ev	evaporation
$F^{(2)}_{\max}$	point of maximum of the second derivative of the Boltzmann factor (or the EGD function)
f	fire; flame
g	gas
hp	heat-up
in	initial; injection
ind	induction
infl	inflection
L	compartment walls surrounding fire lower (cold) layer
lim	limit

max	maximum
min	minimum
p	plume
rad	radiative
s	surface; symmetry; solid
spr	spray
st	stirrer
vap	vaporization
U	compartment walls surrounding fire upper (smoke) layer
V	vent
v	volume
w	wet-bulb temperature; wall
0	ambient; initial; reference
1	fire lower (cold) layer
2	fire upper (smoke) layer

1

Introduction

1.1 Informal Description of Thermal Explosion

The notion of explosion is familiar to everyone. Explosions for high-rise building demolition, accidental domestic gas explosions, volcano eruptions and nuclear explosions are just few examples.

In everyday's perception, explosion power and its consequences are primary questions of interest. Most noticeable explosions are indeed accompanied by sound, thermal and/or shock waves, emission of electromagnetic impulses, radiation, etc., causing significant destruction.

Scientific perception is, however, different. In scientific investigation, the cause and mechanism of explosion and the analysis of conditions that lead to explosion take the center stage.

It is clear that, essentially, explosion is a quick release of a significant amount of energy. However, the mechanisms of such release may be vastly different. Moreover, phenomena that are intrinsically of explosive nature may not cause significant damage. As such, scientifically, in the notion of explosion, the intrinsic properties of the phenomenon and its mechanisms are of importance, rather than its appearance and consequences.

Thermal explosion is a good example. This term refers to explosion that is progressively driven by energy generation, which occurred at earlier stages of the explosion development. Energy, needed for explosion to progress, must be contained within the system, i.e. no external energy supply is required for such explosion to occur. Moreover, energy release is quickly self-accelerating. Self-acceleration means that the rate at which energy is being generated within the system is increasing as the total amount of energy that has been released increases. This is a very important necessary condition for thermal explosion.

Such self-acceleration may only occur in strongly non-linear physical or chemical systems. It should be noted, though, that before the rapid self-accelerating stage there may exist the so-called induction period of thermal explosion (in some cases, very long) during which energy accumulation in the system is very slow.

The present book is confined to the analysis of the very specific mechanism of energy generation in a thermal explosion, that is to the release of energy via exothermic chemical reactions. Heat generated in such reactions causes the reaction rate to grow rapidly, providing a positive feedback mechanism and causing reaction self-acceleration. The exact mechanism of such reaction rate self-acceleration (Arrhenius dependence of the reaction rate on the mixture temperature) will be clarified below.

It should be understood that, of course, in such a process, energy would also be lost from any real system. However, under certain circumstances, the rate of energy dissipation would become less

Thermal Explosion: Theory and Application, First Edition. Vasily B. Novozhilov.

than the rate of energy accumulation. Under such conditions, which are called critical conditions of thermal explosion, the progressive accumulation of energy would proceed.

Thermal explosion may be observed in laboratory conditions, in relatively small-scale vessels containing mixtures of reacting gases. Therefore, the damage caused by such explosions may in some cases be minimal.

This is not to say that the opposite cannot occur. Thermal explosions can be very severe. Some of these occasions have been well documented.

Such was the Texas City disaster that occurred on April 16, 1947 (Fire Prevention and Engineering Bureau of Texas and The National Board of Fire Underwriters 1947). It is believed to be the deadliest industrial accident in the United States history and one of the largest non-nuclear explosions in history with 581 deaths and more than 8,000 victims. The accident, which was a chain of different fires and explosions, started with a detonation of 2,300 tons of ammonium nitrate in the ship (S.S. GRANDCAMP) cargo due to thermal explosion.

Destructions were overwhelming. S.S. GRANDCAMP exploded violently, and there were reports of secondary explosions. An immense tidal wave flooded the vicinity of the explosion. The area was quickly filled with dense smoke from burning chemical facilities and oil refineries. Shock waves were easily felt 10 ml away. Shocks were able to shake buildings and shatter glass windows over wide areas.

Apart from fatalities and casualties, there was widespread property damage. The most extensive damage was sustained by Monsanto Chemical Company and Texas City Terminal Railway Company; however, many more businesses were affected.

Figure 1.1 is an artist's impression of the Texas City Disaster.

Another example is the thermal explosion of dinitolmide (or zoalene) which was used as a poultry feed additive, in a dryer at King's Lynn (Health and Safety Executive 1976) on 27 June, 1976. There was one death and extensive damage to the plant and adjacent buildings.

Figure 1.1 An artist's impression of the Texas City Disaster on April 16, 1947.

The two above-mentioned examples represent a traditional risk of thermal explosion associated with processing, storage or transportation of large amounts of particular chemicals. Recently, less obvious areas of thermal explosion hazards emerged, for example, solid biomass fuels (Koseki, H. 2011) and lithium-ion batteries (British Broadcasting Corporation 2013) industries.

1.2 Historical Remarks and Terminology

The discovery of the theory of thermal explosion was a logical consequence of all the preceding studies on combustion.

A scientific study of combustion began in the eighteenth century, along with the rapid development of chemistry. It was at that time when Lomonosov and Lavoisier rejected false beliefs of their contemporaries that substances capable of combustion must contain a specific element, the so-called phlogiston. In 1773, Lavoisier finally arrived at the correct conclusion that combustion is a process of a chemical combination (reaction) of substances with the oxygen in the air. The developments of the latter half of the eighteenth century (such as the discoveries of nitrogen, hydrogen, carbon monoxide and dioxide) supported Lavoisier's theory.

In the early nineteenth century, Davy conducted some experimental research on combustion, in particular, he measured flame temperatures, studied the effects of dilution and discovered catalytic combustion. He also succeeded in practical inventions related to combustion and is known for Davy's safety lamp, which had flame surrounded by a metallic mesh screen, considerably reducing the risk of its use in mining.

A significant progress in understanding the laws of chemical kinetics was made in the second half of the nineteenth century. Guldberg and Waage discovered the law of mass action, which establishes the direct proportionality of the reaction rate to the product of the concentrations of the reactants. Further, van't Hoff in 1884 and, based on his work, Arrhenius in 1889 quantified the dependence of the reaction rate on mixture temperature. Moreover, Arrhenius uncovered the kinetic essence of Van't Hoff's law, relating a number of active molecules, capable of chemical reaction, to considerations of kinetic theory of gases. The final result of this work, the so-called Arrhenius equation for the reaction rate constant, is a cornerstone of the modern combustion theory and is immensely important for understanding and predicting thermal explosion. The laws of chemical kinetics are reviewed in Section 1.3 below.

A correct understanding of the chemical nature of combustion at this early stage of modern science laid the foundation for the development of power engineering and thermodynamics, as the process of combustion was the major supplier of high-temperature gases and energy. At the end of nineteenth century, intensive investigations were conducted on chemical-thermodynamic equilibria in the system $H_2 - H_2O - CO - CO_2 - C - O_2$, which determines, in the first approximation, the combustion temperature of hydrocarbons.

The application of combustion process continued alongside its scientific investigation. Knowledge gained in laboratories warranted a great contribution of combustion to Industrial Revolution, steadily expanding its application from primitive use in fireplaces and outdoor gas lamps to sophisticated power generation in steam and internal combustion engines (Figure 1.2).

At the early stages of the combustion theory development, the names of Michelson, Berthelot, Vieille, Jouguet, Taffanel, Daniell and Chapman are worth mentioning.

In 1928, Russian scientist Nikolay Nikolaevich Semenov published a paper (Semenov 1928a), which laid the foundation of thermal explosion and, in fact, the whole modern combustion theory.

Figure 1.2 Some applications of combustion.

Semenov's paper considered only the absolutely essential elements of thermal explosion, the features without which autoignition simply cannot exist. He took into account the Arrhenius law of heat generation, using the simplest zeroth-order reaction model, and Newton's law of heat dissipation driven by the difference between the mixture temperature (assumed uniform) and the background one.

Yet, this simplest formulation crystallized a powerful insight into the very nature of autoignition as progressive thermal self-acceleration in the absence of comparable heat dissipation. The same process is at the base of all combustion phenomena; for example, it is occurring on a continuous self-supporting basis in a travelling combustion wave. The explosive self-acceleration of a chemical reaction is at the centre of the modern definition of combustion: 'Combustion is a chemical reaction proceeding under conditions of progressive self-acceleration ...' (Frank-Kamenetskii 2016). In other words, reaction is classified as combustion based on the characteristic features of the chemical process, rather than on the nature of substances involved. From this point of view, one can speak, for example, about the flame of ozone decomposition $2O_3 \rightarrow 3O_2$, although oxygen is released in this reaction rather than being consumed. One can equally speak about combustion processes in the complete absence of oxygen, such as combustion of hydrogen in chlorine.

Semenov considered the theory of thermal explosion as one of the his three major scientific achievements, along with the theory of thermal dielectric breakdown and the theory of chain reactions (Semenov 2004).

The latter theory also contributed greatly to the progress of combustion science. About the same time when Semenov's first paper on thermal explosion appeared, it had become apparent that not all explosive behaviours in chemical processes could be explained by thermal self-acceleration. An alternative self-acceleration mechanism, chain-branching reaction, was predicted by Christiansen and Kramers in 1923, who, however, did not elaborate on the idea further. Semenov demonstrated that such a mechanism indeed explains observed facts (a peculiar combustion behaviour of phosphorus vapour/oxygen mixtures studied by Chariton and Valta in 1926). It is this discovery that eventually resulted in Semenov and Sir Cyril Hinshelwood, who conducted similar studies at Oxford, being awarded the Nobel Prize in Chemistry in 1956 'for their researchers into the mechanism of chemical reactions' (Ölander 1956). Notably, Semenov and Hinshelwood's award was closely related to the very first Nobel Prize in Chemistry given in 1901 to Van't Hoff for '..."the discovery of the laws of chemical dynamics", i.e. the velocity of chemical reactions' (Ölander 1956).

Both thermal explosion and chain reaction studies have been significantly advanced by Semenov and his research school (Berlin and Golubkov 2021; Azatyan 2021) at the Institute of Chemical Physics, the Russian Academy of Sciences, founded by Semenov in 1931. This institute is presently known as the N.N. Semenov Federal Research Centre for Chemical Physics, the Russian Academy of Sciences (Figures 1.3 and 1.4).

Further significant progress in the thermal explosion theory before the Second World War was made by Frank-Kamenetskii (invited by Semenov to work at the Institute of Chemical Physics) who proposed his own formulation of the problem, dropping the assumption of mixture temperature uniformity.

Up to this day, Semenov and Frank-Kamenetskii approaches remain a solid foundation for theoretical and experimental studies on thermal explosion.

Semenov became a full member of the Academy of Science of the Soviet Union in 1932, and he was also elected as a member of several foreign academies.

The final remark must be made on terminology. The terms 'thermal explosion', 'thermal runaway' and 'autoignition' have become synonyms and are used as such throughout the book. In the literature (especially in relation to liquid and solid fuels), the term 'spontaneous ignition' is also being used in the same sense. The latter term emphasizes the absence of any external energy source as a trigger for thermal explosion (as opposed to 'piloted ignition', using various pilots, e.g. spark).

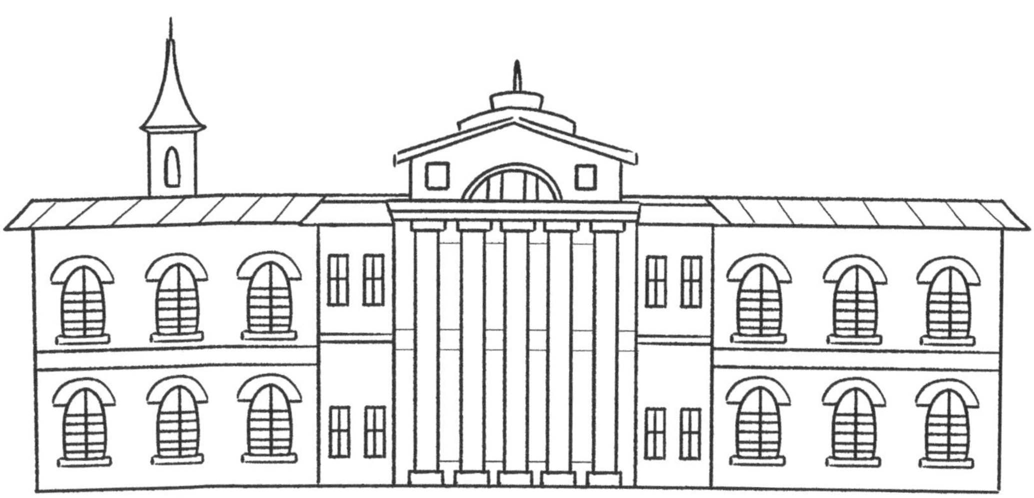

Figure 1.3 Artist's impression of the main building of the N.N. Semenov Federal Research Centre for Chemical Physics, the Russian Academy of Sciences in Moscow.

Figure 1.4 N.N. Semenov monument in front of the main building of the N.N. Semenov Federal Research Centre for Chemical Physics, the Russian Academy of Sciences in Moscow. *Source:* Photo by the author.

1.3 Fundamentals of Chemical Kinetics

The present section presents a brief overview of chemical kinetics, limited to the facts used directly in subsequent chapters.

A general one-step irreversible chemical reaction of arbitrary complexity may be represented by the following formal equation:

$$\sum_{i=1}^{N} \upsilon_i M_i \rightarrow \sum_{i=1}^{N} \mu_i M_i \tag{1.1}$$

Here M_i are molecular or radical species, υ_i are reactants stoichiometric coefficients, and μ_i are products stoichiometric coefficients. The stoichiometric coefficient υ_i or μ_i may be zero if the species M_i is missing in the list of reactants or products, respectively.

The sum of reactants stoichiometric coefficients

$$\sum_{i=1}^{N} \upsilon_i = m \tag{1.2}$$

is called the order of chemical reaction.

Formally, notation (1.1) may be applied to both elementary reactions, expressing the results of direct collision between a number of molecules and radicals, and global reactions, describing the overall chemical transformation of molecular species and consisting of a number of elementary steps (Kuo 1986).

In the former case, kinetic parameters discussed later may be, theoretically, derived from first principles (Gardiner 1972; Benson 1960). In the latter case kinetic parameters, obtained by fitting experimental data, have the meaning of their effective values.

Depending on how many molecular species appears on the LHS of Eq. (1.1), chemical reactions are classified into monomolecular, bimolecular, trimolecular, and so on.

The rate of reaction transformation (1.1) is given by the law of mass action, discovered experimentally by Guldberg and Waage in 1864 (Kuo 1986; Penner 1955)

$$\tilde{w} = \tilde{k}(\tilde{T}) \prod_{i=1}^{N} (\tilde{C}_i)^{\upsilon_i} \tag{1.3}$$

where $\tilde{C}_i$ is the molar concentrations of reactants M_i, and $\tilde{k}$ is the temperature-dependent proportionality factor called the specific reaction rate constant.

Net rate of change of the species M_i is given, according to Eqs. (1.1) and (1.3), by

$$\frac{d\tilde{C}_i}{d\tilde{t}} = (\mu_i - \upsilon_i)\tilde{w} \tag{1.4}$$

Since the concentration of the reactants is proportional to the mixture pressure (given that they obey stoichiometric restriction (1.1)), the law (1.3) implies that the reaction rate is proportional to pressure raised to the power equal to reaction order m. Order of the reaction has, therefore, a distinctive meaning of exponent in power dependence of the reaction rate on pressure.

Specific reaction rate constant, as its name suggests, is independent of concentrations and depends on the mixture temperature $\tilde{T}$ only. The collision frequency analysis of molecules (Kuo 1986) shows that it has the following structure:

$$\tilde{k}(\tilde{T}) = \tilde{A}(\tilde{T}) \exp\left(-\frac{\tilde{E}}{\tilde{R}\tilde{T}}\right) \tag{1.5}$$

where $\tilde{A}(\tilde{T})$ is called the pre-exponential factor, $\tilde{E}$ is the activation energy of chemical reaction, and $\tilde{R}$ is the universal gas constant. Equation (1.5) is called the Arrhenius (essentially, Arrhenius in (1.5) made use of Boltzmann factor) law of temperature dependence, or the Arrhenius function.

Activation energy is a critical kinetic energy level of colliding particles required to break molecular bonds. Its value is typically of the order of bond energy and is specific for a particular reaction. As such, it is an analogy of a potential barrier between two potential wells (upper part of Figure 1.5). For an exothermic reaction, the supply of activation energy to reacting molecules results in the liberation of an amount of energy, exceeding the heat of reaction by an amount exactly equal to activation energy (bottom part of Figure 1.5). The exponential factor in (1.5), arising from Maxwell–Boltzmann distribution, describes the fraction of colliding molecules having energies greater than the activation energy.

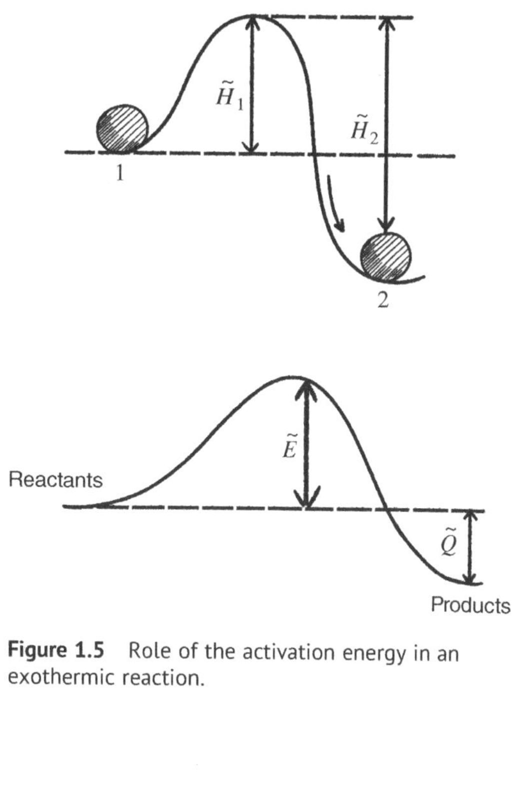

Figure 1.5 Role of the activation energy in an exothermic reaction.

Pre-exponential factor is a weak function of temperature

$$\tilde{A}(\tilde{T}) = \tilde{A}_0 \tilde{T}^{\alpha} \tag{1.6}$$

with the exponent α is typically between 0 and 1 (Kuo 1986). As Arrhenius function, in contrast, is extremely strongly temperature-dependent (see Problem 1), weak temperature dependence in (1.5) is typically neglected in combustion problems, in comparison with the temperature dependence of the Boltzmann factor.

Due to specific form of the Arrhenius function (1.5), temperature scale playing a very important role in thermal explosion problems (and in combustion problems in general), is the so-called characteristic temperature interval $\frac{\tilde{R}\tilde{T}_0^2}{\tilde{E}}$, where $\tilde{T}_0$ is the initial temperature of the reactant (potentially undergoing thermal explosion) mixture.

It is equal to the product of the initial temperature and the so-called Arrhenius number, $Ar = \frac{\tilde{R}\tilde{T}_0}{\tilde{E}}$. The latter property may be considered as the inverse of non-dimensional activation energy. While this number is an essential fundamental factor in combustion kinetics, it is important to realize that in most combustion problems the Arrhenius number is a small parameter.

For typical values of activation energies (Borman and Ragland 1998) and initial temperatures close to room levels, the typical range of Arrhenius numbers is $Ar = 1.2 \cdot 10^{-2} - 2.0 \cdot 10^{-2}$.

For small values of Arrhenius number, it follows immediately from (1.4) that temperature increase, above the initial level, by one characteristic temperature interval results in the reaction constant (and correspondingly, at the same values of reactants concentrations, reaction rate) increase by approximately the factor of e (Problem 2). As will be shown in the next section, this circumstance plays a pivotal role in defining the necessary conditions for thermal explosion.

Correspondingly, non-dimensional temperature in thermal explosion studies is often introduced using the characteristic temperature interval as the natural scale. Specifically, non-dimensional temperature corresponding to the real physical temperature $\tilde{T}$ at given location and time may be defined as $\theta = \frac{\tilde{T} - \tilde{T}_0}{\left(\tilde{R}\tilde{T}_0^2/\tilde{E}\right)}$.

With the introduction of a non-dimensional temperature, exponential factor in the Arrhenius function may be transformed to the following equivalent form:

$$\begin{aligned}
\exp\left(-\frac{\tilde{E}}{\tilde{R}\tilde{T}}\right) &= \exp\left(-\frac{\tilde{E}}{\tilde{R}\tilde{T}_0}\right)\exp\left(\frac{\tilde{E}}{\tilde{R}}\left(\frac{1}{\tilde{T}_0} - \frac{1}{\tilde{T}}\right)\right) \\
&= \exp\left(-\frac{\tilde{E}}{\tilde{R}\tilde{T}_0}\right)\exp\left(\frac{\tilde{E}}{\tilde{R}\tilde{T}_0}\left(\frac{\tilde{T}-\tilde{T}_0}{\tilde{T}}\right)\right) \\
&= \exp\left(-\frac{\tilde{E}}{\tilde{R}\tilde{T}_0}\right)\exp\left(\frac{\tilde{E}(\tilde{T}-\tilde{T}_0)}{\tilde{R}\tilde{T}_0^2}\,\frac{\tilde{T}_0}{\left(\tilde{T}_0 + \frac{\tilde{R}\tilde{T}_0^2}{\tilde{E}}\theta\right)}\right) \\
&= \exp\left(-\frac{\tilde{E}}{\tilde{R}\tilde{T}_0}\right)\exp\left(\frac{\theta}{(1+Ar\theta)}\right)
\end{aligned} \tag{1.7}$$

and consequently, the rate of reaction to the form

$$\tilde{w} = \tilde{k}_0(\theta) \exp\left(\frac{\theta}{(1 + Ar\theta)}\right) \prod_{i=1}^{N} (\tilde{C}_i)^{\nu_i} \tag{1.8}$$

with $\tilde{k}_0(\theta)$ being the value of the Arrhenius function at an initial temperature

$$\tilde{k}_0(\tilde{T}) = \tilde{A}(\theta) \exp\left(-\frac{\tilde{E}}{\tilde{R}\tilde{T}_0}\right) = \exp\left(-\frac{1}{Ar}\right)\tilde{A}(\theta) \tag{1.9}$$

The abbreviated notation $\tilde{k}_0 \stackrel{def}{=} \tilde{k}_0\big|_{\theta=0}$ will also be used.

A very important class of chemical reactions is represented by chain reactions. Chain reactions consist of a number of elementary steps (1.1) involving radical species.

In a simplest form (non-branching chain reaction) radicals are produced (chain-initiation) and react with molecular species in such a way that after radical/molecular collision the number of radicals is preserved (chain-propagation). There are also chain-terminating elementary steps where stable molecules are formed upon radicals collision. It should be noted that chain length, which is a number of molecules of the final product per one originally produced radical, may reach hundreds of thousands and even millions.

The number of radicals may also increase as a result of the elementary reaction step in the chain, in which case branching of the chain occurs, and the reaction is classified as chain-branching. This effect is demonstrated in the mechanism presented in Figure 1.6. Here, chain-branching occurs upon the radical/molecule collisions (r_1, M_2) and (r_4, M_1).

Semenov and his school demonstrated that significant number of chemical reactions propagate by the chain-branching mechanism.

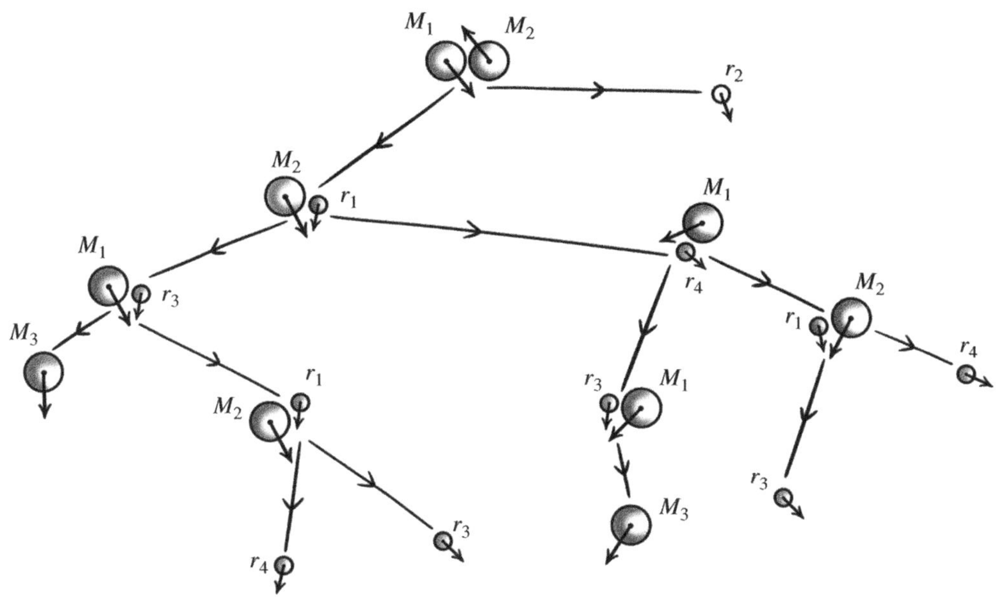

Figure 1.6 An example of the chain-branching reaction with a global kinetic scheme $M_1 + M_2 \rightarrow M_3$. M_i – molecular species; r_i – radical species.

Chain-branching leads to the progressive accumulation of intermediate products (i.e. free radicals) in the mixture, leading potentially to the explosive reaction behaviour. An explosion of this type is called chain explosion, in contrast to thermal explosion. The chain-branching mechanism is responsible for the explosion of hydrogen–oxygen mixtures. All three observable explosion limits are explained based on chain-branching and chain-terminating kinetics (Kuo 1986; Turns 2000).

Before the chain reaction mechanism became known, some chemical phenomena could not be explained. It is exactly the phenomenon of this type (existence of a lower autoignition limit in the reaction between phosphorus and oxygen, discovered by a group of physicists in Leningrad, USSR in 1925) that led to the experimentally-verified discovery of chain-branching reactions. A comprehensive theory of such reactions had been subsequently developed by Semenov in 1920s–1930s.

1.4 Definition of Thermal Explosion

As is the case with any phenomenon, thermal explosion, for its proper study, must be conceptually separated from other phenomena, including similar ones, by some kind of clear and unambiguous definition.

Thermal explosion is distinguished by the fact that the primary effect controlling its development is the reaction rate's dependence on temperature, rather than concentrations. In contrast, in chain explosion, or explosions of mixed chain-thermal type (Barzykin 2004), dependence on radical concentrations is equally important, as these concentrations are affected by chain-branching.

The best definition of thermal explosion was given by the founder of its theory. In his Nobel Lecture Semenov stated:

> 'According to Arrhenius' law, the heat generated by the reaction itself increases the velocity of reaction. The existence of such retroaction is to the greatest possible extent characteristic of most phenomena of the combustion process.
>
> As a result of the reciprocal effect of the reaction on the generation of heat in the mixture on the one hand and the increase in reaction velocity due to this greater heat, as already mentioned, on the other, a self-accelerating surge of heat is set up which we term auto-ignition'. (Semenov 1956).

More than that, Semenov went on explaining that thermal explosion requires specific circumstances for its occurrence.

> 'The dissipation of heat on the walls of the vessel renders the development of a heat surge possible only under certain conditions, which are therefore all-important to the temperature of this autoignition. This temperature, which is a relative value (and by no means a constant of the substance, as was previously assumed by some authorities), is a function of the kinetic and thermal parameters, namely of the order of reaction, the activation energy, the density and composition of the combustible mixture, the heat of reaction, and the dissipation of heat on the vessel walls' (Semenov 1956).

In the space of parameters describing the state (condition) of the system, there are regions corresponding to both non-explosive and explosive reaction regimes, divided by a boundary. Conditions at the boundary are conventionally called critical conditions of thermal explosion.

Semenov's definition makes it possible to formulate necessary conditions for thermal explosion in a quantitative, mathematical form.

The rapid self-acceleration of a chemical reaction requires a noticeable change in the reaction rate upon small temperature changes. As has been discussed in Section 1.3, this occurs for small values of Arrhenius number (reaction rate increases approximately by a factor of e over one characteristic temperature interval). Therefore, the first necessary condition is written in the following form:

$$Ar = \frac{\tilde{R}\tilde{T}_0}{\tilde{E}} \ll 1 \tag{1.10}$$

In addition, of course, the temperature increase by one characteristic interval should not consume too large amounts of reactants, leading to the degeneration of the explosion.

If η is a fraction of substance that has reacted, the temperature increase of the mixture is represented through the following equation:

$$\tilde{T} - \tilde{T}_0 = \frac{\tilde{Q}\eta}{\tilde{c}_p} \tag{1.11}$$

The requirement of $\eta \ll 1$ at $\tilde{T} = \tilde{T}_0 + \dfrac{\tilde{R}\tilde{T}_0^2}{\tilde{E}}$ brings up the second necessary inequality

$$\frac{\tilde{R}\tilde{T}_0^2}{\tilde{E}} \frac{\tilde{c}_p}{\tilde{Q}} \ll 1 \tag{1.12}$$

Inequalities (1.10) and (1.12) form necessary conditions for thermal explosion (Novozhilov 1980).

Once they are fulfilled, the consumption of a small amount of reactants raises the mixture temperature by such a degree that the reaction self-acceleration becomes noticeable. This progressive self-acceleration leads to thermal explosion.

The self-acceleration of a chemical reaction means that the reactive system cannot be in the state of thermal equilibrium (until reactants have been fully consumed). It follows from here that the definition of thermal explosion opens two ways of developing the thermal explosion theory: steady-state and nonsteady. Under the first approach, thermal explosion is equivalent to the impossibility of thermal equilibrium in the system. Under the second, the temperature evolution of the system may be followed until a certain point that, in some sense, corresponds to the total mixture involvement in combustion.

While the mathematical interpretation of the steady-state approach is conceptually straightforward (it amounts to investigating equilibrium points of the system), the nonsteady approach requires some criterion against which system temperature evolution may be judged, qualitatively or quantitatively. It is also clear that such criterion must be related to the inherent properties of the system.

There are a number of possibilities that should be discussed in this regard. The unboundedness (infinite growth) of the system temperature could serve as a qualitative criterion. Unfortunately, it is not suitable for the general (and practically relevant) case $Ar > 0$ for the following reason. The reaction rate function (1.8) and (1.9) has (neglecting pre-exponential temperature dependence and, in line with the necessary condition for explosion, concentration variation) a finite limit at $\theta \to \infty$ ($\tilde{T} \to \infty$). On the other hand, rate of conduction heat dissipation from the system is proportional, in the first approximation, to the difference between the average temperature of the system (Semenov model, in fact, assumes temperature uniformity throughout the mixture) and the temperature of surroundings. This means that the combined contribution of the rate of reaction and the rate of heat dissipation to overall energy balance causes the system temperature to be, strictly speaking, bounded at all temperatures and infinite growth of solution cannot be used as a precise mathematical criterion for explosion for any positive Arrhenius numbers.

Intrinsic properties driving the reactive system's behaviour essentially amounts to the properties of the reaction rate (which is controlled by the Boltzmann factor, neglecting pre-exponential temperature dependence and reactants consumption) and energy dissipation functions. The latter, as mentioned, is a linear function of temperature.

A natural characteristic point of the Boltzmann factor function is its inflection point. (For any criteria which involves the derivatives of the order higher than one linear energy dissipation function may be taken out of consideration.) However, an attempt to relate reaction progression to this point with thermal explosion is equally unsatisfactory.

An inflection point of the Boltzmann factor (which is also an inflection point of the function describing net energy accumulation) is found trivially:

$$\theta_{\inf l} = \frac{1}{Ar}\left[\frac{1}{2Ar} - 1\right] \tag{1.13}$$

It is not difficult to see that physical temperatures corresponding to this point, for Arrhenius numbers in the range $Ar = 1.2 \cdot 10^{-2} - 2.0 \cdot 10^{-2}$, are unrealistically high for combustion reactions, of the order of $\tilde{T}_{\inf l} \approx 5 \cdot 10^3 - 2.0 \cdot 10^{4^\circ}$ K.

The alternative proposed in the present book is to use the point of maximum of the second derivative of the Boltzmann factor as a characteristic point describing the onset of thermal explosion. This point corresponds (in temperature space) to the maximum acceleration of the chemical reaction, which is consistent with the criterion proposed by some other researchers (Barzykin 2004).

The point where the second derivative of the Boltzmann factor reaches its maximum is found by calculating its third derivative. It turns out that it corresponds to the least root of the equation:

$$\begin{aligned} &6\hat{\theta}^2 - 6\hat{\theta} + 1 = 0 \\ &\hat{\theta} = Ar(1 + Ar\theta) \end{aligned} \tag{1.14}$$

which is equal to

$$\hat{\theta}_1 = \frac{1}{2} - \frac{\sqrt{3}}{6}$$

At this point, $\theta_{F^{(2)}_{\max}} \overset{def}{=} \frac{1}{Ar}\left(\frac{\hat{\theta}_1}{Ar} - 1\right)$, third derivative is equal to zero, and changes its sign from positive to negative.

Physical temperatures, in the relevant range of Arrhenius numbers, corresponding to the point $\theta_{F^{(2)}_{\max}}$ are presented in Table 1.1. Their range, $\tilde{T}_{F^{(2)}_{\max}} \approx 3150 - 5250^\circ$ K correlates much better with the upper limit of combustion temperatures ($\tilde{T} \approx 300 - 4000^\circ$ K; Zeldovich *et al.* 1980).

The point of maximum of the second derivative of the Boltzmann factor will be used as the criterion for the onset of autoignition in the non-steady theory of thermal explosion, throughout the book.

Table 1.1 Point of maximum of the second derivative of the Boltzmann factor for different values of Arrhenius number.

Arrhenius number, *Ar*	Point of maximum value of second derivative of the Boltzmann factor, non-dimensional temperature $\theta_{F^{(2)}_{max}}$	Point of maximum value of second derivative of the Boltzmann factor, absolute temperature, °K
0.012	1,375	5,250
0.015	867	4,200
0.020	475	3,150

1.5 Similarities and Differences with Other Phenomena

Thermal autoignition is by no means the only phenomenon exhibiting intrinsic self-acceleration.

In fact, Semenov developed the theory of thermal explosion using analogy with his earlier work on thermal dielectric breakdown (Inge *et al.* 1925). Similar ideas on the latter phenomenon were developed by Fock (1928).

Processes, analogous to thermal explosion, have been noticed in other phenomena associated with the release of energy. Such phenomena include, for example, thermonuclear fusion (Nastoyashii 1972; Nastoyashii and Shevchenko 1972), electrical and optical discharges (Selezneva 1972; Rizer 1973; Makhviladze and Myshenkov 1977), thermal extinguishment of luminescence (Ramsden and Savic 1964; Makshantsev and Finkelburg 1974), formation of solid solutions (Bloshenko *et al.* 1970), baking of metallic powders (Lidorenko and Sidyakin 1972), inductive heating of some bodies, decay of supercooled liquids, and others (Merzhanov and Rumanov 1987).

Best understood is an analogy between thermal explosion and self-heating occurring in flows, or under deformation, of some viscous polymers (Schapery 1964; Gruntfest *et al.* 1964; Korobov and Ratner 1965; Bostanjiyan *et al.* 1965; Barenblatt *et al.* 1965, 1966).

Self-heating emerges in a flow of a melted polymer due to the strong dependence of viscosity (in some approximation, Frenkel 1946) on temperature

$$\tilde{\mu} = \tilde{\mu}_0 \exp\left(\frac{\tilde{B}}{\tilde{R}\tilde{T}}\right) \tag{1.15}$$

Bostanjiyan *et al.* (1965) considered a steady-state flow of the melted polymer in a round infinitely long pipe. They reduced, with a certain change of variables, the problem description to exactly the same equation that appears in the Frank-Kamenetskii formulation of the thermal explosion theory, that is, Eq. (2.27) with $m = 1$. Therefore, the critical value of Frank-Kamenetskii parameter (which for this problem has, of course, different definition) equals to two. For supercritical conditions, the progressive dissipation of kinetic energy leads to self-acceleratory fluid heating. Bostanjiyan *et al.* (1965) termed this phenomenon hydrodynamic thermal explosion.

The same effect is predicted for certain regimes of the Couette flow. The existence of hydrodynamic thermal explosion has been confirmed experimentally by using a rotational viscometer (Merzhanov and Stolin 1971).

Mathematical aspects of hydrodynamic thermal explosion and some related problems have been considered by Khudyaev (2003).

There also exist phenomena of explosive nature, which, despite apparent similarity with thermal explosion, proceed by different mechanisms.

The first example, which should be pointed out, is nuclear fission. In contrast to the above-mentioned thermonuclear fusion, which relies on the high thermal energy input (in order to sustain high-temperature plasma), nuclear fission is driven by progressively increasing production rate of neutrons. In order for fission to proceed, the number of neutrons must increase as a result of a single act of neutron capture by nucleus. Moreover, the average number of secondary neutrons, that is, neutrons produced per one neutron capture act, must be significantly more than one. This process is therefore an analogy of the chain-branching mechanism in chemical reactions.

The second example that may be given lies outside both chemistry and physics. This is an explosively rapid, in the form of epidemic or pandemic, virus spread. There have been an attempts to describe these phenomena using both chain and thermal explosion methods (Philippov 2024). However, it seems that the predominant mode of spread is a chain-like mechanism. The progressive accumulation of virus carriers, driven by interaction between individuals, again makes it an analogy of the chain-branching chemistry mechanism. The COVID-19 pandemic is a recent example of such catastrophic spread.

2

Classical Theory of Thermal Explosion

2.1 General Considerations

The present chapter discusses the basic, and historically, the first model of the combustion system where thermal explosion may be comprehended in terms of underlying physical and chemical processes and also described mathematically.

This is a model of chemically reacting gas mixture contained in a closed vessel at a certain, equal to ambient, temperature (Figure 2.1).

Problem formulation and mathematical analysis of this problem were developed in the pioneering works of Semenov (1928a, b), and extended later by Frank–Kamenetskii (1938, 1939).

An essential feature of the system considered in Figure 2.1 is that the gas mixture is capable of a chemical reaction with energy generation. The system is thermodynamically open, that is, energy transfer may occur between the system and its surroundings through the system boundary. Therefore, energy generated by chemical transformations within the system dissipates into an ambient environment via one or several possible mechanisms (conductive, convective, or radiative modes of heat exchange).

Although the system is at ambient temperature initially, this state is generally not the state of thermal equilibrium. The system may not remain at such temperature indefinitely due to the positive heat generation rate associated with a positive chemical reaction rate at an initial temperature. The evolution of the system in time is determined by dynamic balance between the rates of energy generation, energy dissipation, and the amount of energy contained within the system itself at a particular point in time. The system may achieve the state of thermal equilibrium, with a very low value of the reaction rate, or progress rapidly (thermal explosion) to such temperatures that support intensive combustion.

Several assumptions are made in this chapter before proceeding to the formal mathematical analysis of the system presented in Figure 2.1.

A mixture is considered a quiescent medium. The thermophysical properties of the mixture are assumed to be constant, as well as the ambient temperature.

Justification of the assumption regarding thermophysical properties has roots in the fact that the thermal explosion theory is primarily concerned with the question of the principal possibility of occurrence of thermal explosion. This question is usually answered upon analysis of the development of the system in the vicinity of its initial point, that is, in the region of phase space where the influence of the temperature dependence on thermophysical properties is minimal.

Thermal Explosion: Theory and Application, First Edition. Vasily B. Novozhilov.

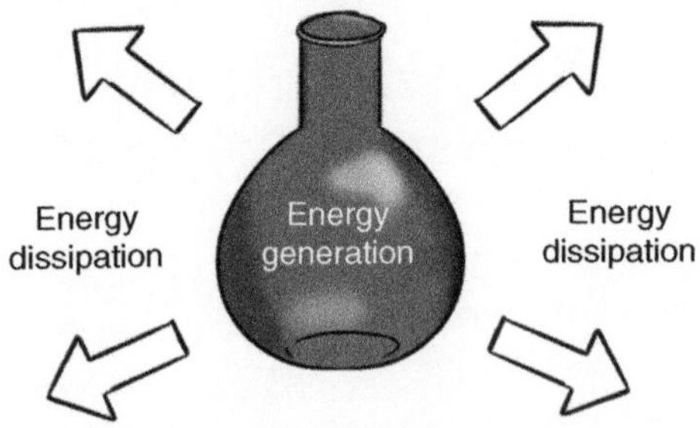

Figure 2.1 Energy evolution in a closed-volume reactive system.

Further, the consumption of reactants is neglected (that is, the reaction rate does not depend on the concentration of the reactants; formally, this corresponds to reaction of the zeroth order $m = 0$). This assumption is justified by the fact that the dependence of the reaction rate on the concentration of the reactants is much weaker than its dependence on temperature (power law compared to Arrhenius function). Investigation of kinetic effects is left to Chapter 4. For a similar reason, the temperature dependence of the pre-exponent is also neglected (i.e. $\alpha = 0$, Equation (1.6), Chapter 1). As follows from the general kinetic theory (Kuo 1986) (see Chapter 1), this dependence follows the power law with the exponent typically between zero and one. This dependence is, therefore, much weaker than the extremely sharp dependence of the Arrhenius function on temperature.

The above neglections are consistent with the general necessary conditions for thermal explosion (inequalities (1.10) and (1.12), Chapter 1), that is, a strong rise in the reaction rate over the characteristic temperature interval, and small consumption of reactants in the course of thermal explosion development.

Overall, all the assumptions single out the most essential features of thermal explosion, the features without which this phenomenon cannot exist. At the same time, they provide a clear and concise mathematical formulation of the problem, which admits relatively simple analysis. This formulation provides the necessary basis on which the investigation of more complicated effects in the thermal explosion theory can be carried out.

Consider now the mathematical formulation of the energy balance for a closed-volume reactive system presented in Figure 2.1. The system is at the initial temperature, $\tilde{T}_0$, and is immersed in the ambient environment with the same temperature.

The two different approaches (owing to Semenov and Frank–Kamenetskii, respectively) to the description of the system in Figure 2.1 will be presented below.

The most basic, Semenov approach assumes spatially uniform temperature $\tilde{T}$ within the vessel. This is a zero-dimensional model, where temperature changes as a function of time only.

The second, more accurate, Frank–Kamenetskii approach considers a three-dimensional (in general case) distributed system with both spatial and time dependence of the system temperature.

By recalling the fundamentals of chemical kinetics presented in Section 1.5 (the Arrhenius dependence of reaction rate on temperature, specifically), the energy rate balance for the Semenov (zero-dimensional) model is written as

$$\tilde{\rho}\tilde{c}_p \frac{d\tilde{T}}{d\tilde{t}} = \tilde{Q}\tilde{A} \exp\left(-\frac{\tilde{E}}{\tilde{R}\tilde{T}}\right) - \tilde{h}\frac{\tilde{S}}{\tilde{V}}(\tilde{T} - \tilde{T}_0) \tag{2.1}$$

Equation (2.1) is a typical energy balance equation of mathematical physics, where the left-hand side (LHS) is the rate of change of the volumetric energy density within the system, and the first term on the right-hand side (RHS) describes chemical source contribution, while the second term on RHS is the energy dissipation rate to surroundings by conduction. It occurs through the surface area of the vessel $\tilde{S}$ and is controlled by the constant (in the present model) heat transfer coefficient.

The following non-dimensional variables are introduced:

$$\theta = \frac{(\tilde{T} - \tilde{T}_0)}{\left(\tilde{R}\tilde{T}_0^2/\tilde{E}\right)}; \quad \tau = \frac{\tilde{Q}\tilde{k}_0\tilde{E}}{\tilde{\rho}\tilde{c}_p\tilde{R}\tilde{T}_0^2}\tilde{t}$$
$$\mathrm{Ar} = \frac{\tilde{R}\tilde{T}_0}{\tilde{E}}; \quad \mathrm{Se} = \frac{\tilde{Q}\tilde{k}_0\tilde{V}\tilde{E}}{\tilde{h}\tilde{S}\tilde{R}\tilde{T}_0^2} \tag{2.2}$$

Here, the excess temperature is scaled by the characteristic temperature interval; *Ar* is the Arrhenius number. The physical meaning of both the latter parameters has been discussed in Chapter 1.

The adopted time scale $\frac{\tilde{\rho}\tilde{c}_p\tilde{R}\tilde{T}_0^2}{\tilde{Q}\tilde{k}_0\tilde{E}}$ is called an adiabatic time scale (Barzykin 2004) since it does not involve any heat transfer parameters (thermal conductivity or heat transfer coefficient) and emerges naturally in the adiabatic version of Equation (2.1) (with the last term on the RHS dropped).

The parameter *Se* is called the Semenov number, or Semenov parameter, and plays a fundamental role in the thermal explosion theory.

By making use of the following evident identities

$$\tilde{T} = \tilde{T}_0 + \left(\frac{\tilde{R}\tilde{T}_0^2}{\tilde{E}}\right)\theta = \tilde{T}_0 + \mathrm{Ar}\tilde{T}_0\theta = (1 + \mathrm{Ar}\theta)\tilde{T}_0$$
$$\tilde{A}\exp\left(-\frac{\tilde{E}}{\tilde{R}\tilde{T}}\right) = \tilde{A}\exp\left(-\frac{\tilde{E}}{\tilde{R}}\frac{1}{(1 + \mathrm{Ar}\theta)\tilde{T}_0}\right)$$
$$= \tilde{A}\exp\left(-\frac{1}{\mathrm{Ar}}\right)\exp\left(\frac{1}{1 + \mathrm{Ar}\theta}\right) = \tilde{k}_0\exp\left(\frac{1}{1 + \mathrm{Ar}\theta}\right) \tag{2.3}$$

where $\tilde{k}_0 = \tilde{A}\exp\left(-\frac{1}{\mathrm{Ar}}\right)$ is (in the view of the above-discussed assumption $\tilde{A} \equiv \tilde{A}_0$) the reaction rate constant at the initial temperature ($\theta = 0$), the following non-dimensional energy rate balance equation

$$\frac{d\theta}{d\tau} = \exp\left(\frac{\theta}{1 + \mathrm{Ar}\theta}\right) - \frac{1}{\mathrm{Se}}\theta \tag{2.4}$$

is obtained from Eq. (2.1).

Therefore, according to the Semenov model, the evolution of the system is governed by the first-order, autonomous ordinary differential equation (ODE)

$$\frac{d\theta}{d\tau} = \dot{G} - \dot{L} = F(\theta) \tag{2.5}$$

with the RHS given by the following function of the non-dimensional temperature

$$F(\theta) = \exp\left(\frac{\theta}{1 + \mathrm{Ar}\theta}\right) - \frac{1}{\mathrm{Se}}\theta \tag{2.6}$$

The function $F(\theta)$, the difference between the rates of heat generation $\dot{G}$ and heat dissipation $\dot{L}$, will be called the energy generation–dissipation (EGD) function. This terminology is consistent with the visualization of the process in Figure 2.1 and reflects the core feature of the thermal explosion phenomenon, which is competition between the rates of energy generation and dissipation.

EGD function, the difference between the rates of energy generation and dissipation, will assume different forms in different problems discussed throughout the book, and generally may also be a function of time.

As discussed earlier, the present chapter considers the simplest initial condition for Equations (2.5) and (2.6)

$$\theta(0) = 0; \quad \tau \geq 0 \tag{2.7}$$

Generally, the initial condition is set as $\theta(0) = \theta_{in}$. All the results, presented in this chapter, are easily generalizable to the case $\theta_{in} \neq 0$ of the initial temperature of the system, which is different from the ambient.

Several plots of the EGD function (2.6), for different values of the Semenov parameter, are presented in Figure 2.2.

It should be noted that the EGD function has an inflection point at $\theta_{\text{inf}\,l} = \dfrac{1}{\text{Ar}}\left(\dfrac{1}{2\text{Ar}} - 1\right)$ and approaches asymptotically $-\infty$ as the temperature θ grows to infinity (the first term is bounded at $+\infty$ by $\exp\left(\dfrac{1}{\text{Ar}}\right)$).

Frank–Kamenetskii formulation differs from Equation (2.1) by the diffusion (heat dissipation) term and replacement of the heat conduction term by the boundary condition. Therefore, temperature generally varies spatially in the system, and consequently, so does the reaction rate. These modifications result in the following distributed (i.e. taking spatial variations of temperature into account) model

$$\tilde{\rho}\tilde{c}_p \frac{\partial \tilde{T}}{\partial \tilde{t}} = \tilde{\lambda}\tilde{\nabla}^2\tilde{T} + \tilde{Q}\tilde{A}\exp\left(-\frac{\tilde{E}}{\tilde{R}\tilde{T}}\right) \tag{2.8}$$

In contrast to the Semenov ODE formulation, the Frank–Kamenetskii model takes the form of a partial differential equation (PDE).

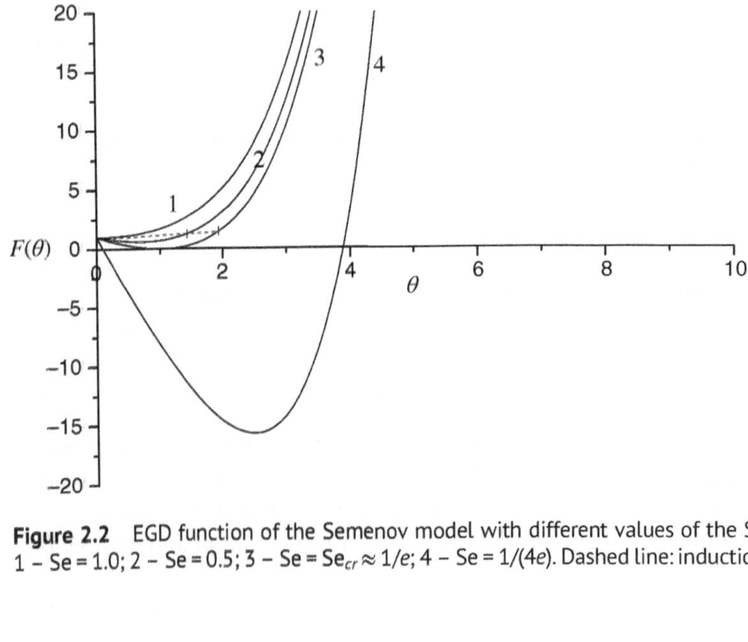

Figure 2.2 EGD function of the Semenov model with different values of the Semenov parameter. Ar = 0.01. 1 – Se = 1.0; 2 – Se = 0.5; 3 – Se = $\text{Se}_{cr} \approx 1/e$; 4 – Se = 1/(4e). Dashed line: induction period span over phase space.

Introducing non-dimensional coordinate in a three-dimensional space as $\vec{\xi} = \frac{\vec{\tilde{x}}}{\tilde{r}}$ (where $\tilde{r}$ is the characteristic size of the vessel), and using again the non-dimensional parameters (2.2), supplemented by one additional parameter

$$\delta = \frac{\tilde{Q}\tilde{k}_0\tilde{E}\tilde{r}^2}{\tilde{\lambda}\tilde{R}\tilde{T}_0^2} \tag{2.9}$$

Equation (2.8) is being transformed to the following form:

$$\frac{\partial\theta}{\partial\tau} = \frac{1}{\delta}\nabla^2\theta + \exp\left(\frac{\theta}{1 + \text{Ar}\theta}\right) \tag{2.10}$$

The parameter δ, emerging in consideration of a distributed model, is called the Frank–Kamenetskii parameter.

It may be worthwhile to note that the two formulations, written in the form of Equations (2.4) and (2.10), possess a certain type of 'symmetry' in a sense that the inverse of the relevant parameter (Semenov or Frank–Kamenetskii, respectively) appear at the front of the terms describing energy dissipation in the system.

Again, the simplest initial and first-type (Dirichlet) boundary conditions are considered in the present chapter

$$\theta\left(0, \vec{\xi}\right) \equiv 0; \quad \theta|_{\partial\Omega} = \theta_s\left(\vec{\xi}_s, \tau\right) = 0; \quad \tau \geq 0 \tag{2.11}$$

where $\partial\Omega$ is the boundary (i.e. the surface of the vessel) of the three-dimensional region Ω.

Both the initial and the first-type boundary conditions may also be non-trivial, so that $\theta\left(0, \vec{\xi}\right) = \theta_{in}\left(\vec{\xi}\right)$ and $\theta|_{\partial\Omega} = \theta_s\left(\vec{\xi}_s, \tau\right)$ are not identically zero.

In general case, other types of boundary conditions are possible (Vladimirov 1971; Volpert and Khudyaev 1975). Such may be the cases of the second-type (Neumann), mixed type (Dirichlet/Neumann conditions on disjoint parts of the boundary), and third-type (linear combination of the Dirichlet and Neumann) conditions. From the point of view of physics of real combustion systems, the latter two types of conditions have much less significance.

Neumann boundary conditions are considered in Chapter 3. The same chapter considers, also, a very specific and important type of boundary conditions, that is, oscillating boundary conditions, of both the Dirichlet and Neumann types.

In accordance with the definition of thermal explosion, presented in Chapter 1, the two methods of investigating thermal explosion possibility in a specific system can be considered.

First, thermal explosion may be considered as impossibility for the system to achieve a state of equilibrium in a certain region of phase space (at low, far below normal combustion, temperatures). In this case, a set of states of equilibrium of the system needs to be described and investigated. If this set turns out to be empty, thermal explosion may occur. Such an approach leads to the steady-state theory of thermal explosion.

Alternatively, the system evolution in time may be followed, in order to see whether it progresses until, at least, the point where the second derivative of the Arrhenius function reaches its maximum. This approach leads to the non-steady theory.

It may be strictly proven that, within the assumptions of the present chapter (in particular, the neglect of chemical kinetics and trivial initial condition), both the steady-state and unsteady approaches are equivalent, that is, they predict the same critical conditions for thermal explosion. This point is discussed in more detail at the end of Section 2.3.

In both the above approaches, the occurence or the impossibility of thermal explosion generally depends on the system control parameters. Transition between these two possibilities occurs at the so-called critical conditions (or critical boundary), which are described by the manifold in the space of control parameters. Traverse of the critical boundary in the space of control parameters changes the scenarios of the system evolution from attaining equilibrium to explosion, and vice versa.

The steady-state theory is considered in Sections 2.2 and 2.3, and the non-steady theory is considered in Section 2.4.

Consider the steady-state theory first. Mathematically, this analysis amounts to investigating the existence of steady-state solutions of the respective (the Semenov or the Frank–Kamenetskii) evolution equations.

2.2 Steady-State Semenov Theory

The points of the phase space, where the RHS $F(\theta)$ turns into zero, and only such points (Arnold 1992) are the equilibrium points of Equation (2.5). These points correspond to the steady-state solutions of the Semenov equation.

Therefore, the locations of equilibrium points are determined by the condition

$$F(\theta) = 0 \tag{2.12}$$

or

$$\exp\left(\frac{\theta}{1 + \mathrm{Ar}\theta}\right) - \frac{1}{\mathrm{Se}}\theta = 0 \tag{2.13}$$

(for $\mathrm{Se} > 0$).

It is clear that at least one such point exists in the view of the behaviour of the function $F(\theta)$ at $\theta \to +\infty$, discussed earlier. This, the largest, value of the equilibrium temperature is located far beyond the inflection point of the Arrhenius function (and thus does not correspond to any realistic combustion temperature). This point is, therefore, irrelevant and is excluded from the following discussion in this chapter. It may be ascribed physical meaning upon taking into account consumption of reactants (Chapters 4 and 5).

In the vicinity of the initial temperature, there can exist zero, one, or two additional roots of Equation (2.13). These possibilities are presented in Figure 2.3.

Starting from the initial (ambient) temperature $\theta(0) = 0$, the system evolution may be controlled by EGD functions of one of the three types, corresponding to the curves 1, 2 or 3 (Figure 2.3), respectively. Sign of $F(\theta)$ determines the direction of the system evolution. Similarly, such evolution may be easily followed for any initial condition $\theta(0) = \theta_{in} \neq 0$. Further, signs of the derivative $dF(\theta)/d\theta$ in the vicinity of each of the equilibrium points determine the stability of each of the steady-state solutions. As follows trivially from the consideration of the latter derivative, the equilibrium point with the least value is asymptotically stable, while the next two points (in the order of increased values) are unstable, including the point where the curves $F(\theta)$ and $\theta = 0$ are tangent. This point corresponds to the curve 2, Figure 2.3.

System control parameters corresponding to the latter point have a particular significance. Upon small variation of these parameters, the state of the system changes between subcritical and supercritical. In the former case, the system evolves towards thermal equilibrium (the case

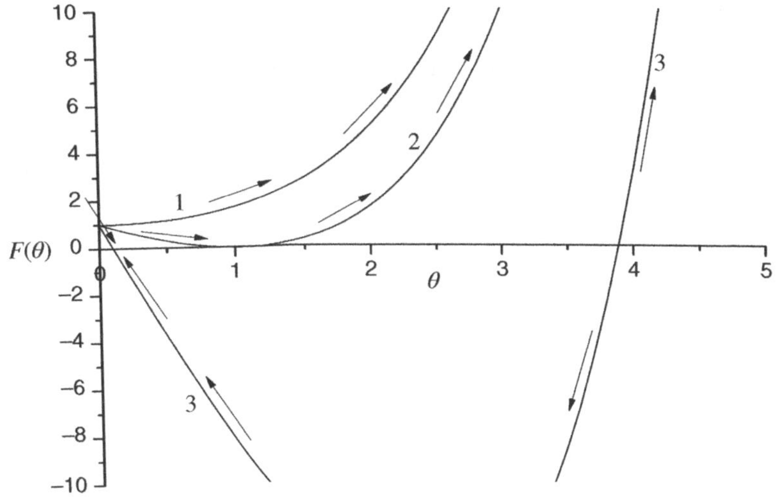

Figure 2.3 EGD function in the vicinity of equilibrium points and the system evolution. Ar = 0.01. 1 – Se = 1.0; 2 – Se = $Se_{cr} \approx 1/e$; 3 – $Se = 1/(4e)$. Arrows – directions of system evolution in different regions of phase space.

of the curve 3, Figure 2.3), in the latter case, thermal explosion develops (the case of the curve 1, Figure 2.3).

Parameters corresponding to the curve 2, Figure 2.3, are therefore critical parameters of thermal explosion and describe the critical boundary between different regimes of the system behaviour.

Semenov formulation contains the only control parameter, that is, the Semenov number Se. Explosion critical conditions must, therefore, be described in terms of this parameter.

It is remarkable that critical conditions in the Semenov model are related to and can be described in terms of quite a general pure mathematical construction called the Legendre transformation (Arnold 1988). This fact has never been noticed before. This procedure (Figure 2.4) generates, from the arbitrary continuous convex function $f(x)$, another function $g(p)$, called the Legendre transformation of $f(x)$. (The Legendre transformation may be defined under conditions for the function $f(x)$ less restrictive, than those given above. The conditions specified here are sufficient for the present

Figure 2.4 Legendre transformation.

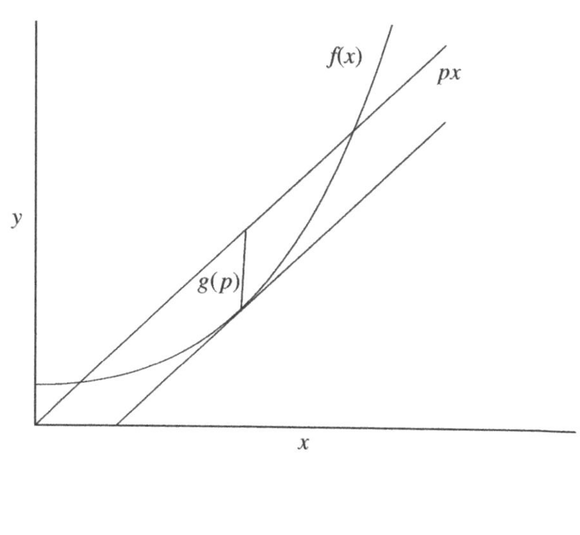

discussion in the view of the fact that EGD function of the Semenov model is continuous and convex.)

Specifically, for the function $f(x)$ and the line with the slope p (Figure 2.4), the function $g(p)$ is given by

$$g(p) = \sup_x (px - f(x)) \tag{2.14}$$

If the function $f(x)$ is strictly convex (i.e. $d^2f(x)/dx^2 > 0$) and its derivative defines a diffeomorphism mapping $\mathbb{R}^1 \to \mathbb{R}^1$, then the Legendre transformation $g(p)$ is also strictly convex and the supremum is achieved at a unique point where $df(x)/dx = p$ (Arnold 1988).

Given the fact that the EGD function of the Semenov model is written as a difference between the Arrhenius (convex) function and the linear function of temperature, it is straightforward to prove (and understand by comparing Figures 2.3 and 2.4) that in terms of the Legendre transform the critical condition of thermal explosion in the Semenov model is given by

$$g\left(\frac{1}{\mathrm{Se}}\right) = 0 \tag{2.15}$$

The requirement (2.15) translates into the following set of equations:

$$\begin{aligned} &\frac{dF}{d\theta}(\theta_{cr}) = 0 \\ &F(\theta_{cr}) = 0 \end{aligned} \tag{2.16}$$

for the unknown temperature θ_{cr}, called the critical temperature.

Since Equation (2.16) contain the Semenov number Se as a parameter (the value of the Arrhenius number Ar is assumed to be fixed), this is effectively the set of two equations for the two unknowns θ_{cr} and Se_{cr}, which must be satisfied simultaneously. The latter value, Se_{cr}, once found, provides the description of the critical conditions in terms of the control parameter, the Semenov number.

It appears that the critical value Se_{cr} is found from Equation (2.16) uniquely.

The set of Equation (2.16) is re-written as

$$\begin{aligned} &\frac{1}{(1+\mathrm{Ar}\theta_{cr})^2}\exp\left(\frac{\theta_{cr}}{1+\mathrm{Ar}\theta_{cr}}\right) - \frac{1}{\mathrm{Se}_{cr}} = 0 \\ &\exp\left(\frac{\theta_{cr}}{1+\mathrm{Ar}\theta_{cr}}\right) - \frac{1}{\mathrm{Se}_{cr}}\theta_{cr} = 0 \end{aligned} \tag{2.17}$$

from where

$$\frac{\theta_{cr}}{(1+\mathrm{Ar}\theta_{cr})^2} = 1 \tag{2.18}$$

It is evident that in the limit $\mathrm{Ar} \to 0$

$$\theta_{cr} \to 1; \ \mathrm{Se}_{cr} \to \frac{1}{e} \tag{2.19}$$

This limit is very important in view of small values of Arrhenius number in real combustion systems (see an estimation in Chapter 1). Note that Equation (2.4) cannot be formally considered for $\mathrm{Ar} = 0$, since at this value the non-dimensional temperature definition (2.2) is not applicable.

Therefore, the correct estimation of the critical parameter Se_{cr} at very small values of the Arrhenius number requires the consideration of the mathematical limit $Ar \to 0$.

In precise terms, for any finite value $Ar > 0$, Equation (2.18) has the two solutions

$$\theta_{cr}^{(1)} = \frac{1 - 2Ar - \sqrt{1 - 4Ar}}{2Ar^2}; \quad \theta_{cr}^{(2)} = \frac{1 - 2Ar + \sqrt{1 - 4Ar}}{2Ar^2} \tag{2.20}$$

and correspondingly, the set of Equations (2.17) is solvable for any $Ar > 0$.

The largest of these two roots corresponds to conditions beyond the inflection point of $F(\theta)$, and as such to unrealistically high temperatures (well above common temperatures of combustion). Therefore,

$$\theta_{cr} = \frac{1 - 2Ar - \sqrt{1 - 4Ar}}{2Ar^2}; \quad Se_{cr} = \frac{(1 + Ar\theta_{cr})^2}{\exp\left(\dfrac{\theta_{cr}}{1 + Ar\theta_{cr}}\right)} \tag{2.21}$$

Correction to conditions (2.21) in the limit $Ar \to 0$ may be easily found upon the expansion of the formula (2.21) for θ_{cr} with respect to Ar for $Ar \ll 1$

$$\theta_{cr} = 1 + 2Ar + 5Ar^2 + o\left(Ar^2\right) \tag{2.22}$$

In the first, linear, approximation

$$\theta_{cr} = 1 + 2Ar; \quad Se_{cr} = \frac{1 + 2Ar}{\exp\left(\dfrac{1 + 2Ar}{1 + Ar}\right)} \tag{2.23}$$

This correction is very small; for $Ar = 0.02$, the relative difference with the estimated ($Ar \to 0$) value $Se_{cr} \approx 1/e$ is below 2%.

It should be pointed out that the result $\theta_{cr} = 1 + o(Ar)$ (i.e. $\theta_{cr} \approx 1$ in the first approximation) has a particular significance. It shows that the pre-explosion heat-up is of order of just one characteristic temperature interval, which confirms the significance of inequality (1.8) as a necessary condition for thermal explosion.

The obtained Se_{cr} corresponds to the value of the Semenov parameter, where the curves $F(\theta)$ and $\theta = 0$ are tangent. It is easy to see that for smaller values $Se < Se_{cr}$, the EGD function has two roots (similar to curve 3, Figure 2.3), the least of which is a stable, low-temperature equilibrium point. On the other hand, for larger values $Se > Se_{cr}$, the EGD function has no roots for temperatures below the inflection point of the Arrhenius function (similar to curve 1, Figure 2.3), and consequently, thermal explosion develops. Therefore, the equality $Se = Se_{cr}$ indeed describes the explosion critical conditions.

2.3 Steady-State Frank–Kamenetskii Theory

The general, non-steady, formulation of the Frank–Kamenetskii model is given by Equation (2.10). As pointed out before, this model accounts for temperature distribution over the volume of the reacting system and its spatial evolution due to the heat conduction process.

Then, the steady-state form of Equation (2.10) is

$$\nabla^2\theta = -\delta \exp\left(\frac{\theta}{1 + Ar\theta}\right) \tag{2.24}$$

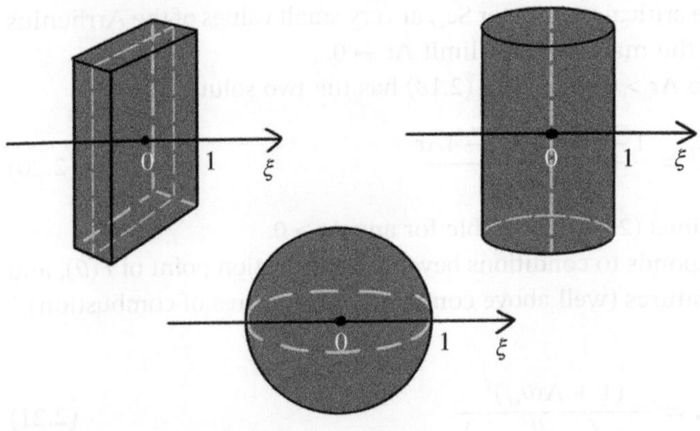

Figure 2.5 Symmetrical cases considered in the Frank–Kamenetskii steady-state theory: planar, cylindrical, and spherical symmetries.

The several cases where Equation (2.24) may be investigated analytically, and which have been considered in the classical theory of thermal explosion, are those of reacting systems possessing certain types of symmetry.

These three cases are shown schematically in Figure 2.5. These are the cases of planar, cylindrical, and spherical symmetries.

In the first case (left upper corner of Figure 2.5), the medium between the two infinite parallel walls, placed at the distance $2\tilde{r}$ apart, is considered. Considering the symmetry plane as parallel to the walls and located half-way between them, temperature distribution is assumed to be invariant with respect to in-plane reflection and any translations parallel to the plane. Therefore, the temperature varies in the direction normal to the walls only, and it is sufficient to know its distribution between the symmetry plane and the wall. In terms of the non-dimensional coordinate system introduced in Figure 2.5 (with its origin located at the symmetry plane), the temperature variation $\theta(\xi)$ must be considered on the interval $\xi \in [0, 1]$.

The second is the case of the medium inside the infinite cylinder (right upper corner of Figure 2.5), with the temperature distribution invariant with respect to rotations around the vertical axis of symmetry of the cylinder and translations parallel to that axis. The origin of the coordinate system is located at the symmetry axis, again the temperature distribution has to be found on $\xi \in [0, 1]$.

The third case (medium inside the sphere, bottom of Figure 2.5) possesses temperature distribution invariance with respect to any rotations around the sphere's centre. Consequently, the temperature changes along the radial direction only. The centre of the coordinate system is at the sphere's centre, the temperature distribution on $\xi \in [0, 1]$ is considered.

Using explicit expression for the Laplacian, Equation (2.24) is re-written for the three discussed symmetry cases as the one-dimensional equation

$$\frac{d^2\theta}{d\xi^2} + \frac{m}{\xi}\frac{d\theta}{d\xi} = -\delta \exp\left(\frac{\theta}{1 + \mathrm{Ar}\theta}\right) \tag{2.25}$$

where m is the symmetry index ($m = 0$ for planar symmetry; 1 for cylindrical symmetry, and 2 for spherical symmetry); $0 \leq \xi \leq 1$.

In the following analysis of the Frank–Kamenetskii model (both the steady-state in the present section and the non-steady in Section 2.4), the so-called Frank–Kamenetskii exponent transformation will be used.

The Frank–Kamenetskii transformation is being performed (Frank–Kamenetskii 1938, 1939, 2016) by expanding the argument in the Arrhenius factor $\exp\left(\frac{\theta}{1+\mathrm{Ar}\theta}\right)$

$$\frac{\theta}{(1+\mathrm{Ar}\theta)} = \theta(1 - \mathrm{Ar}\theta + o(\mathrm{Ar}\theta)) \tag{2.26}$$

for $\mathrm{Ar}\theta \ll 1$, and retaining the leading term. The result of such an operation is the modification of the Arrhenius factor $\exp\left(\frac{\theta}{1+\mathrm{Ar}\theta}\right) \mapsto \exp(\theta)$.

As can be seen from Figures 2.2 and 2.3, traverse of the critical boundary occurs in the region, where the inequality $\mathrm{Ar}\theta \ll 1$ holds (the typical values of Arrhenius number are $\mathrm{Ar} = 1.2 \cdot 10^{-2}$ – $2.0 \cdot 10^{-2}$, see Chapter 1). This fact makes the Frank–Kamenetskii transformation a quite powerful tool for investigating the behaviour of the system in the vicinity of its initial temperature.

As this transformation removes the unknown temperature from the denominator, the non-linear differential equation (2.25) becomes much more tractable. Indeed, it will be demonstrated below that the exact solutions of Equation (2.25) can be obtained at least for the two cases, the planar and the cylindrical symmetry, respectively.

Formally, the Frank–Kamenetskii transformation corresponds to setting $\mathrm{Ar} = 0$ in the Arrhenius factor. This correspondence is due to the trivial fact that $\mathrm{Ar}\theta \ll 1$ for $\mathrm{Ar} = 0$. However, due to the same reason mentioned already in Section 2.2, Equations (2.24) and (2.25) cannot be considered for $\mathrm{Ar} = 0$, as definition of the non-dimensional temperature requires $\mathrm{Ar} > 0$. On the other hand, expansion of the argument on the RHS of the respective equation, and retaining a leading term is a valid mathematical operation. It modifies the Arrhenius factor; however, the correction is rather small. In the vicinity of one characteristic temperature interval around the temperature where the expansion is being made, the relative error of the Frank–Kamenetskii modification does not exceed 5% (Zeldovich *et al.* 1980).

Accuracy of this transformation was also verified by Parks (1961) using the numerical simulations of thermal explosion problem with the exact form of the Arrhenius function. He found the accuracy of the Frank–Kamenetskii transformation very satisfactory for the values of the Arrhenius number $\mathrm{Ar} < 5 \cdot 10^{-2}$. Again, according to the estimation presented in Chapter 1, this inequality is fulfilled for most combustion systems.

Frank–Kamenetskii transformation is the universal approach in the combustion theory which, in some cases, allows either exact analytical solutions to be obtained or values of certain integrals containing the Arrhenius exponent to be estimated. It is being used, for example, in the Zeldovich–Frank–Kamenetskii theory of the laminar flame spread (Zeldovich *et al.* 1980) and the Zeldovich–Novozhilov (ZN) theory of solid propellant combustion (Novozhilov and Novozhilov 2020). It should be noted, however, that the choice of the base temperature, which is being used as a point of expansion in the Frank–Kamenetskii transformation method, is different depending on a particular application. In contrast to thermal explosion problem, where this temperature is an initial temperature of the mixture, in the theory of laminar flame spread, as well as in the ZN theory; for example, this temperature is taken as an adiabatic combustion temperature.

The equation takes the form

$$\frac{d^2\theta}{d\xi^2} + \frac{m}{\xi}\frac{d\theta}{d\xi} = -\delta\exp(\theta) \tag{2.27}$$

$m = 0,1$, or 2; $0 \le \xi \le 1$.

Due to assumed symmetry, the boundary conditions that must be satisfied are

$$\frac{d\theta}{d\xi}(0) = 0; \quad \theta(1) = 0 \tag{2.28}$$

2.3.1 Planar Symmetry

General solution of Equation (2.27) for this case is

$$\theta = -\ln\left[\frac{\delta}{2C_1^2}ch^2(C_1\xi + C_2)\right] \tag{2.29}$$

(Polyanin and Zaitsev 2002), with $C_{1,2}$ being integration constants.

It is easy to check that the first of the boundary conditions (2.28) implies (since $C_1 \neq 0$) that

$$C_2 = 0 \tag{2.30}$$

Then, the following equation for C_1 is obtained:

$$ch(C_1) = \pm\sqrt{\frac{2}{\delta}}C_1 \tag{2.31}$$

Note that only positive values of C_1 need be considered, since at $C_2 = 0$ the transformation $C_1 \mapsto -C_1$ leaves the solution (2.29) invariant.

Taking the positive sign in (2.31), it is seen easily that the solution exists for the values $\delta \leq \delta_{cr}$ only, where the critical value of the Frank–Kamenetskii parameter is determined, in addition to (2.31), by equality of the tangents of the both sides of that equation

$$sh\left(C_1^{cr}\right) = \sqrt{\frac{2}{\delta_{cr}}} \tag{2.32}$$

or

$$C_1^{cr} = \text{Arsh}\left(\sqrt{\frac{2}{\delta_{cr}}}\right) \tag{2.33}$$

With the help of the basic hyperbolic functions identity, Equations (2.31) and (2.32) can be transformed to the single equation that determines δ_{cr}

$$\text{Arsh}\left(\sqrt{\frac{2}{\delta_{cr}}}\right) = \frac{\sqrt{1 + \frac{2}{\delta_{cr}}}}{\sqrt{\frac{2}{\delta_{cr}}}} \tag{2.34}$$

The solution of this equation gives $\delta_{cr} \approx 0.88$; the value was obtained first by Frank–Kamenetskii (1938, 1939, 2016).

For $\delta < \delta_{cr}$, Equation (2.31) has two solutions. Only the smaller of these two corresponds to the stable solution (2.29). This will be demonstrated in Chapter 3.

Various steady-state temperature profiles, including the critical one, are presented in Figure 2.6.

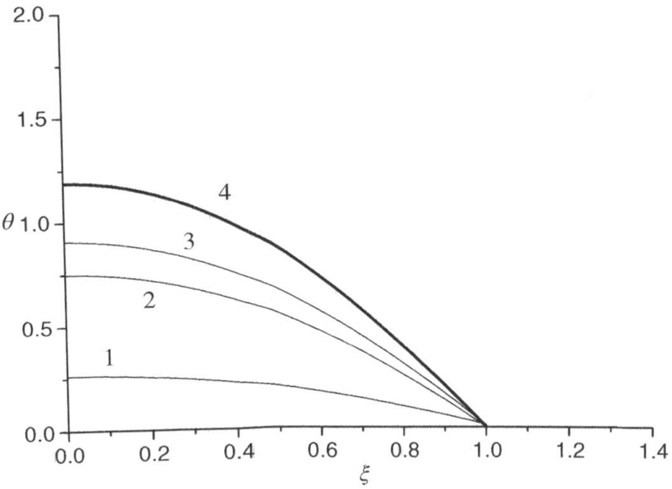

Figure 2.6 Steady-state temperature distributions. Frank–Kamenetskii model. Planar symmetry. Analytical solutions (2.29). 1 – $\delta = 0.40$; 2 – $\delta = 0.80$; 3 – $\delta = 0.84$; 4 – $\delta = \delta_{cr} \approx 0.88$.

2.3.2 Cylindrical Symmetry

The analytical solution for this case

$$\theta = \ln\left(\frac{8-b}{\delta}\right) - 2\ln\left(a\xi^{k_1} + \frac{1}{a}\xi^{k_2}\right) \tag{2.35}$$

where $k_{1,2}$ are the roots of the equation

$$k^2 - 2k + \frac{b}{8} = 0 \tag{2.36}$$

a,b are the integration constants, was given by Lemke (1913).

The restriction (2.36) implies

$$k_1 + k_2 = 2; \;\; k_1 k_2 = \frac{b}{8} \tag{2.37}$$

Finiteness of the solution (2.35) at $\xi = 0$ requires $k_{1,2}$ to be non-negative. Further, the first derivative of the solution contains terms proportional to $k_i\xi^{k_i-1}$; the second derivative contains terms proportional to $k_i(k_i-1)\xi^{k_i-2}$. This excludes the possibility that both k_1 and k_2 are positive, except the case $k_1 = k_2 = 1$. In this case, the symmetry boundary condition (2.28) cannot be satisfied, as the derivative

$$\frac{d\theta}{d\xi} = -2\frac{\left(ak_1\xi^{k_1-1} + \frac{1}{a}k_2\xi^{k_2-1}\right)}{\left(a\xi^{k_1} + \frac{1}{a}\xi^{k_2}\right)} \tag{2.38}$$

does not vanish at $\xi = 0$.

Therefore, the only remaining possibility (to within the renumbering of the roots) is

$$k_1 = 2; \;\; k_2 = b = 0 \tag{2.39}$$

The solution (2.35) turns into

$$\theta = \ln\left(\frac{8}{\delta}\right) - 2\ln\left(a\xi^2 + \frac{1}{a}\right) \tag{2.40}$$

Then, the first (the symmetry) boundary condition (2.28) is satisfied as

$$\frac{d\theta}{d\xi} = -\frac{4a\xi}{\left(a\xi^2 + \frac{1}{a}\right)} \tag{2.41}$$

The remaining constant of integration a is determined from the second of the boundary conditions (2.28), which turns into the equation

$$\left(a + \frac{1}{a}\right)^2 = \frac{8}{\delta} \tag{2.42}$$

It follows immediately that the critical value of the Frank–Kamenetskii parameter is ($a + \frac{1}{a} \geq 2$ for any $a > 0$)

$$\delta_{cr} = 2 \tag{2.43}$$

Solution exists if and only if $\delta < \delta_{cr}$. In this case, there are two solutions of the form (2.40) corresponding to

$$a_1 = \sqrt{\frac{2}{\delta}} + \sqrt{\frac{2}{\delta} - 1}; \quad a_2 = \sqrt{\frac{2}{\delta}} - \sqrt{\frac{2}{\delta} - 1} \tag{2.44}$$

As is the case for the planar symmetry, it will be demonstrated in Chapter 3 that only solution with the smaller value a_2 is stable.

2.3.3 Spherical Symmetry

Increasing the symmetry index in Equation (2.27) further, to $m = 2$ (spherical symmetry)

$$\frac{d^2\theta}{d\xi^2} + \frac{2}{\xi}\frac{d\theta}{d\xi} = -\delta\exp(\theta) \tag{2.45}$$

makes the analysis much less trivial.

Equation (2.45) is the so-called Emden equation (von Kamke 1977; Emden 1907; Lemke 1913), which has also being used in the theory of polytropic gas spheres (e.g. stars). There is no known analytical solution for this case.

Critical conditions are established by the following analysis.

First, the following change of variables (Barzykin and Merzhanov 1958)

$$x = \xi^2\exp(\theta); \quad y = \xi\frac{d\theta}{d\xi} \tag{2.46}$$

is performed.

This transformation turns Equation (2.45) into

$$\frac{dy}{dx} = -\frac{(y + \delta x)}{(y + 2)x} \tag{2.47}$$

Any solution of Equation (2.45) on $\xi \in [0, 1]$ with the boundary conditions (2.28) (if exists) maps, due to the new variables definition (2.46), onto the integral curve of Equation (2.47) exiting from the point (0, 0) and ending at the line $x = 1$.

Integral curves (for different values of the parameter δ) of Equation (2.47) with the initial condition $y(0) = 0$ behave in the following manner for $x > 0$. They have infinite derivative on the line $y = -2$, and zero derivative on the line $y = -\delta x$. Integral curve is spiralling onto the intersection of these two lines, that is the point $\left(\frac{2}{\delta}, -2\right)$. The spiralling nature of integral curves for the Emden equation is discussed in more detail by von Kamke (1977), Emden (1907), and Lemke (1913).

Due to the first of the formulae (2.46), in order for the outer (at $\xi = 1$) boundary condition (2.28) to be satisfied, the integral curve must reach the line $x = 1$, and the variable mapping (2.46) $\xi \mapsto x: [0, 1] \to [0, 1]$, must be such that $\xi(1) = 1$.

Conversely, if integral curve of Equation (2.47) intersects with the line $x = 1$, then the mapping $\xi \mapsto x: [0, 1] \to [0, 1]$ may be constructed in such a way that the further mapping $\xi \mapsto x \mapsto y(x)$, where $y(x)$ satisfies (2.47), determines the curve $\left(\theta(\xi), \frac{d\theta}{d\xi}\right)$ in the parametric form along $\xi \in [0, 1]$.

It is proved by direct calculation (using (2.47)) that along such curve $\frac{d\theta}{d\xi}$ and $\theta(\xi)$ obey the relation (2.45) and, therefore, a solution of Equation (2.45) on $\xi \in [0, 1]$ is constructed.

This solution must also satisfy the boundary conditions (2.28). This is achieved in the following way.

First, setting the requirement $\xi(1) = 1$ for the mapping $\xi \in [0, 1] \to x \in [0, 1]$ forces, due to (2.46), $\theta(1) = 0$. Therefore, the outer boundary condition (2.28) is satisfied.

Local mapping behaviour at the other end of the interval, $\xi = 0$, must force the symmetry boundary condition to be fulfilled.

Since, from (2.46),

$$\xi = \frac{\sqrt{x}}{\sqrt{\exp(\theta)}} \tag{2.48}$$

then any mapping $\xi \in [0, 1] \to x \in [0, 1]$ such that $\frac{x}{\xi^2}$ has finite limit at $\xi \to 0$ results in $\theta(\xi)$ being finite at $\xi \to 0$, and the solution can be continuously extended to a finite value $\theta(0)$. Therefore, upon such process, bounded solution of Equation (2.45) is constructed on the whole interval [0, 1].

The boundary value $\frac{d\theta}{d\xi}(0)$ of such a solution depends, in view of the second formula (2.46), on the asymptotic behaviour of $\frac{y}{\sqrt{x}}$ at $x, y \to 0$ upon approaching the solution starting point (0, 0).

At $y \to 0$ $y(x)$ approaches the solution of the equation

$$\frac{dy}{dx} = -\frac{1}{2}\left(\frac{y}{x} + \delta\right) \tag{2.49}$$

The latter has the solution $y(x) = -\frac{\delta}{3}x$ with the initial condition $y(0) = 0$. For this solution $\frac{y(x)}{\sqrt{x}} = -\frac{\delta}{3}\sqrt{x} \xrightarrow[x \to 0]{} 0.$

Therefore,

$$\frac{d\theta}{d\xi} = \frac{y}{\xi} = \sqrt{\exp(\theta(x))}\frac{y}{\sqrt{x}} \underset{\xi \to 0}{\longrightarrow} 0 \tag{2.50}$$

since the finiteness of $\theta(x)$ has been ensured at the previous construction step.

The above considerations show that solution of Equation (2.46) with the boundary conditions (2.28) (steady-state temperature distribution in the spherically symmetrical case) exists whenever the integral curve of Equation (2.47) with the initial condition $y(0) = 0$ intersects the line $x = 1$. The question now is when, and how many such intersections exist, depending on the value of the parameter δ.

First of all, it is clear that the explosion critical condition corresponds to the situation when the integral curve is tangent with the line $x = 1$ (one intersection), while no intersections occur at larger values of δ.

This case is illustrated in Figure 2.7, where such a critical integral curve is shown. The critical value of the parameter δ has to be found by numerical integration of Equation (2.45) and turns out to be $\delta_{cr} \approx 3.32$.

For the values $\delta > \delta_{cr}$, integral curves lie below the critical one and do not cross the line $x = 1$.

For the values $\delta < \delta_{cr}$, multiple intersections occur, and their number may be different. Therefore, the solution of the BVP (2.45) and (2.28) is not unique. An example is demonstrated in Figure 2.8 for $\delta = 2$. This figure shows a beginning segment of the integral curve. For each of the points B of intersection between the integral curve and the line $x = 1$ a separate solution of the problem (2.45) and (2.28) can be constructed. In fact, for this particular case, the number of such intersections and correspondingly the number of solutions are infinite.

The identification of the stable that is physically realizable solutions within the whole set of formal solutions is carried out in Chapter 3.

Critical values of the Frank–Kamenetskii parameter δ for the three symmetrical cases (planar, cylindrical, and spherical) are summarized in Table 2.1.

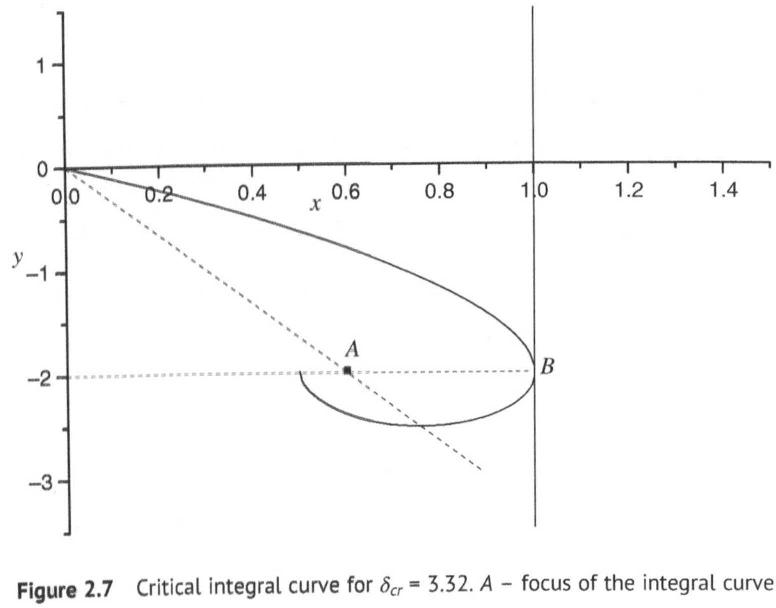

Figure 2.7 Critical integral curve for $\delta_{cr} = 3.32$. A – focus of the integral curve at the location $\left(\frac{2}{\delta}, -2\right)$. B – point of the intersection with the line $x = 1$.

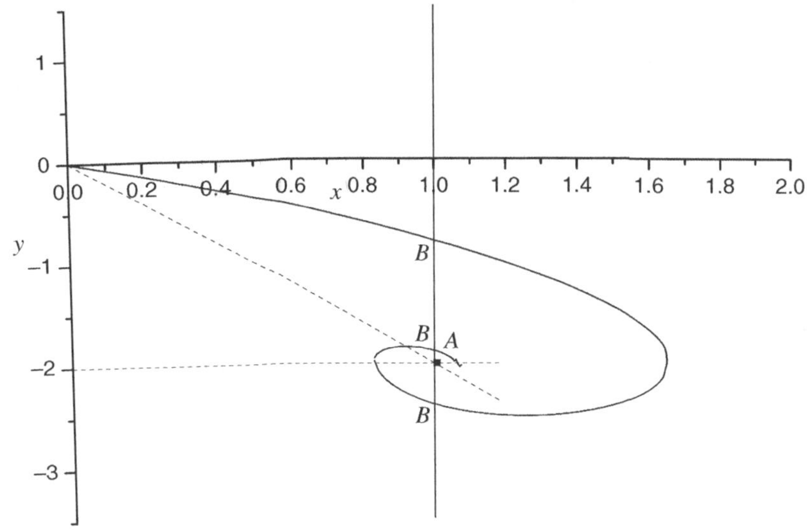

Figure 2.8 Critical integral curve for $\delta = 2.0$. A – focus of the integral curve at the location $\left(\frac{2}{\delta}, -2\right)$. B – various points of the intersection with the line $x = 1$.

Table 2.1 Critical values of the Frank–Kamenetskii parameter for different symmetrical cases.

Symmetry index, m	Critical Frank–Kamenetskii parameter, δ_{cr}
0	0.88
1	2
2	3.32

It may be noted that an alternative method of analysis of the symmetrical cases considered in Sections 2.3.1–2.3.3 may be developed using consideration of the envelope of the family of solutions corresponding to possible different temperature distributions in the system (Zeldovich *et al.* 1980).

An important note must be made at the conclusion of this section. Sections 2.3.1–2.3.3 established (in the Frank–Kamenetskii formulation, and for different types of symmetries) the values δ_{cr} such that positive (on the semi-open interval $0 \leq \xi < 1$) steady-state solution of the Frank–Kamenetskii equation exists for $\delta \leq \delta_{cr}$. *A 'priori* that does not mean, however, that thermal explosion occurs for $\delta > \delta_{cr}$. However, the latter fact is indeed true in the view of the following important theorem.

Theorem 2.1 (Volpert and Khudyaev 1975).
Solution of the Frank–Kamenetskii problems (2.10) and (2.11) (modified with the Frank–Kamenetskii transformation) is a growing (for any $0 \leq \xi < 1$) function of τ, which is bounded at $\tau \rightarrow \infty$ if and only if a positive (on $0 \leq \xi < 1$) solution of the BVP problem (2.27) exists.

Moreover, $\lim_{\tau \rightarrow \infty} \theta\left(\tau, \vec{\xi}\right)$ is the least positive solution of the problems (2.10) and (2.11).

As such, the absence of positive solutions of the BVP problem (2.27) (for $\delta > \delta_{cr}$) implies infinite growth of temperature, that is, thermal explosion.

The theorem also implies the equivalence of the steady-state and non-steady Frank–Kamenetskii approaches of investigating critical conditions. The critical conditions, predicted by both methods, are identical.

Identical theorem and equivalence hold also for the Semenov model.

2.4 Non-steady Theory

An alternative (and under certain restrictions equivalent) approach to investigating behaviour of the system (evolution towards stable equilibrium or developing thermal explosion) is a solution of non-steady governing equations, such as Equation (2.4) (Semenov model) or Equation (2.10) (Frank–Kamenetskii model).

Evolution of the system temperature for the simplest, the Semenov case is presented in Figure 2.9. Equation (2.4) is being solved numerically with the trivial initial condition $\theta(0) = 0$. Correspondingly, the system evolution trajectories start from the origin of the coordinate system and evolve according to the two scenarios: they either grow rapidly until inflection point of the Arrhenius function (curves 1–4) or approach asymptotically some low value, which corresponds to thermally stable equilibrium (curves 5, 6).

Generalization to the case of non-trivial initial condition $\theta(0) \neq 0$ does not represent any difficulty and may be investigated by the same numerical technique.

A simple analytical solution of the problem can be obtained in the case of adiabatic explosion, that is, for the system that does not lose energy to surroundings.

For this adiabatic case, Equation (2.4) assumes the simplified form

$$\frac{d\theta}{d\tau} = \exp\left(\frac{\theta}{1 + \mathrm{Ar}\theta}\right) \qquad (2.51)$$
$$\theta(0) = 0$$

It is easy to see that in the limit, $\mathrm{Ar} \to 0$ solution of Equation (2.51) approaches asymptotically the solution of the equation

$$\frac{d\theta}{d\tau} = \exp(\theta) \qquad (2.52)$$
$$\theta(0) = 0$$

that is

$$\theta(\tau) \sim -\ln(1-\tau) \quad (\mathrm{Ar} \to 0) \qquad (2.53)$$

Solution (2.53) is presented in Figure 2.9 by curve 1 and is in fact indistinguishable from the numerical solution of the problem (2.51) with $\mathrm{Ar} = 0.01$. This indistinguishable difference between the two solutions may be expected in view of the very small value of Arrhenius number.

It should be noted that the analytical solution (2.53) may provide quite reasonable description of thermal explosion in some practical systems, if these systems operate in nearly adiabatic conditions.

An important characteristic of thermal explosion is an induction period. Informally speaking, an induction period is the phase of thermal explosion development where the process develops slowly, that is, the temperature of the system remains close to the initial one and consumption of reactants

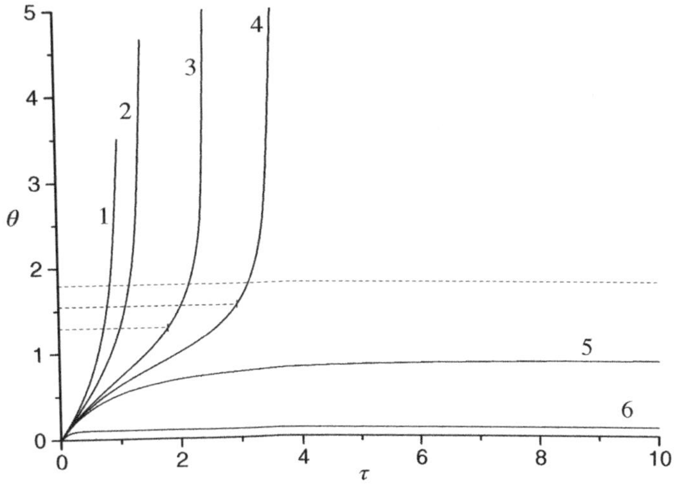

Figure 2.9 Non-steady solutions. Semenov model. Ar = 0.01. 1 – adiabatic (numerical); analytical adiabatic asymptotic (2.53) Ar $\rightarrow$ 0; 2 – Se = 1.0; 3 – Se = 0.50; 4 – Se = 0.43; 5 – Se = $Se_{cr} \approx 1/e$; 6 – Se = 1/(4*e*). Dashed line: induction period span.

is negligible. It is followed by the phase of extremely rapid, progressive accumulation of energy and temperature growth, that is, by the phase normally associated with thermal explosion as such.

The moment in time associated with the end of induction period divides, therefore, the process into two very distinctive stages.

Depending on the system parameters, the induction period may be quite prolonged. For example, the thermal autoignition of storage of biofuels or waste materials (Chapter 7) may take months to develop. Even lengthier induction periods may occur in such systems if thermal explosion occurs eventually as a result of several cycles of seasonal ambient temperature variations (Chapter 3).

Induction period is a very important parameter. In particular, the existence of lengthy period may result in a false perception of a system as being in a safe steady-state (or quasi- steady-state) thermal equilibrium, while in reality a system is a hazard and may develop disastrous thermal instability later on. The identification and accurate estimation of induction period is, therefore, a matter of great practical importance from the safety point of view.

On the other hand, induction period, if properly identified, presents an opportunity to relieve the system from conditions leading to thermal explosion, by some external action (Chapters 8 and 9).

Induction period may be defined as a threshold where chemical reaction reaches a relatively small percentage of its maximum rate (5%, 10%, or similar) in the course of thermal explosion development (Zeldovich *et al.* 1980). This definition is ambiguous.

Here, a more invariant definition is considered, which is based on the geometrical shape of the EGD function. Another advantage of the present definition is that it is applicable to the reaction of the zeroth order, that is, when all the kinetic effects are neglected, as is the case in the present chapter. The results based on the kinetic threshold definition (Zeldovich *et al.* 1980) are discussed briefly in Chapter 4.

For $Se_{cr} < Se < 1$, the interval between the points where the function $F(\theta)$ assumes the values $F(0)$ is considered. Figure 2.2 demonstrates the two such intervals, for EGDs corresponding to $Se = 0.5$ and $Se = Se_{cr} \approx 1/e$. The span of the induction period over the phase space (i.e. over the system temperature range) is the distance from the axis $\theta = 0$ to the point where the dashed line ends on the curve 2 or 3, respectively.

Such a definition is motivated by the two facts:

1) the EGD function remains, within the specified period, bounded by its initial value $F(\theta) \le F(0)$ (Figure 2.2), thus preventing excessively rapid growth of temperature, and
2) this is the region where the influence of the negative (heat dissipation) term of the EGD function is the most profound (in relative terms).

Due to the reasons (1) and (2), the specified region of the phase space corresponds to the slow, induction stage of thermal explosion.

Within the specified region, the EGD function is approximated by the quadratic polynomial

$$F^a(\theta) = \frac{1}{\alpha_o\alpha_1}(\theta-\alpha_o)(\theta-\alpha_1) + \frac{F(\alpha_o)}{\alpha_o(\alpha_o-\alpha_1)}\theta(\theta-\alpha_1) + \frac{F(\alpha_1)}{\alpha_1(\alpha_1-\alpha_0)}\theta(\theta-\alpha_0) \tag{2.54}$$

connecting the points $(0, F(0))$, $(\alpha_0, F_{\min})$ and $(\alpha_1, F(0))$

where $\alpha_1 > 0$ is the point at which $F(\theta)$ assumes its initial value, that is, $F(\alpha_1) = F(0)$, $0 < \alpha_0 < \alpha_1$ is the point of minimum of the function $F(\theta)$ on the interval $[0, \alpha_1]$.

Such a polynomial has the form

$$F^a(\theta) = 1 - \beta_1\theta + \beta_2\theta^2 \tag{2.55}$$

where

$$\begin{aligned} \beta_1 &= \frac{(\alpha_o+\alpha_1)}{\alpha_o\alpha_1} + \frac{\alpha_1}{\alpha_o(\alpha_o-\alpha_1)}F(\alpha_o) + \frac{\alpha_0}{\alpha_1(\alpha_1-\alpha_0)}F(\alpha_1) \\ \beta_2 &= \frac{1}{\alpha_o\alpha_1} + \frac{1}{\alpha_o(\alpha_o-\alpha_1)}F(\alpha_o) + \frac{1}{\alpha_1(\alpha_1-\alpha_0)}F(\alpha_1) \end{aligned} \tag{2.56}$$

Here α_0 is the root of the equation

$$\frac{1}{(1+\mathrm{Ar}\theta)^2}\exp\left(\frac{\theta}{1+\mathrm{Ar}\theta}\right) = \frac{1}{\mathrm{Se}} \tag{2.57}$$

α_1 is non-zero root of the equation

$$\exp\left(\frac{\theta}{1+\mathrm{Ar}\theta}\right) - \frac{1}{\mathrm{Se}}\theta = 1 \tag{2.58}$$

Both the roots exist for $\mathrm{Se}_{cr} < \mathrm{Se} < 1$.

Re-writing Equation (2.5) in the form

$$\frac{d\theta}{F(\theta)} = d\tau \tag{2.59}$$

the induction period is calculated as

$$\int_0^{\alpha_1} \frac{d\theta}{F(\theta)} = \int_0^{\tau_{ind}} d\tau \tag{2.60}$$

from where

$$\tau_{ind} = \int_0^{\alpha_1} \frac{d\theta}{F(\theta)} \approx \int_0^{\alpha_1} \frac{d\theta}{F^a(\theta)} \tag{2.61}$$

The latter integral is easily calculated yielding the result

$$\tau_{ind} \approx \frac{2}{\sqrt{4\beta_2 - \beta_1^2}} \left[\mathrm{Arctg}\left(\frac{2\beta_2\alpha_1 - \beta_1}{\sqrt{4\beta_2 - \beta_1^2}} \right) - \mathrm{Arctg}\left(-\frac{\beta_1}{\sqrt{4\beta_2 - \beta_1^2}} \right) \right] \tag{2.62}$$

Induction period, τ_{ind}, calculated according to the formula (2.61), is shown in Figure 2.9. Similar to Figure 2.2, the induction period is the distance from the axis $\theta = 0$ to the point, where the dashed line ends on curve 3 or 4, respectively. It is infinite for curve 5.

The induction periods are shown in Figure 2.9 for the same two values of the Semenov parameter, as in Figure 2.2, that is, for Se = 0.5 (curve 3) and Se = $\mathrm{Se}_{cr} \approx 1/e$ (curve 5). The latter is infinite, as expected, and is shown in Figure 2.9 by infinite dashed line. Besides, the induction period is also shown for the value in-between the two, for Se = 0.43 (curve 4).

In all cases, the calculated induction periods in Figure 2.9 agree well with the intuitive visual subdivision of temperature time history into relatively slow, and extremely rapid growth stages.

Consider now the Frank–Kamenetskii non-steady formulation (2.10), where the Frank–Kamenetskii transformation has been used, that is the evolution equation

$$\frac{\partial \theta}{\partial \tau} = \frac{1}{\delta} \nabla^2 \theta + \exp(\theta) \tag{2.63}$$

Figure 2.10 shows temperature distributions for the case of planar symmetry, for different values of the Frank–Kamenetskii parameter, and different times. Therefore, the following version of Equation (2.63)

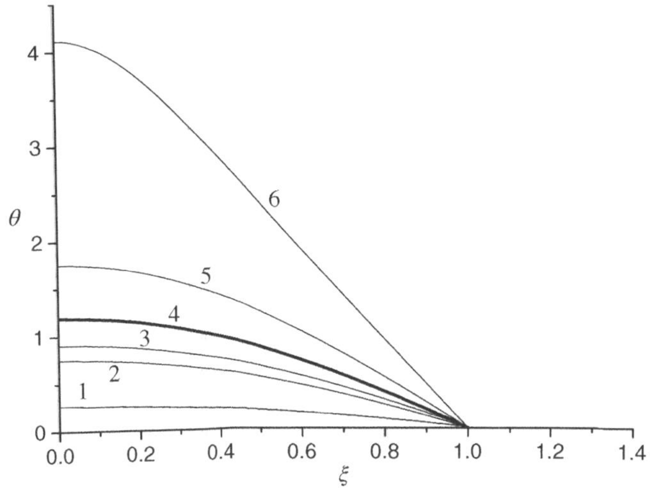

Figure 2.10 Non-steady temperature distributions. Frank–Kamenetskii model. Planar symmetry. 1 – $\delta = 0.40$ analytical solution (2.29) and non-steady numerical solution $\tau \to \infty$; 2 – $\delta = 0.80$ analytical solution (2.29) and non-steady numerical solution $\tau \to \infty$; 3 – $\delta = 0.84$ analytical solution (2.29) and non-steady numerical solution $\tau \to \infty$; 4 – $\delta = 0.88 \approx \delta_{cr}$ analytical solution (2.29) and non-steady numerical solution $\tau \to \infty$; 5 – $\delta = 1.00$, $\tau = 3.00$ non-steady numerical solution; 6 – $\delta = 1.00$, $\tau = 3.50$ non-steady numerical solution.

$$\frac{\partial \theta}{\partial \tau} = \frac{1}{\delta}\frac{\partial^2 \theta}{\partial \xi^2} + \exp(\theta) \tag{2.64}$$

is being solved, with the initial and boundary conditions

$$\theta(\xi, 0) = 0; \quad \frac{\partial \theta}{\partial \xi}(0, \tau) = 0; \quad \theta(1, \tau) = 0$$
$$\xi \in [0, 1]; \quad \tau \in [0, \infty] \tag{2.65}$$

Curves 1–4 correspond to subcritical conditions. The Frank–Kamenetskii parameter for curve 4 is very close to (only slightly below) critical. For these cases, the non-steady solutions approach analytical steady-state profiles (2.29) in the limit of infinitely increasing non-dimensional time. This is a rather expected behaviour, however, it should be noted that this result is obtained here for the trivial initial condition. Generally speaking, the system, despite the fact that its Frank–Kamenetskii parameter is below critical value (obtained from the steady-state theory), may not remain at subcritical conditions if the initial temperature distribution is different from trivial. That is, in the non-steady theory occurrence or non-occurrence of thermal explosion depends, generally, on initial distribution. This consideration becomes important, for example, if initially hot localized region exists within the system (spotted thermal explosion considered in Chapter 4).

Another explosion possibility, with both the trivial initial condition and subcritical value of the steady-state Frank–Kamenetskii parameter, exists in the case of complicated heat exchange regime at the system boundaries. Such are the examples of dynamical regimes of thermal explosion (e.g. thermal explosion with oscillating boundary conditions, Chapter 3) and conjugate thermal explosion (Chapter 4).

For the values of the Frank–Kamenetskii parameter exceeding critical, temperature in the central region of the vessel grows progressively until inflection point of the Arrhenius function (curves 5 and 6).

The temperature at the symmetry boundary, as a function of time, is shown in Figure 2.11. It may be easily seen that despite the fact that the Frank–Kamenetskii formulation results in a distributed

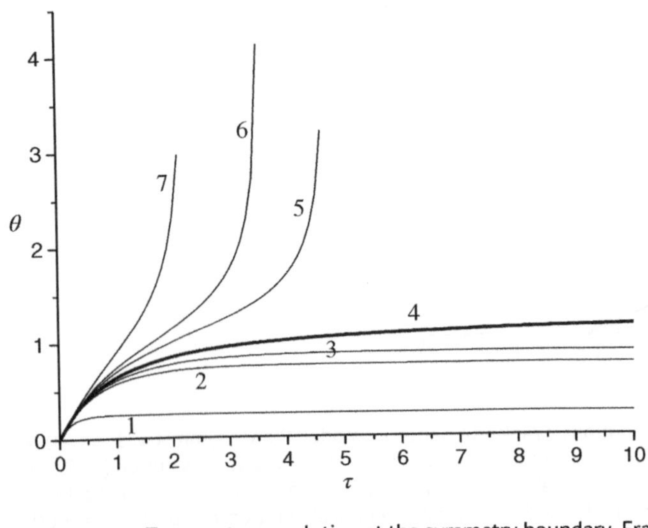

Figure 2.11 Temperature evolution at the symmetry boundary. Frank–Kamenetskii model. Planar symmetry. 1 – $\delta = 0.40$; 2 – $\delta = 0.80$; 3 – $\delta = 0.84$; 4 – $\delta = \delta_{cr} \approx 0.88$; 5 – $\delta = 0.95$; 6 – $\delta = 1.00$; 7 – $\delta = 1.20$.

model (taking into account spatial temperature difference within the system), the temperature behaviour in time (in the vicinity of the symmetry boundary) is, in principal, qualitatively completely similar to the temperature behaviour in the Semenov model (Figure 2.9), which neglects such spatial variations.

In principle, it is possible to define the induction period in the Frank–Kamenetskii model, for example, by following the temperature time history at the centre of the vessel. The theory of such induction period, whatever method is being used to define it, is likely to be rather complicated. In contrast to Semenov formulation, in the Frank–Kamenetskii model, the induction period would depend on the initial condition in quite non-trivial manner. This difficulty is attributable to the fact that the initial condition in this case is represented by a function, belonging to a functional space of infinite dimension. Induction period for the Frank–Kamenetskii model is not considered in the present book.

2.5 Comparison of the Semenov and the Frank–Kamenetskii Formulations

In conclusion of this chapter, it is instructive to compare briefly the two approaches to the thermal explosion theory, that is, the Semenov and the Frank–Kamenetskii formulations.

The most fundamental, the Semenov theory of thermal explosion predicts correctly the qualitative characteristics of this phenomenon, for example, existence of critical conditions, existence of both stable and unstable steady-state solutions, and others.

However, the core assumption of the Semenov theory, that is the uniform distribution of temperature in space, is rather restrictive. Strict fulfilment of this condition requires extremely intensive natural or forced convection in the mixture which, of course, occurs only in very special circumstances. It is also obviously inapplicable to solid reacting systems.

Further, the convective heat transfer coefficient entering the Semenov Equation (2.1) and controlling heat losses from the system is not a precisely determined parameter. Its accurate estimation requires the analysis of the gas dynamics in a boundary layer adjacent to the wall of the vessel and is a complicated task by itself.

Due to the above reasons, only qualitative agreement between the Semenov theory and experiment may be expected.

It is clear that the Frank–Kamenetskii theory, admitting spatial temperature evolution, is more physically sound. It should also be noted that in its simplest form, the Frank–Kamenetskii formulation imposes the same boundary condition as the Semenov one. This is the assumption of constant temperature of the wall. This is equivalent to assuming the infinite thermal conductivity of the material of the wall; the wall temperature is forced to be equal to ambient temperature. Still, the Frank–Kamenetskii theory is more flexible in this respect as it easily admits generalization to other types of boundary conditions (discussed briefly above), which may better suit a particular combustion system.

As such it has been, for a long time, considered as an alternative to the Semenov approach.

It should be noted that, despite considerable differences in underlying assumptions, both approaches agree qualitatively in predicting two fundamentally different types (quasi-steady and growing) of non-steady solutions (shown in Figures 2.9 and 2.11).

More formal reconciliation between formulations of the two theories may be achieved upon the extended interpretation of the Semenov theory. Under such interpretation, the theory is considered as been applied to spatially averaged values of the temperature and reaction rate

$$\langle \tilde{T} \rangle = \frac{1}{\tilde{V}} \int_{\tilde{V}} \tilde{T} d\tilde{V}; \quad \langle \tilde{k}(\tilde{T}) \rangle = \frac{1}{\tilde{V}} \int_{\tilde{V}} \tilde{k}(\tilde{T}) d\tilde{V} \tag{2.66}$$

This extended approach is illustrated here for the steady-state formulations of the two theories. In these variables, the extended version of the Semenov Equation (2.1) is a consequence of the spatially averaged Frank–Kamenetskii Equation (2.8)

$$\tilde{\rho}\tilde{c}_p \frac{1}{\tilde{V}} \int_{\tilde{V}} \frac{\partial \tilde{T}}{\partial \tilde{t}} d\tilde{V} = \frac{1}{\tilde{V}} \tilde{\lambda} \int_{\tilde{V}} \tilde{\nabla}^2 \tilde{T} d\tilde{V} + \frac{1}{\tilde{V}} \tilde{Q} \int_{\tilde{V}} \tilde{k}(\tilde{T}) d\tilde{V} \tag{2.67}$$

or, due to the Gauss–Ostrogradsky theorem

$$\tilde{\rho}\tilde{c}_p \frac{\partial \langle \tilde{T} \rangle}{\partial \tilde{t}} = \frac{1}{\tilde{V}} \tilde{\lambda} \int_{\tilde{S}} \left(\tilde{\nabla} \tilde{T} \cdot \vec{n} \right) d\tilde{S} + \tilde{Q} \langle \tilde{k}(\tilde{T}) \rangle \tag{2.68}$$

With the additional assumptions

$$-\tilde{\lambda} \int_{\tilde{S}} \left(\tilde{\nabla} \tilde{T} \cdot \vec{n} \right) d\tilde{S} = \tilde{h}\tilde{S}(\langle \tilde{T} \rangle - \tilde{T}_0); \quad \langle \tilde{k}(\tilde{T}) \rangle = \tilde{k}\langle \tilde{T} \rangle \tag{2.69}$$

the final result (given that partial and full derivatives over time in this case coincide) is

$$\tilde{\rho}\tilde{c}_p \frac{d\langle \tilde{T} \rangle}{d\tilde{t}} = \tilde{Q}\tilde{k}\langle \tilde{T} \rangle - \tilde{h}\frac{\tilde{S}}{\tilde{V}} (\langle \tilde{T} \rangle - \tilde{T}_0) \tag{2.70}$$

which is identical to the Semenov equation (2.1), written in spatially averaged variables. The Semenov equation preserves its form upon transformation to spatially averaged variables.

The identities (2.69) are the major assumptions required for the formulation and application of the extended Semenov theory. If the mixture temperature is uniform over volume due to, for example, intensive stirring and all the temperature variation occur in infinitely thin layer, adjacent to the vessel walls, then the assumptions (2.69) are fulfilled exactly. Furthermore, the parameter $\tilde{h}$ has physical meaning of a real heat transfer coefficient. In this case, the Semenov model is applied in its original form (2.1). Otherwise, the relations (2.69) are only approximate, and the Semenov theory is applied in the form (2.70). In this sense, the extended Semenov model becomes an approximation of the more accurate Frank–Kamenetskii model. The quality of this approximation may be judged by considering particular examples.

For most real systems, the assumptions (2.69) are fulfilled only to within a certain degree of accuracy. In particular, the second of the assumptions (2.69) may be, in general, quite inaccurate for non-linear functions (such is the case, for example, with the identical assumption in turbulent flames). For the systems described by the Semenov and Frank–Kamenetskii models (uniform quiescent mixtures), however, the latter assumption may be quite reasonable. Zeldovich *et al.* (1980) investigated the accuracy of such extended Semenov approach considering steady-state solutions in the case of planar symmetry. They found that in this case, the assumption $\langle \tilde{k}(\tilde{T}) \rangle = \tilde{k}\langle \tilde{T} \rangle$ is fulfilled with very good accuracy. The relative error is less than 3% at the explosion limit (i.e. for the solution formally corresponding to the critical condition), and even less for sub-critical solutions.

Frank–Kamenetskii (2016) demonstrated that the extended Semenov model (2.70) predicts the same critical conditions as the Frank–Kamenetskii model (in the steady-state formulation, for the planar, cylindrical, and spherical symmetries), upon a replacement of its heat transfer coefficient by the following effective value:

$$\tilde{h} \equiv \tilde{h}_{eff} = \frac{\tilde{\lambda}\tilde{V}e}{\tilde{S}\tilde{r}^2}\delta_{cr} \tag{2.71}$$

Here, the critical values δ_{cr} are given in Table 2.1.

Equation (2.71) allows an important relation between the critical Frank–Kamenetskii number and an eigenvalue of the Laplace heat transfer operator to be established (Frank–Kamenetskii 2016).

It must be emphasized that the analysis developed below provides an estimation only, that is, predicts an approximate value of the critical Frank–Kamenetskii parameter.

Effective heat transfer coefficient in the extended Semenov model may be estimated by neglecting the chemical reaction term and considering the problem of the inert cooling of heated mixture within a given volume Ω of arbitrary shape.

The corresponding 'exact' HTE for the excess temperature $\tilde{\vartheta} = \tilde{T} - \tilde{T}_0$ is (which is identical to the truncated form of the Frank–Kamenetskii equation) is

$$\frac{\partial\tilde{\vartheta}}{\partial\tilde{t}} = \tilde{\kappa}\tilde{\nabla}^2\tilde{\vartheta} \tag{2.72}$$

Consider Equation (2.72) with an appropriate initial (e.g. uniformly elevated temperature of the mixture) and homogeneous boundary condition (i.e. temperature at the enclosing surface is equal to ambient; $\tilde{\vartheta} = 0$ at $\partial\Omega$).

The solution of this linear PDE, obtained by separation of variables (Tikhonov and Samarskii 1963), has the form of the series

$$\tilde{\vartheta} = \sum_{n=1}^{\infty} \tilde{C}_n X_n\left(\vec{x}\right) \exp(-\tilde{\eta}_n\tilde{\kappa}\tilde{t}) \tag{2.73}$$

Here, $\tilde{\eta}_n, n \in \mathbb{N}$ are the eigenvalues of the corresponding Sturm–Liouville problem for the Laplace (i.e. steady-state heat transfer) operator

$$\begin{aligned} &\tilde{\nabla}^2 X + \tilde{\eta}X = 0 \\ &\qquad X = 0, \vec{x} \in \partial\Omega \end{aligned} \tag{2.74}$$

where $\partial\Omega$ is the surface enclosing given three-dimensional region Ω, and X_n are corresponding eigenfunctions (i.e. non-trivial solutions of (2.74)) of spatial coordinates forming an orthonormal system in $L^2(\Omega)$. The expansion coefficients $\tilde{C}_n$ are uniquely determined by the initial condition.

According to the general theory of PDEs, (Tikhonov and Samarskii 1963; Volpert and Khudyaev 1975), the eigenvalues of the problem (2.74) are all positive and form monotonically increasing sequence $0 < \tilde{\eta}_1 \leq \ldots \leq \tilde{\eta}_k \leq \tilde{\eta}_{k+1} \leq \ldots; \tilde{\eta}_k \to \infty, k \to \infty$. Due to their positiveness, the eigenvalues may also be written in the form $\tilde{\eta}_k = \tilde{\mu}_k^2,\ \ k = 1, \ldots, \infty$.

Due to the latter property, the following asymptotic equality (Volpert and Khudyaev 1975) holds

$$\tilde{\vartheta} \simeq \tilde{C}_1 X_1\left(\vec{x}\right) \exp(-\tilde{\eta}_1\tilde{\kappa}\tilde{t}),\ \ \tilde{t} \to \infty \tag{2.75}$$

Therefore, in the first approximation

$$\frac{d\tilde{\vartheta}}{d\tilde{t}} = \frac{\partial\tilde{\vartheta}}{\partial\tilde{t}} = -\tilde{C}_1\tilde{\eta}_1\tilde{\kappa}X_1\left(\vec{x}\right) \exp(-\tilde{\eta}_1\tilde{\kappa}\tilde{t}) = -\tilde{\eta}_1\tilde{\kappa}\tilde{\vartheta} \tag{2.76}$$

On the other hand, the extended Semenov Equation (2.70) provides for inert cooling

$$\frac{d\langle\tilde{\vartheta}\rangle}{d\tilde{t}} = -\frac{\tilde{h}}{\tilde{\rho}\tilde{c}_p}\frac{\tilde{S}}{\tilde{V}}\langle\tilde{\vartheta}\rangle \tag{2.77}$$

The solution (2.76) still depends on spatial coordinates as temperature in (2.72) and (2.76) is non-uniform. In order for both solutions (2.76 and 2.77) to behave identically at a particular characteristic point in the domain (e.g., at the location of the maximum temperature), the heat transfer coefficient in the Semenov model must have the effective value

$$\tilde{h} \equiv \tilde{h}_{eff} = \tilde{\eta}_1\tilde{\lambda}\frac{\tilde{V}}{\tilde{S}} \tag{2.78}$$

The formulae (2.71) and (2.78) give the suggested value of the effective heat transfer parameter in the extended Semenov model. By comparing these formulae, it may be concluded that the critical Frank–Kamenetskii parameter is approximately proportional to the least eigenvalue of the steady-state heat transfer operator

$$\delta_{cr} \approx \tilde{\eta}_1\frac{\tilde{r}^2}{e} \tag{2.79}$$

Frank–Kamenetskii (2016) slightly adjusted the relation (2.79) to the 'exact' form

$$\delta_{cr} = \gamma\tilde{\eta}_1\frac{\tilde{r}^2}{e}, \; \gamma \approx 0.93 \tag{2.80}$$

by applying (2.79) to the two symmetrical cases, with cylindrical and spherical symmetry, and choosing the median value of the correction factor γ.

In this form, the relation (2.80) has proven to be a helpful way to predict approximate values of the critical Frank–Kamenetskii parameter for bodies of arbitrary shapes. In particular, this method is used in Section 5.3.

There are also more accurate space-averaging methods with various weighting functions (Khudyaev 1964; Volpert and Khudyaev 1975; Khudyaev 2003). These methods may be efficiently used for constructing solutions of approximate problems of the theory of thermal explosion in vessels of arbitrary shapes.

3

Extended Mathematical Theory of Thermal Explosion

3.1 Generalized Boundary Conditions

The present section considers several examples of thermal explosion problems where boundary conditions are more complicated, compared to those considered in Chapter 2.

The first of such problems is thermal explosion description with the boundary conditions of the third type, specifically Newton's law of cooling.

The problem, using the Frank-Kamenetskii model, is formulated in the following way (Barzykin and Merzhanov 1958)

$$\frac{d^2\theta}{d\xi^2} + \frac{m}{\xi}\frac{d\theta}{d\xi} = -\delta \exp(\theta) \tag{3.1}$$

$m = 0$, 1, or 2; $0 \le \xi \le 1$.

This equation is completely identical to Eq. (2.27) in Chapter 2.

The boundary conditions, however, now take the form

$$\frac{d\theta}{d\xi}(0) = 0; \ \frac{d\theta}{d\xi}(1) = -Bi\theta\,(1) \tag{3.2}$$

The second of this conditions, on the right boundary, is Newton's law of cooling (heat flux from reacting system to ambient medium is proportional to the temperature difference between system boundary and surroundings), taking note that non-dimensional ambient temperature $\theta(\infty) = 0$, and the Biot number in Eq. (3.2) is defined as $Bi = \frac{\tilde{h}\tilde{r}}{\tilde{\lambda}}$.

Values of the Biot number $0 < Bi < \infty$ describe different heat transfer regimes. If $Bi \to \infty$, that is heat exchange is so intensive that an increase in the surface temperature is prevented, the problem turns into the Frank-Kamenetskii one (as considered in Chapter 2) $\theta(1) \to 0$. The opposite limit $Bi \to 0$ corresponds to adiabatic case (no energy loss to surrounding medium).

In the formulations (3.1) and (3.2) the critical Frank-Kamenetskii parameter is a function of the Biot number. Determination of this dependence, $\delta_{cr} = \delta_{cr}(Bi)$, is of primary interest in the present problem.

General forms of solutions of Eq. (3.1) for different geometries are discussed in Section 2.3, Chapter 2.

For the planar symmetry ($m = 0$), due to the first (symmetry at $\xi = 0$) of conditions (3.2), solution has the form (2.29) with $C_2 = 0$

$$\theta = -\ln\left[\frac{\delta}{2C_1^2} ch^2(C_1\xi)\right] \tag{3.3}$$

Thermal Explosion: Theory and Application, First Edition. Vasily B. Novozhilov.

As already noted in Chapter 2, it is sufficient to consider positive values of C_1 only.

The integration constant $C_1 > 0$ must be found from the second boundary condition, which translates into the following equation

$$\frac{2C_1^3}{\delta} \frac{sh(2C_1)}{\ln\left[\frac{\delta}{2C_1^2} ch^2(C_1)\right] ch^2(C_1)} = Bi \tag{3.4}$$

$(\frac{\delta}{2C_1^2} ch^2(C_1) \neq 1$; otherwise it is easy to see that the required boundary condition cannot be satisfied for any value of C_1).

It is equally easy to see that, for a fixed value of Bi, sufficiently large values of the parameter δ will make the left-hand side (LHS) of Eq. (3.4) strictly limited by Bi. As such, no solution on the interval $0 < C_1 < \infty$ exists for sufficiently large δ, and therefore there exists a critical value $\delta_{cr} = \delta_{cr}(Bi)$ permitting the solution of Eq. (3.4).

The function $\delta_{cr} = \delta_{cr}(Bi)$ for the planar case $m = 0$ will be presented later.

For the cylindrical symmetry ($m = 1$), again due to the symmetry boundary condition, the solution assumes the form (2.40)

$$\theta = \ln\left(\frac{8}{\delta}\right) - 2\ln\left(a\xi^2 + \frac{1}{a}\right) \tag{3.5}$$

The constant of integration $a > 0$ is to be found from the equation

$$\frac{4a}{\left(a + \frac{1}{a}\right)} = Bi\left[\ln\left(\frac{8}{\delta}\right) - 2\ln\left(a + \frac{1}{a}\right)\right] \tag{3.6}$$

Critical conditions correspond to the situation where Eq. (3.6) has single positive root, and this happens when the tangents of the curves on the LHS and the RHS of (3.6) at some point are identical.

Equality of the derivatives

$$\frac{d}{da}\left\{\frac{4a}{\left(a + \frac{1}{a}\right)}\right\} = \frac{8a}{(a^2 + 1)^2} \tag{3.7}$$

and

$$\frac{d}{da}\left\{Bi\left[\ln\left(\frac{8}{\delta}\right) - 2\ln\left(a + \frac{1}{a}\right)\right]\right\} = 2\,Bi\frac{(1 - a^2)}{a(1 - a^2)} \tag{3.8}$$

gives the value of the integration constant corresponding to critical conditions

$$a_{cr} = \sqrt{\sqrt{\frac{4}{Bi^2} + 1} - \frac{2}{Bi}} \tag{3.9}$$

For the case of cylindrical symmetry, Formula (3.9) allows the dependence $\delta_{cr}(Bi)$ to be found explicitly. Indeed, substituting the critical value (3.9) back to Eq. (3.6), one finds the relation between the Frank-Kameneskii parameter and the Biot number at critical conditions

$$\delta_{cr} = \frac{8}{\left(a_{cr} + \frac{1}{a_{cr}}\right)^2} \exp\left[-\frac{4a_{cr}^2}{(a_{cr}^2 + 1)Bi}\right] \tag{3.10}$$

Finally, consider the case of a spherical symmetry $m = 2$. As discussed in Section 2.3.3, Chapter 2, the equation, with the help of new variables (2.46)

$$x = \xi^2 \exp(\theta);\ y = \xi \frac{d\theta}{d\xi} \tag{3.11}$$

may be transformed to the form (2.47)

$$\frac{dy}{dx} = -\frac{(y + \delta x)}{(y + 2)x} \tag{3.12}$$

The latter equation, with appropriate boundary conditions derived from (3.2), may be easily integrated numerically. The dependence $\delta_{cr}(Bi)$ may be constructed by obtaining (approximately) the point of touch of the integral curve of Eq. (3.12) with the curve $y = -Bi \ln(x)$, obtained from the second of the boundary conditions (3.2).

An important discovery made by Barzykin and Merzhanov (1958) is that there exists a universal function approximating (to within the accuracy of $1 - 2\%$) the dependence $\delta_{cr}(Bi)$ for all the three symmetries, that is

$$\frac{\delta_{cr}}{\delta_\infty} = \varphi\,(Bi);\ m = 0, 1, 2 \tag{3.13}$$

where

$$\delta_\infty = \delta_{cr}(Bi \to \infty)$$

The form of the function $\varphi(Bi)$ is apparent from the case of cylindrical symmetry, where the exact result (3.10) has been obtained, and the fact that $\delta_{\infty,m=1} = 2$ (Table 2.1)

$$\varphi(Bi) = \frac{4\left(\sqrt{\frac{4}{Bi^2} + 1} - \frac{2}{Bi}\right)}{\left(\sqrt{\frac{4}{Bi^2} + 1} - \frac{2}{Bi} + 1\right)^2} \exp\left[-\frac{4\left(\sqrt{\frac{4}{Bi^2} + 1} - \frac{2}{Bi}\right)}{\left(\sqrt{\frac{4}{Bi^2} + 1} - \frac{2}{Bi} + 1\right)Bi}\right] \tag{3.14}$$

This function is shown in Figure 3.1. As discussed earlier, the limit $Bi \to \infty$ gives the classical Frank-Kamenetskii problem. Accordingly, $\varphi(Bi) \to 1$ at infinity, and Eq. (3.13) ensures the required result. In the opposite limit $Bi \to 0$, $\varphi(Bi) \to 0$ as well, and Eq. (3.13) correctly predicts the adiabatic thermal explosion for any $\delta > 0$.

Grigoriev *et al.* (1966) considered the conjugate problem where explicit heat conduction equation for energy dissipation into semi-infinite ambient space is solved, coupled (through the boundary condition) with the Frank-Kamenetskii problem within the reacting mixture.

The precise formulation of the problem, $m = 0, 1, 2$ being the symmetry index as usual, is the following

$$\gamma \frac{\partial \theta}{\partial \tau} = \varphi(\eta) \exp\left(\frac{\theta}{1 + Ar\theta}\right) + \frac{1}{\delta}\left(\frac{\partial^2 \theta}{\partial \xi^2} + \frac{m}{\xi}\frac{\partial \theta}{\partial \xi}\right); 0 \le \xi \le 1$$

$$\omega_\kappa \gamma \frac{\partial \theta}{\partial \tau} = \frac{1}{\delta}\left(\frac{\partial^2 \theta}{\partial \xi^2} + \frac{m}{\xi}\frac{\partial \theta}{\partial \xi}\right); 1 \le \xi < \infty$$

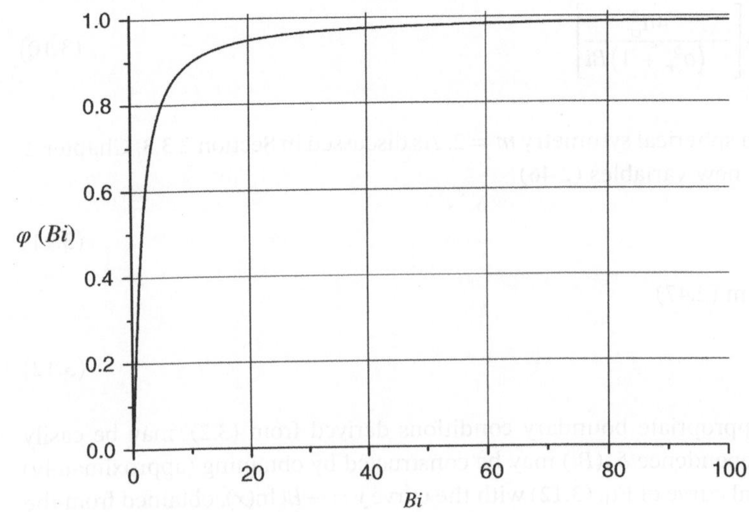

Figure 3.1 Universal dependence (3.14).

$$\frac{\partial \eta}{\partial \tau} = \varphi(\eta) \exp\left(\frac{\theta}{1 + Ar\theta}\right)$$
$$\theta = 0; \eta = 0; \tau = 0 \tag{3.15}$$
$$\frac{\partial \theta}{\partial \xi}(0, \tau) = 0; \theta(\infty, \tau) = 0;$$
$$\theta(1-0, \tau) = \theta(1+0, \tau)$$
$$\omega_\lambda \frac{\partial \theta}{\partial \xi}(1-0, \tau) = \frac{\partial \theta}{\partial \xi}(1+0, \tau)$$

Here the time scale different from the one in Chapter 2 is used, resulting in the appearance of the parameters

$$\gamma = \frac{\tilde{c}_p \tilde{\rho} \tilde{R} \tilde{T}_0^2}{\tilde{Q} \tilde{E} \tilde{C}_0}; \ \omega_\kappa = \frac{\tilde{\kappa}}{\tilde{\kappa}_1} \tag{3.16}$$

in front of time derivatives in the first two equations. (The subscript “1” here and below refer to ambient space $\xi \geq 1$).

The first two equations are energy balances in the reacting system and ambient space, respectively. The third is the kinetic equation for the progress variable η, the degree (the ‘depth’) of transformation. $\eta = 0$ corresponds to a completely fresh mixture, and $\eta = 1$ to the fully reacted mixture. Kinetic effects will be considered in detail in Chapter 4. Grigoriev *et al.* (1966) considered first-order kinetics

$$\varphi(\eta) = 1 - \eta \tag{3.17}$$

Solutions for the problems in the reacting mixture and ambient space are coupled through the conditions at the boundary of the system (i.e. at the wall of the vessel), that is the last two equations of the set (3.15). These describe the continuity of the temperature field and the heat flux through the boundary respectively. The parameter

$$\omega_\lambda = \frac{\tilde{\lambda}}{\tilde{\lambda}_1} \tag{3.18}$$

is the ratio of thermal conductivities in the two contacting media.

The problem (3.15) may be reduced to BVP on the interval $0 \le \xi \le 1$. Indeed, the 'outer' HTE (for the ambient space $1 \le \xi < \infty$) is a linear one, and therefore readily tractable.

This problem may be rewritten in the following form

$$\frac{\partial \theta}{\partial \tau} = \frac{1}{\delta_1}\left(\frac{\partial^2 \theta}{\partial \xi^2} + \frac{m}{\xi}\frac{\partial \theta}{\partial \xi}\right);\ 1 \le \xi < \infty$$

$$\theta = 0;\ \tau = 0 \tag{3.19}$$

$$\theta(1+0,\tau) = \theta_s(\tau);\ \theta(\infty,\tau) = 0$$

where

$$\delta_1 = \omega_\kappa \gamma \delta.$$

The function $\theta_s(\tau)$ (temperature at the boundary of the reacting system) is unknown at the moment, and will have to be found from solution of the 'inner' problem, as reformulated below.

The solution of the problem (3.19), for arbitrary continuous function $\theta(\tau)$, may be written analytically, for example, by Laplace or Fourier Transform methods. Correspondingly, the value of the derivative $\frac{\partial \theta}{\partial \xi}(1+0,\tau)$ may also be found analytically and used as a second (together with symmetry) boundary condition for the "inner" problem.

As a result, the "inner" problem may be written in the following form

$$\gamma\frac{\partial \theta}{\partial \tau} = \varphi(\eta)\exp\left(\frac{\theta}{1+Ar\theta}\right) + \frac{1}{\delta}\left(\frac{\partial^2 \theta}{\partial \xi^2} + \frac{m}{\xi}\frac{\partial \theta}{\partial \xi}\right);0 \le \xi \le 1$$

$$\frac{\partial \eta}{\partial \tau} = \varphi(\eta)\exp\left(\frac{\theta}{1+Ar\theta}\right)$$

$$\theta = 0, \eta = 0, \tau = 0 \tag{3.20}$$

$$\frac{\partial \theta}{\partial \xi}(0,\tau) = 0$$

$$\frac{\partial \theta}{\partial \xi}(1,\tau) = -\frac{\sqrt{\delta}}{\sigma}\frac{d}{d\tau}\int_0^\tau \frac{\theta_s(z)}{\sqrt{\tau - z}}dz, \sigma = \sqrt{\frac{\pi\omega_\lambda^2}{\gamma\omega_\kappa}}; m = 0$$

or

$$\frac{\partial \theta}{\partial \xi}(1,\tau) = -\frac{4}{\pi^2\omega_\lambda}\frac{d}{d\tau}\int_0^\tau \theta_s(\tau - z)\int_0^\infty \exp\left(-\frac{u^2\sigma^2 z}{\pi\delta\omega_\lambda^2}\right)$$

$$\times \frac{du}{u\left[J_0^2(u) + Y_0^2(u)\right]}dz; m = 1$$

or

$$\frac{\partial \theta}{\partial \xi}(1,\tau) = -\frac{\sqrt{\delta}}{\sigma}\frac{d}{d\tau}\int_0^\tau \frac{\theta_s(z)}{\sqrt{\tau - z}}dz - \frac{\theta_s(\tau)}{\omega_\lambda}; m = 2$$

Here J_0 and Y_0 are the Bessel functions of the zeroth order of the first and second kinds, respectively.

The solution of the BVP (3.20) does not exist for arbitrary function $\theta_s \equiv \theta(1, \tau)$ (temperature at the boundary of the reacting system), but only for a specific function, which is found as a part of the solution of the problem (3.20).

The initial conjugate problem (3.15) has been reduced, therefore, to an equivalent BVP (3.20) for the temperature of the reacting mixture on the interval $\xi \in [0, 1]$.

Boundary conditions at $\xi = 1$ for the problem (3.20) or, equivalently, solutions of the problem (3.19), clearly show that steady-state solutions do not exist for the conjugate problem (3.15) for any type of symmetry. (From (3.20), it is clear that derivatives $\frac{\partial \theta}{\partial \xi}(1, \tau) \neq const$ for any $\theta_s(\tau) = const$).

Further analysis of solutions of the "outer" problem (3.19) shows, however, an important difference between the planar and cylindrical cases $m = 0, 1$, on the one side, and the spherical case $m = 2$ on the other. This difference has profound effect on the existence of critical conditions.

In the case of bounded (i.e. subcritical) solution $\frac{\partial \theta}{\partial \xi}(1 + 0, \tau) \to 0$ as $\tau \to \infty$ for the former two cases. Therefore, temperature gradient on the "outer" (ambient) side of the boundary flattens out, and the rate of heat transfer away from the reacting mixture approaches zero at infinite time. In this case thermal explosion will always occur unless there is reactant consumption. In other words, critical conditions do not exist for the reaction of zeroth order (no reactant depletion, thermal explosion occurs for any set of input parameters); for the reaction of positive order critical conditions exist as energy generation rate decreases during the induction period due to reactant consumption. Depending on the relation between different controlling parameters, both explosive and quasi steady-state regimes are possible, and these will be divided by critical conditions.

The situation in the latter, spherical symmetry, case is different. For the solution bounded in time, the gradient $\frac{\partial \theta}{\partial \xi}(1 + 0, \tau)$ will approach finite limit at $\tau \to \infty$. Therefore, a minimum rate of energy dissipation from the system exists at any time. Correspondingly, critical conditions will also exist for the reaction of any order (including zero).

The conjugate problem formulation (3.15) means that the critical value of the Frank-Kamenetskii number is a function of four parameters $\delta_{cr} = \delta_{cr}(Ar, \gamma, \omega_\kappa, \omega_\lambda)$. The parameters Ar and γ have weak influence on the critical value of Frank-Kamenetskii parameter δ (Parks 1961; Barzykin *et al.* 1963). Taking note of this fact, and making use of the parameter σ introduced in Eq. (3.20), the critical conditions of thermal explosion in the conjugate problem (3.15) may be written in the form

$$
\begin{aligned}
\delta_{cr} &= \delta_{cr}^{(0)}(\sigma); m = 0 \\
\delta_{cr} &= \delta_{cr}^{(1)}(\sigma, \omega_\lambda); m = 1 \qquad (3.21) \\
\delta_{cr} &= \delta_{cr}^{(2)}(\sigma, \omega_\lambda); m = 2
\end{aligned}
$$

where the upper index of critical functions is identical to symmetry index. For planar symmetry, the critical Frank-Kamenetskii number is a function of single parameter σ, in the view of exact form of the boundary condition (at $\xi = 1$) (3.20).

The critical conditions $\delta_{cr}^{(m)}$ may be obtained by the numerical solution of the problem (3.20). Examples of the functions (3.21) are presented by Barzykin and Merzhanov (1958). They demonstrated also that the numerical solutions of the problems (3.15) and (3.20) coincide within a good degree of accuracy.

Some other studies also considered the effect of generalized boundary conditions.

For example, Adler (1975) studied, in two-dimensional formulation, the effect of small spatial oscillations of boundary temperature. The problem is formulated in non-dimensional variables, using exactly the same scaling as in Chapter 2, in the following way.

Consider again the reacting system as a slab shown in the left upper corner of Figure 2.5. Recalling translation invariance assumptions of Section 2.3, assume now invariance to translations in just one direction, parallel to the symmetry plane. Such formulation leads to the two-dimensional analogue of Eq. (2.27) ($m = 0$)

$$\frac{d^2\theta}{d\xi^2} + \frac{d^2\theta}{d\varsigma^2} = -\delta\exp(\theta)$$
$$\xi \in [0,1], \varsigma \in]-\infty, \infty[\tag{3.22}$$

The following boundary conditions are imposed

$$\frac{d\theta}{d\xi}(0,\varsigma) = 0; \theta(1,\varsigma) = \varepsilon\cos(\omega\varsigma); 0 < \varepsilon \ll 1 \tag{3.23}$$

The first is the standard symmetry condition, as in (2.28), the second condition imposes spatial oscillations in ς–direction with small amplitude ε and wave number ω.

The Equation (3.22), the so-called Liouville equation, can be integrated considering an unknown function of two real variables $\theta(\xi, \varsigma)$ as a function of a complex variable $z = \varsigma + i\xi$, i.e. $\theta = \theta(z)$ (Stuart 1967; Adler 1975). One of the possible solutions (as it is easy to verify directly) is

$$\theta(\xi,\varsigma) = -2\ln\left[\frac{1}{2}(A+B)ch(\omega\xi) + \frac{1}{2}(A-B)\cos(\omega\varsigma)\right] \tag{3.24}$$

where A and B are real integration constants, arbitrary to within the relation

$$AB = \frac{\delta}{2\omega^2} \tag{3.25}$$

Consider now A, B determined from the following set of equations

$$(A+B)ch(\omega) + (A-B) = 2\exp\left(-\frac{1}{2}\varepsilon\right)$$
$$(A+B)ch(\omega) - (A-B) = 2\exp\left(\frac{1}{2}\varepsilon\right) \tag{3.26}$$

that is

$$A = \frac{ch\left(\frac{1}{2}\varepsilon\right)}{ch(\omega)} - sh\left(\frac{1}{2}\varepsilon\right); B = \frac{ch\left(\frac{1}{2}\varepsilon\right)}{ch(\omega)} + sh\left(\frac{1}{2}\varepsilon\right) \tag{3.27}$$

and the solution (3.24) becomes

$$\theta(\xi,\varsigma) = -2\ln\left[ch\left(\frac{1}{2}\varepsilon\right)\frac{ch(\omega\xi)}{ch(\omega)} - sh\left(\frac{1}{2}\varepsilon\right)\cos(\omega\varsigma)\right] \tag{3.28}$$

Asymptotically, at $\varepsilon \to 0$ $\theta(1, \varsigma) \sim \varepsilon\cos(\omega\varsigma)$, thus satisfying the boundary condition (3.23). The constants A, B still have to obey restrictions (3.25), that is,

$$\frac{\delta}{2\omega^2} = \frac{ch^2\left(\frac{1}{2}\varepsilon\right)}{ch^2(\omega)} - sh^2\left(\frac{1}{2}\varepsilon\right) \tag{3.29}$$

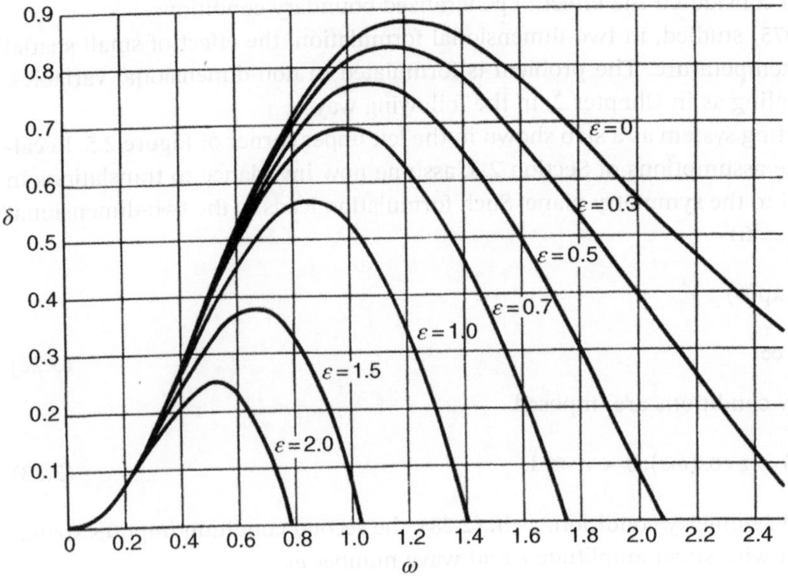

Figure 3.2 Solutions $\delta = \delta(\omega)$ of Eq. (3.30) for various values of oscillation amplitude ε. *Source:* Reproduced with permission from Adler (1975) / Elsevier.

This can be rearranged as

$$sh^2\left(\frac{1}{2}\varepsilon\right) = \frac{1}{th^2(\omega)}\left[\text{sech}^2(\omega) - \frac{\delta}{2\omega^2}\right] \tag{3.30}$$

As, for given amplitude ε and wave number ω, the Frank-Kamenetskii parameter δ increases, the critical conditions of thermal explosion correspond to the point $\delta = \delta_{cr}$ of disappearance of the root of Eq. (3.30).

Solutions $\delta = \delta(\omega)$ of Eq. (3.30) for various values of amplitude ε are presented in Figure 3.2. Critical values δ_{cr} are given by the maximum value of δ on each of the curves.

For $\varepsilon = 0$ (the absence of oscillations making the notion of wave number irrelevant) where boundary conditions (3.23) reduce to conventional Frank-Kamenetskii conditions (2.28) Eq. (3.30) (or Figure 3.2) give $\delta_{cr} \approx 0.88$, in full agreement with classical Frank-Kamenetskii theory (Table 2.1). For $\delta \leq \delta_{cr}$ solution (3.28) turns into solution (2.29).

For a finite amplitude $\varepsilon > 0$ and subcritical values for $\delta < \delta_{cr}$, there are exactly two different types of temperature oscillations (with two different values of wave number ω) for which constructed steady-state solution (3.28) exists. An important comment that has to be made at this point is that Adler's (1975) study leaves open the question whether a steady-state solution for BVP (3.22 and 3.23) exists, for $\delta < \delta_{cr}$, for any wave number $\omega > 0$. By its construction, solution (3.28) satisfies the BVP (3.22 and 3.23) but, as has been discussed just above, only for a discrete set of wave numbers. The possibility, therefore, exists that thermal explosion develops from small surface temperature oscillations with certain wave lengths even if the Frank-Kamenetskii parameter is below its critical value derived by Adler (1975). If this is indeed the case, then critical conditions for the considered thermal explosion problem must be re-formulated. This question remains unsettled.

Shouman and Donaldson (1975) considered the effect of unsymmetrical Dirichlet boundary conditions in the thermal explosion problem for the infinite (in two dimensions) slab (Figure 2.5) with a unidirectional heat conduction, in the direction orthogonal to slab walls. They reduced the PDE problem to integral equation and solved it numerically. Explosion critical conditions appear as critical thickness of reacting media which turns out to be a function of the two unequal temperatures, maintained at both boundaries of the slab.

3.2 Dynamical Regimes

Constant temperature of the environment surrounding the reacting system (and corresponding constant-value Dirichlet boundary condition) is not the only case of interest.

In many cases, ambient temperature changes according to the known law. For example, for underground construction an explosive (charge) may need be lowered to a very deep well. As charge moves down, temperature in the well will change. At the depth of 8 Km this temperature may reach ~350° C, and exceed 400° C in more deep wells. It is important to know how would explosive respond to such heating. Thermal explosion may occur before actually planned charge application, thwarting the course of work. Similarly, thermal explosion is not acceptable in kinetic experimental studies such as TG-DTA where sample (e.g., explosive or propellant) is subjected to gradual heating. Another notable case is the high-temperature synthesis of various materials in thermal explosion regime. Such synthesis is typically conducted by enforcing a linear temperature rise on the surface of the sample. Apart from the linear law of change, the oscillating behaviour of ambient temperature presents significant interest for a certain class of reacting systems experiencing daily and/or seasonal temperature variations. This case will also be discussed in detail later.

Linear temperature variation at the boundary is the simplest case of a dynamical regime, which has been studied extensively by a number of researchers (Merzhanov and Strunina 1965; Strunina *et al.* 1965; Barzykin 1973, 2004; Merzhanov *et al.* 1977, 1996; Sânchez-Rodriguez *et al.* 2017).

It is clear that for any constant rate of temperature rise $\tilde{\psi} > 0$, for the reaction of the zeroth order (no reactant consumption), critical conditions do not exist. Under such conditions, thermal explosion inevitably occurs. Critical conditions in the discussed problem can only relate to the kinetic effects of reactant consumption. In the latter formulation, that is, taking into account kinetic effects, the problem is considered in Chapter 4. Nevertheless, it is clear that at sufficiently high rate of heating thermal explosion occurs regardless of any details of finite-rate chemical kinetics (such as reaction order).

Typical features of thermal explosion development may be, however, considered for the case of zeroth-order reaction. In principle, both the boundary conditions of the first or the second type may be imposed. Taking, for concreteness, the Neumann boundary condition and neglecting reactant consumption, the problem is written in the following form:

$$\begin{gathered}\frac{\partial \theta}{\partial \tau} = \frac{1}{\delta}\left(\frac{\partial^2 \theta}{\partial \xi^2} + \frac{m}{\xi}\frac{\partial \theta}{\partial \xi}\right) + \exp\left(\frac{\theta}{1 + Ar\theta}\right) \\ \xi \in [0, 1]; \tau \in [0, \infty) \\ \frac{\partial \theta}{\partial \xi}(0, \tau) = 0; \frac{\partial \theta}{\partial \xi}(1, \tau) = -Bi(\theta - \psi\tau)\end{gathered} \tag{3.31}$$

where

$$\psi = \frac{\tilde{\rho}\tilde{c}_p}{\tilde{Q}\tilde{k}_0}\tilde{\psi}$$

Thermal explosion development for the planar symmetry case $m = 0$ is illustrated in Figure 3.3.

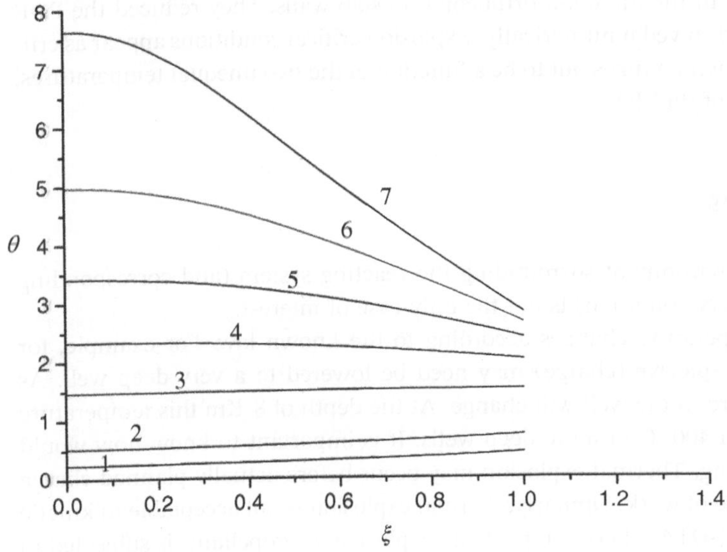

Figure 3.3 Thermal explosion development in linear dynamical regime. Zeroth order reaction. Planar symmetry case $m = 0$. $Ar = 0.01$, $Bi = 15.0$, $\psi = 3.0$. 1 – $\tau = 0.15$; 2 – $\tau = 0.25$; 3 – $\tau = 0.50$; 4 – $\tau = 0.65$; 5 – $\tau = 0.75$; 6 – $\tau = 0.80$; 7 – $\tau = 0.82$.

A distinctive feature of the linear dynamical regime is a heat-up stage. During this stage, the temperature at the boundary grows according to the prescribed law, outpacing the temperature growth at the centre of the vessel. The latter follows in essentially inert regime (chemical reaction rate is negligible). For the set of parameters in Figure 3.3, the heat-up stage extends from $\tau = 0.0$ to $\tau \sim 0.50$ (curves 1–3). Once the chemical reaction rate picks up, the temperature profile across the mixture levels off (curve 4).

At the following stage thermal explosion develops in a way qualitatively similar to the classical nonsteady Frank-Kamenetskii model, with the maximum temperature observed at the centre location and rapidly increasing in an accelerating manner. In Figure 3.3 this stage lasts from $\tau \sim 0.65$ to $\tau \sim 0.82$ (curves 4–7).

It should be noted that at very high rates of ambient temperature growth rate $\psi \gg 1$ the boundary $\xi = 1$ will become a location of temperature runaway. The self-ignition regime with maximum of temperature at the centre of the vessel will turn into the ignition regime with the maximum located at the boundary of the system.

Sânchez-Rodriguez *et al.* (2017) revisited, relatively recently, the problem of thermal explosion under linear heating conditions and applied theoretical analysis to thin films. They came to the conclusion that the idea, put forward earlier by some researchers, of a low-temperature synthesis of oxide films via thermal explosion is essentially unviable.

A more complicated problem arises in considering laws of background temperature change, which are significantly different from the linear, for example, oscillating ambient temperature behaviour (Novozhilov 2016). Such a problem is of interest and importance, for example, in assessing fire and explosion hazards associated with certain classes of combustion systems. These include large-scale storages of chemically active solid materials, such as various kinds of biomass, coal, military munitions, solid propellants, pyrotechnics or similar substances. These systems are likely to be subjected to temperature variations of daily (especially significant in regions with desert-like climate) or yearly (seasonal) nature. Such variations may, in principal, cause significant deviations from the system's thermal equilibrium and lead to thermal explosion (fire).

It is qualitatively clear that, in contrast to the problem with linearly growing ambient temperature, critical conditions in the case of oscillating temperature exist for reaction of any order, including zero. Indeed, consider a linear ambient temperature growth at such a high rate that leads to thermal explosion. Behaviour of ambient temperature beyond the induction period (i.e. after explosion) is irrelevant. Taking such a sinusoid profile that majorizes the linear profile over the induction period, one can immediately see that thermal explosion occurs for such ambient temperature variation as well. Therefore, thermal runaway may occur under oscillations of ambient temperature. On the other hand, if the system is at subcritical state, sufficiently small oscillations of ambient temperature cannot produce supercritical conditions. It follows from the above arguments that critical conditions for thermal explosion under the oscillating behaviour of the background temperature do exist. Since oscillation temperature variations are bounded, these conditions exist for reaction of any order, including zero with no influence of kinetics.

The mathematical model applicable to this situation is essentially identical to Eq. (3.31), with the difference in the boundary condition at the external boundary of the system. Both the conditions of the first and the second type are sensible from the physical point of view. Correspondingly, the problem, with the boundary conditions of either the first or the second type, may be written in the following form (Novozhilov 2016)

$$\frac{\partial \theta}{\partial \tau} = \frac{1}{\delta}\left(\frac{\partial^2 \theta}{\partial \xi^2} + \frac{m}{\xi}\frac{\partial \theta}{\partial \xi}\right) + \exp\left(\frac{\theta}{1 + Ar\theta}\right)$$

$$\xi \in [0, 1]; \tau \in [0, \infty) \tag{3.32}$$

$$\theta(0, \tau) = 0; \tau \in [0, \infty)$$

$$\frac{\partial \theta}{\partial \xi}(0, \tau) = 0; \theta(1, \tau) = A_0 \sin(\omega\tau)$$

or

$$\frac{\partial \theta}{\partial \xi}(0, \tau) = 0; \ \frac{\partial \theta}{\partial \xi}(1, \tau) = -Bi(\theta - A_0 \sin(\omega\tau))$$

where $A_0 = \frac{\tilde{A}_0}{\left(\tilde{R}\tilde{T}_0^2/\tilde{E}\right)}$ and $\omega = \frac{\tilde{\rho}\tilde{c}_p\tilde{R}\tilde{T}_0^2}{\tilde{Q}\tilde{k}_0\tilde{E}}\tilde{\omega}$ are non-dimensional amplitude and frequency of oscillations, respectively.

The behaviour of the system is described by the following parameters: Frank-Kamenetskii number, Arrhenius number, oscillation amplitude, oscillation frequency and, in the case of boundary conditions of the second type, the Biot number. The Frank-Kamenetskii parameter is supposed to have a subcritical value for the static boundary conditions ($A_0 = 0$), and describes the degree of subcriticality of the system. The latter is a measure of how much lower is the actual parameter δ compared to its critical value δ_{cr} at static ambient conditions.

Therefore, the following sets of control parameters are considered:

(δ, Ar, A_0, ω) for the boundary conditions of the first type;
(δ, Ar, A_0, ω, Bi) for the boundary conditions of the second type.

Numerical solution of BVP (3.32) allows critical conditions of thermal explosion with respect to various parameters to be found. The planar symmetry case $m = 0$ is discussed below.

Figures 3.4–3.8 show critical dependencies $A_0(\omega)$, with all the other parameters being fixed. Thermal explosion develops if the point (A_0, ω) is above a particular curve. Regions below the curves correspond to parameters resulting in bounded system temperature variations.

Figures 3.4 and 3.5 refer to the boundary conditions of the first type. Figure 3.4 demonstrates the effects of the Frank-Kamenetskii parameter, in the subcritical range of its change, on the critical boundaries $A_0(\omega)$. Critical amplitude, at fixed frequency, decreases with increase in the Frank-Kamenetskii number. This trend is expected as with increasing the Frank-Kamenetskii parameter the state of the system approaches the critical one under steady ambient conditions. It is evident that, except for the range of low frequencies $0 < \omega \leq \approx 1$, critical dependencies of the amplitude on frequency are very close to linear. Figure 3.5 shows the effect of Arrhenius number on critical dependencies $A_0(\omega)$ for fixed values of the Frank-Kamenetskii parameter. Again, all the dependencies shown in Figure 3.5 are essentially linear. Critical amplitude decreases (at fixed frequency) with

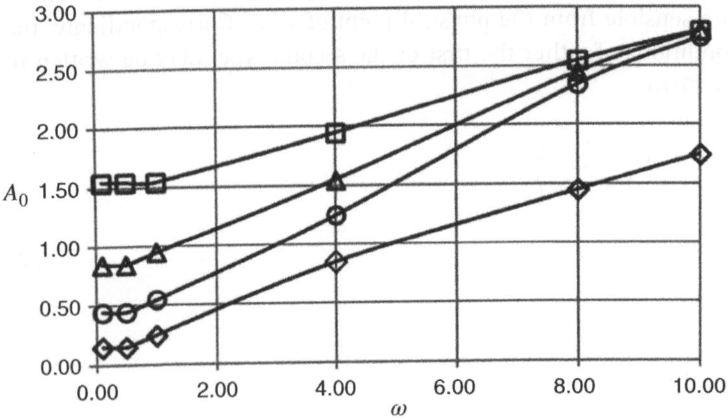

Figure 3.4 Critical dependencies $A_0(\omega)$. Boundary conditions of the first type. $Ar = 0$. □ – $\delta = 0.2$; Δ – $\delta = 0.4$; ○ – $\delta = 0.6$; ◇ – $\delta = 0.8$. *Source:* Novozhilov (2016) / Springer Nature / CC BY 4.0.

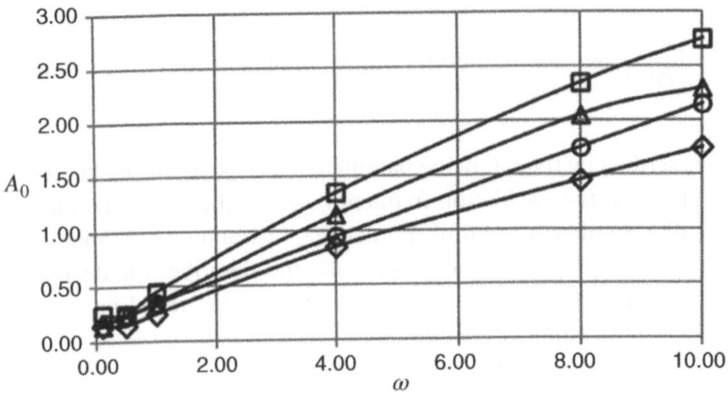

Figure 3.5 Critical dependencies $A_0(\omega)$. Boundary conditions of the first type. $\delta = 0.8$. □ – $Ar = 0.06$; Δ – $Ar = 0.04$; ○ – $Ar = 0.025$; ◇ – $Ar = 0.0$. *Source:* Novozhilov (2016) / Springer Nature / CC BY 4.0.

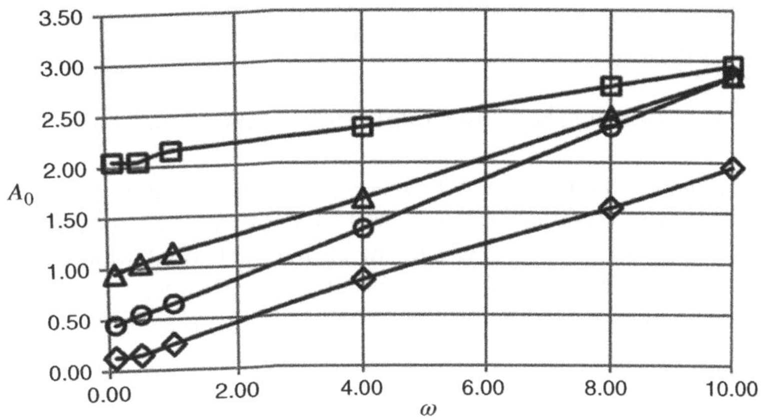

Figure 3.6 Critical dependencies $A_0(\omega)$. Boundary conditions of the second type. $Ar = 0$; $Bi = 15.0$. □ – $\delta = 0.1$; Δ – $\delta = 0.3$; ○ – $\delta = 0.5$; ◇ – $\delta = 0.7$. *Source:* Novozhilov (2016) / Springer Nature / CC BY 4.0.

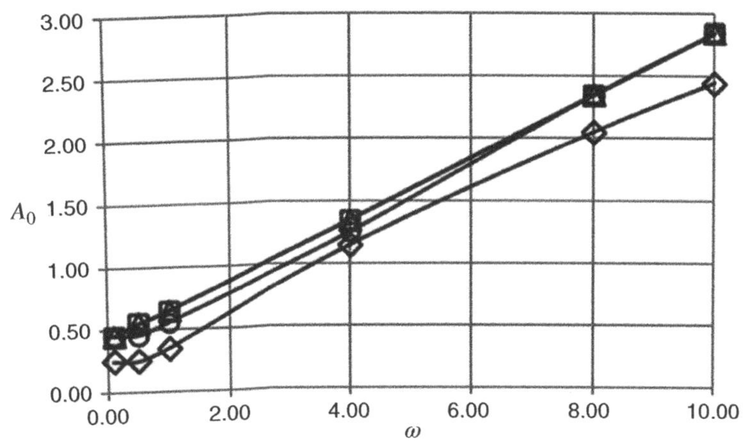

Figure 3.7 Critical dependencies $A_0(\omega)$. Boundary conditions of the second type. $Ar = 0$; $\delta = 0.5$. □ – $Bi = 20$; Δ – $Bi = 15.0$; ○ – $Bi = 10.0$; ◇ – $Bi = 5.0$. *Source:* Novozhilov (2016) / Springer Nature / CC BY 4.0.

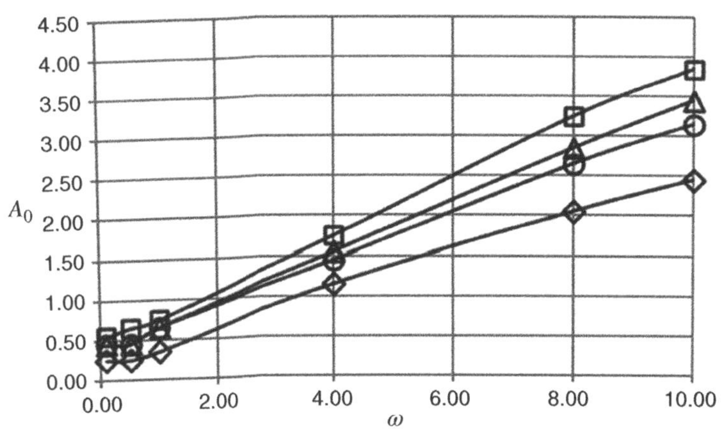

Figure 3.8 Critical dependencies $A_0(\omega)$. Boundary conditions of the second type. $\delta = 0.5$; $Bi = 10.0$. □ – $Ar = 0.06$; Δ – $Ar = 0.04$; ○ – $Ar = 0.025$; ◇ – $Ar = 0.0$. *Source:* Novozhilov (2016) / Springer Nature / CC BY 4.0.

the decrease in the Arrhenius number as it should be, considering the form of the chemical reaction term in Eq. (3.32).

The explanation of the overall behaviour of critical boundaries $A_0(\omega)$ in both the Figures 3.4 and 3.5 is rather obvious. For a fixed frequency ω thermal explosion occurs for sufficiently large amplitude A_0 since the time spent by system in a "hot" environment during each oscillation cycle becomes larger than the induction period. On the other hand, for a fixed amplitude, the system moves to subcritical conditions with increasing frequency as the latter time becomes shorter than the induction period of thermal explosion.

Figure 3.4 also illustrates an important and essential feature of the considered problem: system that is thermally stable under constant ambient temperature may develop instabilities (thermal explosion) if ambient conditions vary in oscillating manner. Indeed, the values of the Frank-Kamenetskii parameter in Figure 3.4 are below the critical one at static conditions ($\delta_{cr} \approx 0.88$, Table 2.1).

Figures 3.6–3.8 present similar results on critical boundaries $A_0(\omega)$ for boundary conditions of the second type. The additional parameter that appears in this case is the Biot number. Each of the Figures 3.6–3.8 has two parameters from the set (δ, Ar, Bi) fixed, while the remaining parameter is allowed to vary. Figure 3.6 is an analogue of Figure 3.4 showing the same trends in dependence of critical boundary $A_0(\omega)$ on the Frank-Kamenetskii parameter. Figure 3.7 demonstrates that dependence of critical conditions on the Biot number is in fact very weak. Finally, Figure 3.8 shows the influence of the Arrhenius number. General trends are very similar to the case of boundary conditions of the first type (Figure 3.5) although critical amplitude is overall somewhat higher. Again, all the dependencies in Figures 3.6–3.8 are very close to linear, and their qualitative explanations are the same as those for boundary conditions of the first type.

It is instructive to observe the temporal characteristics of thermal explosion. Consider first temperature distributions across the slab. At low frequency of ambient temperature oscillations the system temperature oscillations at any given point inside the media follow essentially the same frequency (time scale of heat propagation across the sample is larger, but comparable with the time scale of ambient temperature variations). Consequently, the temperature distribution remains monotonic. This is evidenced, for example, by the profiles $1-3$ in Figure 3.9. Note that temperatures at locations deep inside the media somewhat lag behind temperature variations at the external boundary. Deviation from this simple pattern develops as the oscillation frequency grows and the time scale of ambient temperature oscillations becomes much shorter than the time scale of heat propagation across media. This leads to appearance of non-monotonic profiles, such as the curves 4 and 10 in Figure 3.9.

The temperature time history can be illustrated by considering a particular location, for example, the plane of symmetry $\xi = 0$. In the case of static boundary conditions, this location normally corresponds to maximum temperature. For oscillating boundary conditions this is not true anymore (Figure 3.9). Temperature oscillations at the centreplane of the slab are presented in Figure 3.10.

It is interesting to observe the behaviour of the temperature under subcritical conditions. These are oscillations, and for low amplitudes, for example, $A_0 = 1.8$ they are essentially harmonic, shifted up along θ-axis. This shift is caused by contribution from chemical term during the positive semi-period of oscillation. This contribution becomes larger as the amplitude approaches its critical value (Figure 3.10). Not only the amplitude increases considerably during the positive semi-period of oscillations, but the peak of oscillation also exhibits a noticeable delay compared to the curves

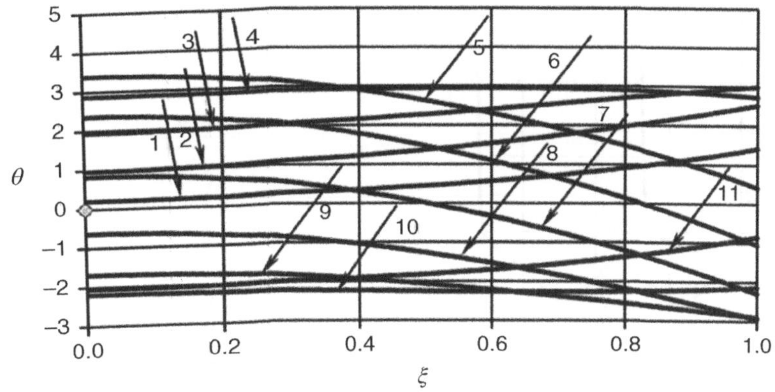

Figure 3.9 Temperature distribution across the medium. Boundary conditions of the first type. $\delta = 0.2$; $Ar = 0.04$; $\omega = 10.0$; $A_0 = 3.0$. 1 – $\tau = 0.05$; 2 – $\tau = 0.10$; 3 – $\tau = 0.15$; 4 – $\tau = 0.20$; 5 – $\tau = 0.30$; 6 – $\tau = 0.35$; 7 – $\tau = 0.40$; 8 – $\tau = 0.45$; 9 – $\tau = 0.50$; 10 – $\tau = 0.55$; 11 – $\tau = 0.60$. *Source:* Novozhilov (2016) / Springer Nature / CC BY 4.0.

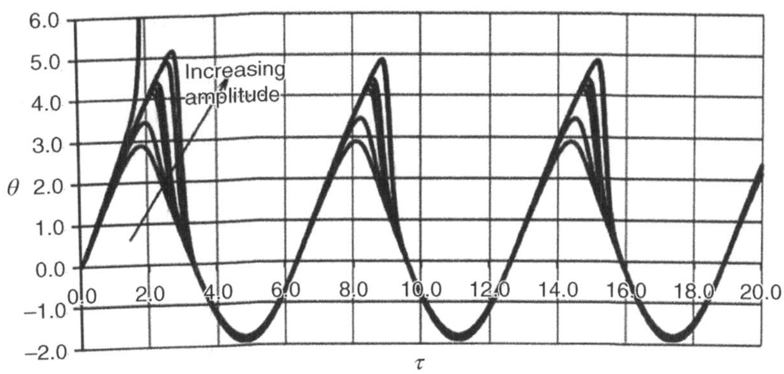

Figure 3.10 Temperature time history at $\xi = 0$. Boundary conditions of the first type. $\delta = 0.2$; $Ar = 0.04$; $\omega = 1.0$; $A_0 = 1.8$. Arrow points in the direction of amplitude increase for the following values: $A_0 = 1.8$, $A_0 = 1.9$, $A_0 = 1.93$, $A_0 = 1.931$, $A_0 = 1.9315$, $A_0 = 1.9318$, $A_0 = 1.931802$. *Source:* Novozhilov (2016) / Springer Nature / CC BY 4.0.

corresponding to lower frequencies. Thermal explosion, in the case of parameters presented in Figure 3.10, occurs upon further increase in the imposed amplitude. This happens early in the process development, that is before the completion of the first period of oscillations corresponding to the subcritical values of the amplitude.

The case presented in Figure 3.10 is not, however, the only possible scenario of thermal explosion. Figure 3.11 demonstrates that the thermal runaway may also occur after a number of oscillations (curve 2, Figure 3.11). In this case, the peak of the amplitude grows slightly during the induction period, from one oscillation period to another, until thermal explosion eventually occurs.

Transition to thermal explosion through temperature oscillations during the induction period is a remarkable unique scenario that is possible under dynamic ambient conditions. Novozhilov (2016) presents the estimation of parameters which show that thermal explosion scenarios described above may well occur in real combustion systems.

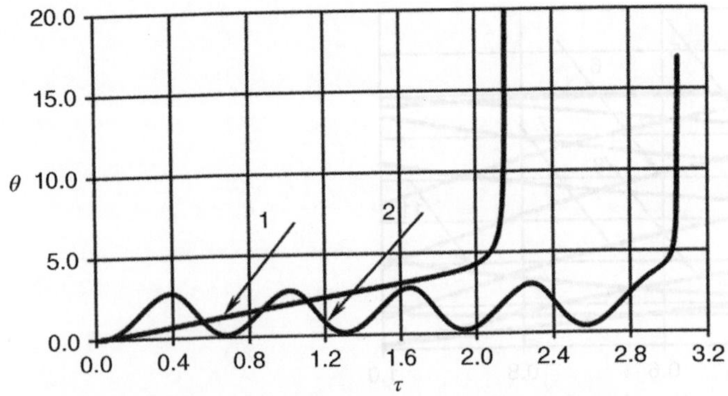

Figure 3.11 Two possible scenarios of thermal explosion development. Temperature time history at $\xi = 0$. 1 – Boundary conditions of the second type. $\delta = 0.2$; $Bi = 15$; $Ar = 0.04$; $\omega = 1.0$; $A_0 = 1.8$. 2 – Boundary conditions of the first type. $\delta = 0.6$; $Ar = 0.0$; $\omega = 10.0$; $A_0 = 2.8$. *Source:* Novozhilov (2016) / Springer Nature / CC BY 4.0.

Data presented in Figures 3.4–3.11 summarizes the influence of all the control parameters on the thermal explosion development under oscillating ambient conditions. Lower oscillation frequencies require lower amplitudes for explosion development as the system stays longer close to the maximum temperature of the oscillations. This fact has important practical implications as low-frequency oscillations are the most difficult to detect. However, the most important outcome of the presented analysis is the fact that oscillations may transfer the system from the state of thermal equilibrium to explosion conditions. At $\omega = 0$ (no oscillations), the parameters of the system are below critical (at static conditions) in Figures 3.4–3.8, however, explosion becomes possible if the oscillations of environment conditions develop.

3.3 Thermal Explosion in a Region of Arbitrary Shape

The establishment of critical conditions of thermal explosion for a region of arbitrary shape (Figure 3.12), where stable solution (if exists) does not possess any symmetry, turns out to be a very interesting and difficult problem (Volpert and Khudyaev 1975; Khudyaev 2003).

In the present section this problem is considered within the framework of the Frank-Kamenetskii theory, therefore, it consists of finding critical Frank-Kamenetskii parameter δ_{cr} for the BVP

$$\nabla^2\theta + \delta\exp(\theta) = 0$$
$$\theta|_{\partial G} = 0; G \subset \mathbb{R}^n, n = 2, 3 \tag{3.33}$$

with the ODE identical to (2.27), but without the symmetry boundary condition (the first of the conditions (2.28)).

For the boundary condition of the first type, δ_{cr} depends on the region only $\delta_{cr} = \delta_{cr}(G)$, so that the problem becomes purely geometrical.

As follows from the results of Section 2.5, for boundary conditions of the first type and for dimensions 2 or 3 of the region G, the critical parameter δ_{cr} in the BVP (3.33) and the first eigenvalue $\tilde{\eta}_1$ of the BVP

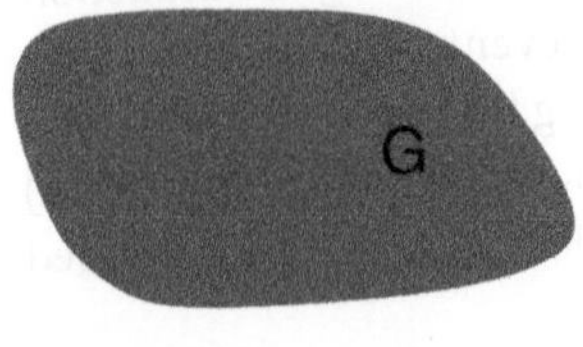

Figure 3.12 An arbitrary space region.

$$\nabla^2 X + \eta X = 0$$
$$X|_{\partial G} = 0; G \subset \mathbb{R}^n, n = 2, 3 \tag{3.34}$$

depend identically on the linear scale of the region G (Eq. (2.80); $\eta_1 = \frac{\tilde{\eta}_1}{\tilde{r}^2}$). Therefore, the ratio $\aleph(G) = \frac{\delta_{cr}(G)}{\eta_1(G)}$ turns out to be invariant with respect to self-similar transformations of the region G and depends on its geometrical shape only.

One estimation for $\aleph(G)$ follows from the following general theorem.

Theorem 3.1 (Khudyaev 2003).
Consider the following BVP

$$L_\eta(u) \equiv \nabla^2 u + \eta a\left(\vec{\xi}\right)F(u) = 0$$
$$u|_{\partial G} = 0; G \subset \mathbb{R}^n, n = 2, 3 \tag{3.35}$$
$$\eta > 0, a\left(\vec{\xi}\right) \geq 0, F(u) > 0 \textit{ for } u \geq 0$$

where the region G is bounded.
If the following condition

$$m = \inf_{u>0} \frac{F(u)}{u} > 0 \tag{3.36}$$

is met, and the limit (finite or infinite)

$$\lim_{u\to\infty} \frac{F(u)}{u} \tag{3.37}$$

exists, then for the value η_{cr} in the problem (3.35) the following inequality holds

$$\eta_{cr} \leq \frac{\eta_1}{m} \tag{3.38}$$

where $\eta_1 \equiv \eta_1|_{F(u)=u}$ is the first eigenvalue of the linear problem (3.35) with $F(u) \equiv u$, and the number m is defined by Eq. (3.36).

ODE (3.35) is a generalized form of the ODE (2.27) for the Frank-Kamenetskii model. Critical value η_{cr} has the same meaning as in the Frank-Kamenetskii model, that is, BVP (3.35) has solution for all $\eta < \eta_{cr}$, and has no solution for all $\eta > \eta_{cr}$.

Conditions (3.36) and (3.37) warrant existence of $0 < \eta_{cr} < \infty$; solvability of the problem (3.35) for $\eta = \eta_{cr}$ depends on the form of the function $F(u)$ (Khudyaev 2003). Obviously, for $a\left(\vec{\xi}\right) \equiv 1$ and $F(u) \equiv \exp(u)$ $\eta_{cr} = \delta_{cr}$.

For the case of exponential function $m = e$, and Theorem 3.1 delivers the following general (for any region $G \subset \mathbb{R}^n$, $n = 2, 3$) estimation

$$\aleph(G) \leq \frac{1}{e} \approx 0.368 \tag{3.39}$$

The following approximate method of estimating $\delta_{cr}(G)$ is based on the following conjectured 'isoperimetric-kind' theorem (Khudyaev 1966; Volpert and Khudyaev 1975)

Among all regions with the dimension 3(2) *with fixed value of* η_1 *for the BVP (3.34)* $\delta_{cr}(G)$ *assumes a minimum value on a ball (circle), and maximum value on a strip.*

This conjecture is supported by the fact that in theorems of such kind the proposition of extremality of round-shaped regions usually turns out to be true. The most convincing argument, however, in support of the above conjecture is its fulfilment in all the cases that admit exact analysis. Nevertheless, the proof of this theorem is an open problem (Volpert and Khudyaev 1975; Khudyaev 2003).

Conditional on the proposition above, the following is true for any region $G \subset \mathbb{R}^n$, $n = 2, 3$

$$\aleph_n \leq \aleph(G) \leq \aleph_1 \tag{3.40}$$

where the indices $n = 1, 2, 3$ refer to strip, round cylinder and ball, respectively.

For the latter regions of simplest shape, problems (3.33) and (3.34) are reduced to interval, circle, and ball, respectively.

The corresponding values of η_1 and δ_{cr} are found analytically (Khudyaev 2003; Frank-Kamenetskii 2016), allowing the values of $\aleph(G)$ to be determined as well

$$\begin{aligned} &\eta_1 = \frac{\pi^2}{4}; j^2 \approx 5.783; \pi^2 \\ &\delta_{cr} \approx 0.878; 2.000; 3.322 \\ &\aleph_1 \approx 0.356; \aleph_2 \approx 0.346; \aleph_3 \approx 0.337 \end{aligned} \tag{3.41}$$

The estimations (3.41) show that even the most simple and general estimation (3.39) gives very reasonable approximation for actual values of δ_{cr}. Besides, the comparison of (3.41) with the universal estimation (3.39) supports one's hope on the weak dependence of $\aleph(G)$ on the region.

Developing the arguments behind the estimation (3.40) further, Khudyaev (1966) proposed that, at least for convex bodies of revolution ($n = 3$), the estimation (3.40) may be improved as

$$\aleph_3 \leq \aleph(G) \leq \aleph_2 \tag{3.42}$$

Conjectures leading to estimations (3.40) and (3.42) were first put forward by Khudyaev (1966). A similar idea was used by Frank-Kamenetskii (2016) when calculating δ_{cr} for cube and cylinder of finite height, however, without referring to any mathematical reasoning.

Considering inequalities (3.40) and (3.42), different "interpolating" formulae for $\delta_{cr}(G)$ may be proposed.

Khudyaev (1966) suggests the two ways

$$\delta_{cr}(G) \cong \eta_1(G)\left[\aleph_n \frac{P_0}{P} + \aleph_1\left(1 - \frac{P_0}{P}\right)\right] \tag{3.43}$$

$$\delta_{cr}(G) \cong \eta_1(G)\left[\aleph_n \frac{P_1}{P} + \aleph_1\left(1 - \frac{P_1}{P}\right)\right] \tag{3.44}$$

where n (3,2) is the dimension of the region G, P is its surface area (perimeter), P_0 is the surface area (perimeter) of the ball (circle) of the same volume (area) as G, and P_1 is the surface area (perimeter) of ball (circle) with the same value of first eigenvalue η_1 as G.

Obviously, relations (3.43) and (3.44) define interpolation only if

$$\frac{P_0}{P} \leq 1, \frac{P_1}{P} \leq 1 \tag{3.45}$$

The first of the inequalities (3.45) follows from the classical isoperimetric theorem (Pólya and Szegö 1951): *among all the regions of given volume (area) ball (circle) has the least surface area (perimeter).*

The second one follows from the Faber–Krahn theorem (Pólya and Szegö 1951): *among all the regions of given volume (area) ball (circle) has the least value of* η_1. According to this theorem, the volume of ball with the eigenvalue $\eta_1(G)$ does not exceed the volume of G. Therefore, $P_1 \leq P_0$.

In view of the latter inequality, and the fact that $\aleph_n < \aleph_1$ (estimation (3.41)) the approximation (3.43) always gives lower critical values than the approximation (3.44).

The theorem of Faber and Krahn provides one more way of interpolation. Specifically, if $\overline{\eta}_1$ is the eigenvalue of the ball (circle) of the same volume (area) as G, then $\dfrac{\overline{\eta}_1}{\eta_1} \leq 1$ and one can estimate

$$\delta_{cr}(G) \cong \eta_1(G)\left[\aleph_n \frac{\overline{\eta}_1(G)}{\eta_1(G)} + \aleph_1\left(1 - \frac{\overline{\eta}_1(G)}{\eta_1(G)}\right)\right] \tag{3.46}$$

For convex bodies of revolution ($n = 3$), according to (3.42), one can use the formulae (3.43), (3.44) and (3.46) replacing $\aleph_1$ by $\aleph_2$.

Note that P_0 and $\overline{\eta}_1$ (as being inversely proportional to squared radius; using (3.41)) can easily be expressed in terms of volume (area) V of the region G, and P_1 in terms of η_1

$$\begin{aligned} &P_0 = (4\pi)^{1/3}(3V)^{2/3}; P_1 = \frac{4\pi^3}{\eta_1}; \overline{\eta}_1 = \pi^2\left(\frac{4\pi}{3V}\right)^{2/3}, n = 3 \\ &P_0 = 2\sqrt{\pi V}; P_1 = \frac{2\pi j}{\sqrt{\eta_1}}; \overline{\eta}_1 = \frac{\pi j^2}{V}\ (j^2 \approx 5.783), n = 2 \end{aligned} \tag{3.47}$$

The formulae (3.43), (3.44) and (3.46) take, correspondingly, the form:

For arbitrary three-dimensional region G with volume V and surface area P

$$\delta_{cr}(G) \cong \begin{cases} \eta_1\left[\aleph_1 - (\aleph_1 - \aleph_3)\dfrac{1}{P}(4\pi)^{1/3}(3V)^{2/3}\right] \\ \eta_1\aleph_1 - (\aleph_1 - \aleph_3)\dfrac{4\pi^3}{P} \\ \eta_1\aleph_1 - (\aleph_1 - \aleph_3)\pi^2\left(\dfrac{4\pi}{3V}\right)^{2/3} \end{cases} \tag{3.48}$$

For arbitrary two-dimensional region G with area V and perimeter P.

$$\delta_{cr}(G) \cong \begin{cases} \eta_1\left[\aleph_1 - (\aleph_1 - \aleph_2)\dfrac{2\sqrt{\pi V}}{P}\right] \\ \eta_1\aleph_1 - \dfrac{2\pi j}{P}(\aleph_1 - \aleph_2)\sqrt{\eta_1} \\ \eta_1\aleph_1 - (\aleph_1 - \aleph_2)\dfrac{\pi j^2}{V} \end{cases} \tag{3.49}$$

For convex bodies of revolution

$$\delta_{cr}(G) \cong \begin{cases} \eta_1\left[\aleph_2 - (\aleph_2 - \aleph_3)\dfrac{1}{P}(4\pi)^{1/3}(3V)^{2/3}\right] \\ \eta_1\aleph_2 - (\aleph_2 - \aleph_3)\dfrac{4\pi^3}{P} \\ \eta_1\aleph_2 - (\aleph_2 - \aleph_3)\pi^2\left(\dfrac{4\pi}{3V}\right)^{2/3} \end{cases} \tag{3.50}$$

Table 3.1 Comparison of approximations (3.48–3.50) with "exact" values of critical Frank-Kamenetskii parameter for different two- and three-dimensional regions.

Region, dimension and non-dimensional geometrical parameters	**Region Type/ Formula**	**1**	**2**	**3**	δ_{cr}	$\frac{\eta_1}{e}$
Cylinder, $n = 3$, radius $r = 1$, height $h = 2$	*III* / (3.50)	2.77	2.79	2.78	2.78	3.03
Cube, $n = 3$, side $d = 2$	*I* / (3.48)	2.52	2.53	2.51	2.53	2.73
Ring, $n = 2, \frac{1}{2} \leq r \leq 1$	*II* / (3.49)	13.7	13.8	13.8	13.8	14.3
Sector, $n = 2, \frac{1}{5} \leq r \leq 1$, angle $\frac{\pi}{2}$	*II* / (3.49)	9.43	9.46	9.45	–	10.0

A priori there are no reasons to favour any particular formula in each of the groups (3.48–3.50). From general considerations, one can only be sure that the first formula in each group provides lower estimation than the second one ($P_1 \leq P_0$).

Table 3.1 compares "exact" critical values δ_{cr} with their approximations obtained by formulae (3.48) (Region Type *I*), (3.49) (Region Type *II*), and (3.50) (Region Type *III*). For each of formulae groups *I*, *II* or *III*, respective row lists values delivered by the first, the second and the third formula in that group, respectively. The universal upper estimation $\frac{\eta_1}{e}$, which follows from (3.39), is provided in the last column.

"Exact" values for cube and cylinder were found by Parks (1961), for a ring by Barzykin and Merzhanov (unpublished Report of the Institute of Chemical Physics of the Academy of Sciences of the USSR; reproduction of their result can be found in Khudyaev 2003).

As evidenced in Table 3.1, the choice of any particular formula within each of the groups may be rather arbitrary. All the formulae require the knowledge of the eigenvalue η_1. Third formula in each group requires the knowledge of V, the second one requires the knowledge of P, and first equation requires the knowledge of both the latter geometrical parameters.

Determination of the eigenvalue η_1 is, in general case, a difficult problem. However, for many regions of practical interest this value may be found analytically, by solving BVP problem (3.34). Besides, various methods for approximate calculation of η_1 are available.

3.4 Stability of Thermal Explosion Solutions

In Chapter 2, a number of solutions, at conditions below critical, describing the thermal equilibrium of a reactive system were obtained. The question of stability of such solutions, that is their rigidity in the presence of small perturbations, unavoidable in any real physical systems was not, however, raised.

Meanwhile, in order for any mathematical solution to reflect the physically realizable state of a real system, it must be stable.

The present section concerns with the problem of stability of solutions developed in Chapter 2.

It should be noted, first of all, that in the framework of the Semenov theory (uniform temperature of the reacting system) the respective question is rather trivial, and in fact has been answered in Chapter 2. Specifically, the least root of the EGD function corresponds to stable solution, while the second largest root to unstable (see, e.g. Figure 2.3 for illustration).

The same problem considered for Frank-Kamenetskii solutions (Section 2.3, Chapter 2) is, however, much more involved.

The most logical and comprehensive treatment of this class of problems may be achieved in the framework of rather general mathematical theories of quasi-linear PDEs developed mostly in the monographs by Volpert and Khudyaev (1975) and Khudyaev (2003).

Stability of a steady-state solution is understood here as asymptotic stability in the sense of the definition (Khudyaev 2003) given below.

Consider BVP

$$\begin{aligned}&\nabla^2 u + a\left(\vec{\xi}\right)F(u) = 0\\&u|_{\partial G} = 0; G \subset \mathbb{R}^n, n = 2, 3\\&a(x) \geq 0, F(u) > 0\,for\,u \geq 0\end{aligned} \tag{3.51}$$

in the bounded, connected region G. Function $a\left(\vec{\xi}\right)$ is assumed here and below to be continuous.

Solution of this BVP will be interpreted as a steady-state solution of BVP for unsteady HTE

$$\begin{aligned}&\frac{\partial u}{\partial \tau} = \nabla^2 u + a\left(\vec{\xi}\right)F(u)\\&u|_{\partial G} = 0; G \subset \mathbb{R}^n, n = 2, 3\\&a\left(\vec{\xi}\right) \geq 0, F(u) > 0\,for\,u \geq 0\end{aligned} \tag{3.52}$$

Definition 3.1 (Khudyaev 2003).

Solution $u_0\left(\vec{\xi}\right)$ of BVP (3.51) is called stable, if there exists $\delta > 0$ and bounded positive functions $\psi\left(\vec{\xi}\right)$ and $g\left(\vec{\xi}\right)$ such that for any ε, $|\varepsilon| \leq 1$

(a) in the interval $\tau \in (0, \infty)$ there exists solution $u_\varepsilon\left(\tau, \vec{\xi}\right)$ of the Eq. (3.52) with the initial condition

$$u|_{\tau=0} = u_0\left(\vec{\xi}\right) + \varepsilon\psi\left(\vec{\xi}\right) \tag{3.53}$$

(b) for all $\vec{\xi} \in G$ and $\tau > 0$ the following inequality holds

$$\left|u_\varepsilon\left(\tau, \vec{\xi}\right) - u_0\left(\vec{\xi}\right)\right| \leq |\varepsilon| g\left(\vec{\xi}\right)\exp(-\delta\tau) \tag{3.54}$$

Inequality (3.54) means that the influence of the initial perturbation of the form $\varepsilon\psi\left(\vec{\xi}\right)$ diminishes with time.

Since condition (3.53) admits $\varepsilon = \pm 1$, then for any initial condition of the form

$$u|_{\tau=0} = u_0\left(\vec{\xi}\right) + \varphi\left(\vec{\xi}\right); \left|\varphi\left(\vec{\xi}\right)\right| \leq \left|\psi\left(\vec{\xi}\right)\right| \tag{3.55}$$

respective solution $u\left(\tau, \vec{\xi}\right)$ of the Eq. (3.52) is defined on the whole semi-space $\tau > 0$ and at $\tau \to \infty$ exponentially approaches $u_0\left(\vec{\xi}\right)$.

Indeed, the solutions $u_1\left(\tau, \vec{\xi}\right)$ and $u_{-1}\left(\tau, \vec{\xi}\right)$ of the problem (3.52) and (3.53) at $\varepsilon = \pm 1$ are, respectively, the upper and the lower functions (Khudyaev 2003) of the problem (3.52) and (3.55) for $0 < \tau < \infty$. Therefore, the solution of the latter problem is defined in this interval, and

$$u_{-1}\left(\tau, \vec{\xi}\right) \leq u\left(\tau, \vec{\xi}\right) \leq u_1\left(\tau, \vec{\xi}\right) \tag{3.56}$$

Inequality (3.54), which is satisfied for $u_{\pm 1}\left(\tau, \vec{\xi}\right)$, holds also for $u\left(\tau, \vec{\xi}\right)$, since

$$-g\left(\vec{\xi}\right)\exp(-\delta\tau) \leq u_{-1} - u_0 \leq u - u_0 \leq u_1 - u_0 \leq g\left(\vec{\xi}\right)\exp(-\delta\tau) \tag{3.57}$$

Consider now the case of problem with parameter, especially important for the theory of thermal explosion, that is, the modified version of BVP (3.51)

$$\nabla^2 u + \lambda a\left(\vec{\xi}\right)F(u) = 0$$
$$u|_{\partial G} = 0; G \subset \mathbb{R}^n, n = 2, 3 \tag{3.58}$$

Here parameter λ varies in the interval $0 < \lambda < \lambda_{cr} \leq \infty$. Let $u_\lambda\left(\vec{\xi}\right)$ denote the minimum (in pointwise sense) positive solution of BVP (3.58) for given λ.

Having in mind Definition 3.1, the following fundamental theorem may be proved.

Theorem 3.2 (Volpert and Khudyaev 1975).
If the following conditions

$$\inf a(x) > 0, F(u) > 0, \frac{\partial^2 F(u)}{\partial^2 u} \geq 0, for\ u \geq 0 \tag{3.59}$$

are met, then among positive solutions of problem (3.58) for $0 < \lambda < \lambda_{cr}$ *the minimum positive solution* $u_\lambda\left(\vec{\xi}\right)$, *and only this solution, is stable.*

This theorem is powerful, as it allows all the positive solutions, except just one, to be rejected as unphysical.

It may be also proved (Volpert and Khudyaev 1975, Khudyaev 2003) that at critical condition $\lambda = \lambda_{cr}$ solution of BVP (3.58) is unique and unstable.

Recalling now the results of Section 2.3, Chapter 2 the following conclusions may be drawn.

(a) Planar symmetry ($m = 0$). For any $\delta < \delta_{cr}$ solution (2.29) with the least root of the equation

$$ch(C_1) = \sqrt{\frac{2}{\delta}}C_1 \tag{3.60}$$

is stable; solution with the largest root is unstable.

Existence of the minimum solution follows from the chain of quite delicate theorems proved by Volpert and Khudyaev 1975. These theorems are not repeated here, and the details may be found in the latter monograph. The fact that the minimum solution exists allows the solutions (2.29) with different values of C_1 to be compared at just any one point. It is clear then that the minimum solution corresponds to the least root of Eq. (3.60).

Solution for $\delta = \delta_{cr}$ is unstable.

(b) Cylindrical symmetry ($m = 1$). For any $\delta < \delta_{cr}$ solution (2.40) with the least root

$$a = \sqrt{\frac{2}{\delta}} - \sqrt{\frac{2}{\delta} - 1} \tag{3.61}$$

of the Eq. (2.42) is stable; the solution with the largest root is unstable. In this case, the fact that the minimum solution corresponds to the root (3.61), the least of the two roots (2.44), is checked trivially.

Solution for $\delta = \delta_{cr}$ is unstable.

(c) Spherical symmetry ($m = 2$). For any $\delta < \delta_{cr,}$ the minimum positive solution is stable. According to the remark in point (a), such a minimum solution exists.

Solution for $\delta = \delta_{cr}$ is unstable.

3.5 Interpretation of Thermal Explosion in Terms of Theory of Catastrophes and Control Theory

The theory of thermal explosion has a close connection with more abstract branches of mathematics, in particular with the Theory of Catastrophes and Control Theory.

Catastrophes, formally speaking, are particular cases of a wider notion of bifurcations. Historically, catastrophes emerged as special kinds (versal families, or deformations) of bifurcations of mappings.

In practice (e.g. Arnold 1984), and definitely for the purpose of the present book, the terms "catastrophe" and "bifurcation" may be used interchangeably.

For a parameterised family of mathematical objects bifurcation (catastrophe), in broad sense, refer to a sudden qualitative (topological) change in the object's properties/structure upon small displacement in the parameters space.

A very simple illustration of the essence of the Semenov theory in terms of bifurcations may be developed as follows (Figure 3.13). Increase in the Semenov number corresponds to movement from the bottom to the upper part of the diagram. The bottom part shows two roots of Eq. (2.6), that is, the case $Se < Se_{cr}$ with the EGD function similar to curve 4 in Figure 2.2. The first bifurcation occurs when the two roots merge at $Se = Se_{cr}$, as shown in the middle part (intersection of all three lines) of Figure 3.13. At this point, the EGD function is similar to that shown by curve 3 in Figure 2.2. Upon further increase in the Semenov number the second bifurcation (denoted by dashed line in Figure 3.13) occurs that leads to the disappearance of the roots (in the region of realistic combustion temperatures) of the EGD function and correspondingly to the thermal explosion behaviour of the system at $Se > Se_{cr}$. The upper part of the diagram in Figure 3.13 shows an empty set of roots.

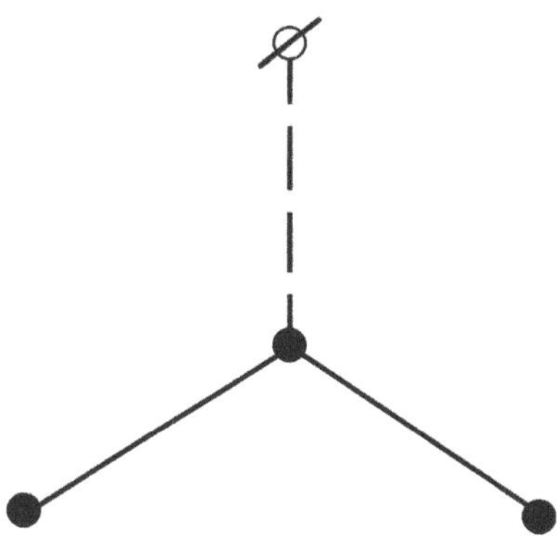

Figure 3.13 Bifurcation diagram for the roots of EGD function (in the region $0 < \theta < \theta_{inf\ l}$). ● – real root; Ø – empty set. Bottom row (two roots) corresponds to $Se < Se_{cr}$, middle row (one root) to $Se = Se_{cr}$, upper row (empty set of roots) to $Se > Se_{cr}$.

In a more comprehensive way, the Semenov theory is interpreted by a bifurcation (fold catastrophe) diagram presented in Figure 3.14. For greater clarity, values of dimensional physical parameter, namely heat transfer coefficient, are shown qualitatively along the horizontal axis. This axis, alternatively, may be thought as depicting values of the nondimensional Semenov number as the latter is inversely proportional to heat transfer coefficient (Chapter 2).

The equilibrium state of the reacting system may be followed as value of the heat transfer coefficient

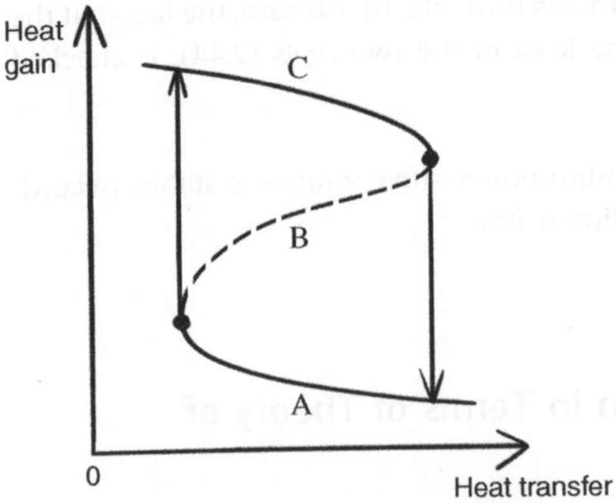

Figure 3.14 Fold catastrophe diagram for Semenov model, Eqs (2.5) and (2.6). Solid lines – stable branches, dashed lines – unstable branches. Arrows – discontinuous changes of state of the system (catastrophes). Upward arrow describes thermal explosion.

changes. For very large values of the heat transfer coefficient, the EGD function (2.6) has just one, stable, root. This situation is seen as the only intersection of vertical (i.e. at fixed value of heat transfer coefficient $\tilde{h}$) line with the curve (*ABC*). Such point of intersection corresponds to the state of the system on the curve's lower stable branch *A*. As heat transfer coefficient decreases, the point depicting the state of the system continues to evolve along the stable branch *A*. During this movement, the region in the parameter space where the EGD function has formally three roots is entered. At a fixed value of the heat transfer coefficient $\tilde{h}$, these three roots may be obtained as a vertical line intersections with the branches *A*, *B* and *C*. The branch *B* corresponds to an unstable middle root (as, e.g. for curve 3 in Figure 2.3) of the EGD function and is, therefore, unstable. The branch *C* is, similar to *A*, stable. It corresponds to the largest (Chapter 2) root of the EGD function, which corresponds to physically irrelevant temperature, but manifests thermal explosion.

As the point is achieved where the lower (*A*) and the middle (*B*) branches merge, the continuous evolution of the system upon further decrease in the heat transfer coefficient becomes impossible. The point depicting state of the system must jump to the upper branch *C*. This is the so-called fold catastrophe (upward arrow, Figure 3.14).

From that point, if heat transfer coefficient decreases further, the only intersection of vertical line (with the fixed value of $\tilde{h}$) with the curve is at the stable branch *C* (condition with completed thermal explosion).

If, on the other hand, the heat transfer coefficient starts to increase again, the system evolves along the branch *C* (formally attained post-explosion temperature starts to decrease) until such a state where the middle and the largest root merge. At this point, another fold catastrophe occurs (downward arrow, Figure 3.14) as upon the further increase in the heat transfer coefficient EGD function has the only root on the bottom branch *A*. Correspondingly, the equilibrium state of the system continues to evolve further to the right along the branch *A*.

In the classical Semenov theory, the state of a system is described by just one phase variable (uniform temperature) and just one parameter (Semenov number).

If the number of parameters is increased by one, then another, more complicated, cusp catastrophe may be observed (Figure 3.15).

Presently, one case of such system behaviour is known in the thermal explosion theory. This occurs in certain models describing compartment fire flashover as thermal runaway (Bishop *et al.* 1993). Here, the fire radius (i.e. essentially fire size) plays a role of the second parameter.

Consider the system behaviour at the fixed value of the heat transfer coefficient first.

At large values of heat transfer coefficient $\tilde{h}$ (i.e. when the rate of heat dissipation is high), increase in the fire radius leads to a continuous, gradual increase in the energy accumulation rate in the compartment (curve 1, Figure 3.15).

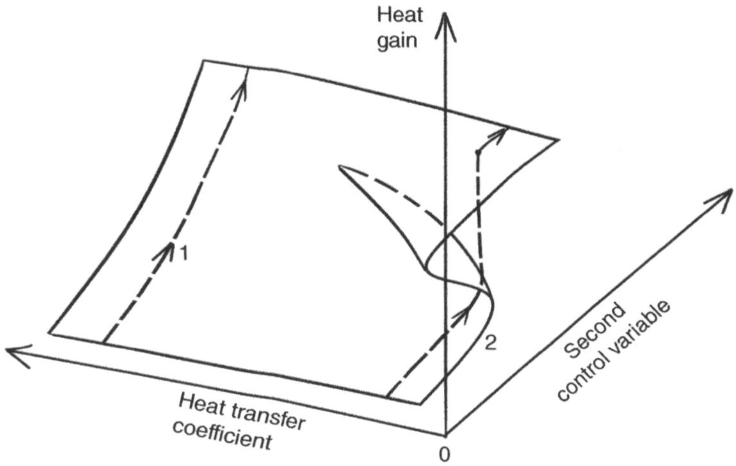

Figure 3.15 Cusp catastrophe diagram describing thermal explosion in a system with two parameters. Dashed curves: 1 – smooth evolution of the system equilibrium state; 2 – catastrophic evolution of the system equilibrium state (thermal explosion).

In contrast, at small values of $\tilde{h}$ behaviour of the system changes dramatically. At a certain point, as the fire radius increases, the state of the system must change dramatically (curve 2, Figure 3.15). This is a point of cusp catastrophe, manifesting thermal explosion.

It should be noted that in a certain range of fixed values of the second parameter (e.g. fire radius) behaviour of the system, upon changing heat transfer coefficient, would follow the fold catastrophe pattern of Figure 3.14. The curve similar to that shown in Figure 3.14 would be obtained by the cut of the cusp control surface (Figure 3.15) by the plane parallel to the ($\tilde{h}$– "Heat Gain") plane.

Models describing the compartment fire flashover as thermal explosion are discussed in Chapter 6.

Considering connection of thermal explosion with the control theory, it should be noted, first of all, that chemically reacting system (Figure 2.1) considered in the framework of the Semenov (or Frank-Kamenetskii) theory possesses an intrinsic mechanism of positive feedback. This mechanism is embodied into positive, monotonically increasing segment of the EGD function (2.6). In the classical thermal explosion theory there is no availability of dynamic control. Nevertheless, the effect of positive feedback may be formally considered within the framework of the control theory as "intrinsic" control. It is this feedback that drives the system towards thermal instability. As the system evolves along positive, monotonically increasing segment of the EGD function, energy release leads to a higher temperature of the system and, in turn, to progressively higher energy release rates. The temperature rise is self-accelerating and the system is caught in a positive feedback loop. This self-accelerating mechanism is presented in Figure 3.16 as a scheme typical for the

Figure 3.16 System with positive feedback.

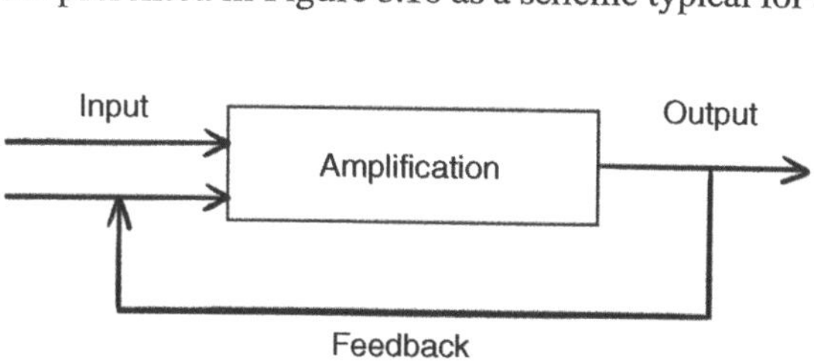

control theory. Here input and output should be interpreted as the energy release rate; the mechanism of amplification is the temperature rise of the reacting mixture.

Further, if external action on the system may be performed, then a control problem in a strict sense of the control theory may be posed. It is clear that such a problem must be the one of preventing thermal explosion. In a simplest formulation, the Semenov number may be considered as control parameter which is changed either in abrupt or continuous way. Along those lines, various control problem formulations are considered in Chapter 8.

3.6 Review of Other Results in Mathematical Theory of Thermal Explosion

Theory of thermal explosion has attracted, over nearly a hundred years of its existence, significant interest from both physicists and mathematicians.

The most comprehensive mathematical treatment of the theory, so far, has been offered by Volpert and Khudyaev (1975) and Khudyaev (1963, 1964, 2003) in the course of more general development of the theory of quasi-linear elliptic PDEs.

Solutions of this type of equations are steady-state solutions of the theory of thermal explosion. Therefore, the study of this class of PDEs allows the questions of existence of solutions describing thermal equilibrium in the system and their stability to be resolved.

Major results of the studies by Volpert and Khudyaev (1975) and Khudyaev (1963, 1964, 2003) may be summarized briefly as follows.

Mixed boundary conditions (of the first and the third type on disjoint parts of the boundary) have been considered. Criterion of solvability has been formulated in terms of the existence of lower and upper functions (belonging to the space $BV^2(G)$) of the problem.

Further, the case of the elliptic BVP with the parameter has been considered. This case is directly related to thermal explosion theory, as the Frank-Kamenetskii number may be chosen as a parameter. For such a case, the existence of a positive critical value of the parameter has been established in the sense of the following alternative. Either elliptic BVP with parameter is solvable for all non-negative values of that parameter, or such a critical positive value of the parameter exists that the problem is solvable for all the values of the parameter less than critical, and not solvable for all the parameter values larger than critical. The criterion for the existence of a critical value has been established. Upper bound estimation, as well as the asymptotic formula for the critical value have been derived.

The stability of solutions, including solutions for the problem with parameter, has been investigated in the framework of the asymptotic stability definition, which generalizes corresponding definition from the theory of ODEs. Several equivalent criteria of stability have been established. Some of the most important corollaries of those criteria (Theorem 3.2, and uniqueness and instability of solution at the critical value of the parameter) have already been discussed above in Section 3.4.

It must be appreciated that all the results discussed above have been obtained under rather mild, natural restrictions on the function describing source term (such as this function being non-negative), that is in much more general form compared to the Frank-Kamenetskii model formulation (exponential source term).

These results have been used in Khudyaev (1995) for explicit analysis of the stability of Frank-Kamenetskii model solutions in the case of spherical symmetry.

Historically, the stability of symmetrical solutions of the Frank-Kamenetskii model has been investigated by far less general methods than those developed by Volpert and Khudyaev (1975) and Khudyaev (2003). A classical work in this regard is due to Istratov and Librovich (1963), where the question of stability was resolved for the first time. These investigators used a standard approach of perturbing steady-state solutions and following perturbation evolution in time. It is worthwhile to note, however, that their approach to construction and investigation of stability of steady-state solutions was based on the invariance of the Frank-Kamenetskii equation under transformations belonging to a certain Lie group. This method was initially introduced to the theory of thermal explosion by Barenblatt, and was first described in the section that he wrote for a more general review by Gelfand (1959).

With regard to the region of arbitrary shape, the result obtained independently by Kaganov (1965) and somewhat later, by different method, by Sivashinskii (1967) is noticeable. The result says that if steady-state solution exists for a particular region, it also exists for any embedded region. This circumstance allowed Boddington *et al.* (1971) to obtain, using appropriate embeddings, estimations of the critical Frank-Kamenetskii parameter for the bodies of various shapes. It may be derived further (Sivashinskii 1967) that stable steady-state solution always exists for a vessel of sufficiently small size, irrespective of its shape. The latter fact is, of course, perfectly clear from physical reasoning.

Thus, Kaganov (1965) and Sivashinskii (1967) proved the existence and stability of solutions "in the small". As important as the latter result may look, it is far less general than the one obtained by Volpert and Khudyaev (1975) and Khudyaev (2003), who established the existence of solutions, for the arbitrary shape of vessel, for all parameter values less than critical. The reproduction of the result by Kaganov (1965) and Sivashinskii (1967) may also be found in Zeldovich *et al.* (1980).

Of considerable interest is the generalization of results of the thermal explosion theory for energy source functions more general than exponential and obeying some natural properties (e.g. non-negativeness, convexity, and the like). Much progress in this direction has been made, in fact, by Volpert and Khudyaev (1975) and Khudyaev (2003) who try to formulate their results in the most general form. Also, Kaganov (1967) addressed the problem of the stability of the thermal explosion theory solutions for an arbitrary energy generation function and an arbitrary region.

Historically, the stability of symmetrical solutions of the Frank-Kamenetskii model has been investigated by far less general methods than those developed by Volpert and Khudyaev (1975) and Khudyaev (2003). A classical work in this regard is due to Istratov and Librovich (1963), where the question of stability was resolved for the first time. These investigators used a standard approach of perturbing steady-state solutions and following perturbation evolution in time. It is worthwhile to note, however, that their approach to construction and investigation of stability of steady-state solutions was based on the invariance of the Frank-Kamenetskii equation under transformations belonging to a certain Lie group. This method was initially introduced to the theory of thermal explosion by Barenblatt, and was first described in the section that he wrote for a more general review by Gelfand (1959).

With regard to the region of arbitrary shape, the result obtained independently by Kaganov (1965) and somewhat later, by different method, by Sivashinskii (1967) is noticeable. The result says that if steady-state solution exists for a particular region, it also exists for any embedded region. This circumstance allowed Boddington et al. (1971) to obtain, using appropriate embeddings, estimations of the critical Frank-Kamenetskii parameter for the bodies of various shapes. It may be derived further (Sivashinskii 1967) that stable steady-state solution always exists for a vessel of sufficiently small size, irrespective of its shape. The latter fact is, of course, perfectly clear from physical reasoning.

Thus, Kaganov (1965) and Sivashinskii (1967) proved the existence and stability of solutions in [illegible] [illegible] [illegible] [illegible] [illegible] [illegible] [illegible] [illegible] [illegible]

4

Thermal Explosion in a Quiescent Medium

4.1 Kinetic Effects

The present chapter considers the effects of reactant consumption in the system.

Consider, first of all, a modified Semenov model assuming, without the loss of generality, monomolecular reaction of the order *m*.

Energy conservation equation, instead of (2.1), assumes now the following form:

$$\tilde{\rho}\tilde{c}_p \frac{d\tilde{T}}{d\tilde{t}} = \tilde{Q}\tilde{A}\exp\left(-\frac{\tilde{E}}{\tilde{R}\tilde{T}}\right)\tilde{C}^m - \tilde{h}\frac{\tilde{S}}{\tilde{V}}\left(\tilde{T} - \tilde{T}_0\right) \tag{4.1}$$

Non-dimensional variables of Chapter 2 are modified in the following way:

$$\begin{aligned} \theta &= \frac{\left(\tilde{T} - \tilde{T}_0\right)}{\left(\tilde{R}\tilde{T}_0^2/\tilde{E}\right)}; \tau = \frac{\tilde{Q}\tilde{k}_0\tilde{C}_0^m\tilde{E}}{\tilde{\rho}\tilde{c}_p\tilde{R}\tilde{T}_0^2}\tilde{t} \\ \mathrm{Ar} &= \frac{\tilde{R}\tilde{T}_0}{\tilde{E}}; \mathrm{Se} = \frac{\tilde{Q}\tilde{k}_0\tilde{C}_0^m\tilde{V}\tilde{E}}{\tilde{h}\tilde{S}\tilde{R}\tilde{T}_0^2} \end{aligned} \tag{4.2}$$

Transformations similar to those in Chapter 2 lead to the following final form of the energy balance equation:

$$\frac{d\theta}{d\tau} = \exp\left(\frac{\theta}{1 + \mathrm{Ar}\theta}\right)\varphi(\eta) - \frac{1}{\mathrm{Se}}\theta \tag{4.3}$$

where

$$\varphi(\eta) = (1 - \eta)^m \tag{4.4}$$

is the so-called kinetic function depending on the progress variable

$$\eta = 1 - \frac{\tilde{C}}{\tilde{C}_0} \tag{4.5}$$

Consider now the kinetic equation

$$\frac{d\tilde{C}}{d\tilde{t}} = -\tilde{w} \tag{4.6}$$

Thermal Explosion: Theory and Application, First Edition. Vasily B. Novozhilov.

assuming

$$i = 1; \quad \tilde{C}_1 \equiv \tilde{C}; \quad \upsilon_1 = m \tag{4.7}$$

Introducing an extra non-dimensional parameter, the so-called Todes number

$$Td = \frac{\tilde{Q}\tilde{E}\tilde{C}_0}{\tilde{\rho}\tilde{c}_p\tilde{R}\tilde{T}_0^2} \tag{4.8}$$

and using basic kinetics formulae (Section 1.3, Chapter 1), Eq. (4.6) may be written in the following non-dimensional form:

$$\frac{d\eta}{d\tau} = \frac{1}{Td}\exp\left(\frac{\theta}{1+\mathrm{Ar}\theta}\right)\varphi(\eta) \tag{4.9}$$

Corresponding equation in the Frank–Kamenetskii formulation (with spatially distributed temperature) may be derived in completely analogous way.

The initial condition for Eqs. (4.3) and (4.9) may be set (in the simplest form) as

$$\theta = 0; \quad \eta = 0 \ \text{ at } \ \tau = 0 \tag{4.10}$$

Even in the Semenov formulation (Eqs. (4.3), (4.9) and (4.10)) accounting for reactant depletion analytically presents significant challenges.

It is easy to see that the only stationary point for Eq. (4.9) is $\eta = 1$ ($\varphi = 0$), which corresponds to full reactant consumption ($\tilde{C} = 0$). Therefore, the set of Eqs. (4.3) and (4.9) has no stationary points corresponding to the finite positive rate of a chemical reaction. Therefore, the steady-state theory of thermal explosion cannot be developed in the same way as it was done for the Semenov model with the zero-order chemical reaction in Chapter 2. Critical conditions must be found from the non-steady theory, applying one of the approximate methods.

Such available methods include the quasi-stationary theory (Merzhanov and Dubovitskii 1958, 1960; Volpert and Khudyaev 1975) and, for dynamical regimes, for example, with a linear increase of ambient temperature, pseudo-isothermal approximation (Barzykin 2004).

Another approximate method that can be easily employed is the numerical integration of the set of Eqs. (4.3), (4.9) and (4.10), combined with applying definition of thermal explosion (temperature growth to the point of maximum of the second derivative of the EGD function) adopted in the present book.

Results of such numerical solution are presented in Figures 4.1 and 4.2. Figure 4.1 shows temperature time histories at different values of the Semenov parameter. Due to reactant consumption temperature of the system rises initially, but then starts to decrease as progress variable reaches its maximum value $\eta = 1$ (reaction is completed), and the only remaining mechanism described by Eq. (4.3) is energy sink (second term on the RHS). Curve 2 (Figure 4.1) has its maximum at around $\theta \approx 1{,}400$, very close to the maximum of the second derivative of the EGD function (without kinetic factor, Table 1.1, Chapter 1). Therefore, curve 2 with Se = 0.42 corresponds, essentially, to critical conditions at Ar = 0.012. Curve 3 in Figure 4.1 corresponds to conditions above critical (descending part of that curve is not shown). Its maximum is well above the point of the maximum of the second derivative of the EGD function (i.e. $\gg \approx 1{,}375$, see Table 1.1). Finally, curve 1 refers to subcritical conditions. It shows a weak temperature rise, followed by its continuous decrease.

Corresponding Figure 4.2 demonstrates the behaviour of the progress variable for the same three cases. The progress variable reaches its maximum $\eta = 1$ for curves 2 and 3, while reaction continues for curve 1 ($\eta < 1$).

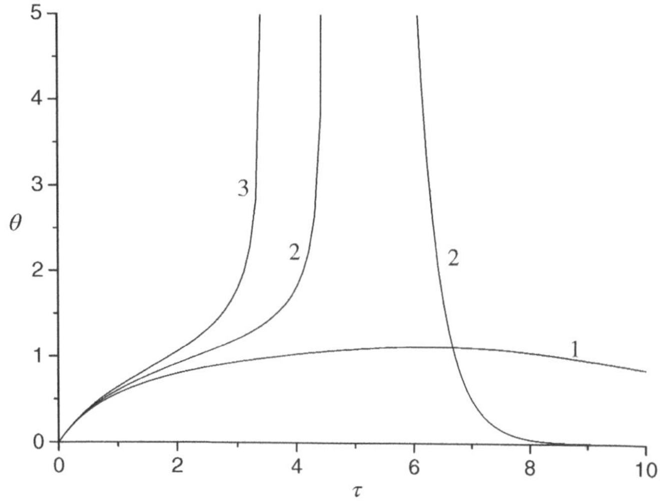

Figure 4.1 Temperature histories for the Semenov model (4.3, 4.9, 4.10) with reactant consumption. Reaction of the first order $m = 1$. Ar = 0.012; Td = 200.0. 1 – Se = 0.40; 2 – Se = 0.42; 3 – Se = 0.45.

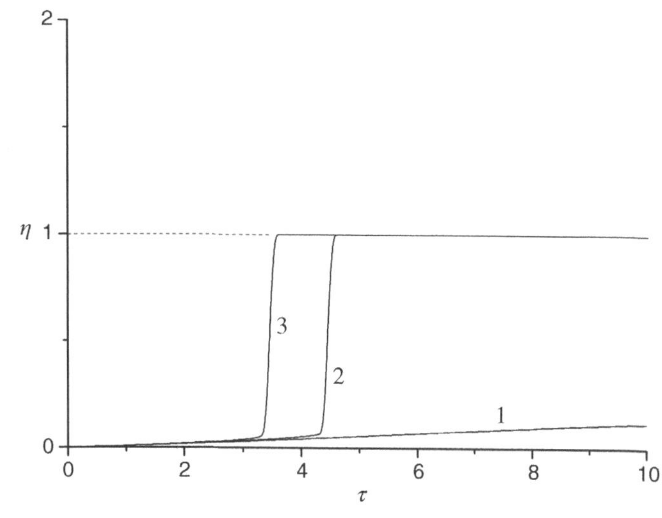

Figure 4.2 Progress variable histories for the Semenov model (4.3, 4.9, 4.10) with reactant consumption. Reaction of the first order $m = 1$. Ar = 0.012; Td = 200.0 1 – Se = 0.40; 2 – Se = 0.42; 3 – Se = 0.45.

Critical values of the Semenov number may be calculated based on the thermal explosion criterion established in Chapter 1, using Table 1.1. It turns out that in the practical range of Arrhenius numbers, $0.010 \leq \text{Ar} \leq 0.020$, dependence of critical Semenov parameter on that number is rather weak (Figure 4.3).

Consideration of the problem with spatially distributed temperature (Frank–Kamenetskii model) demonstrates essentially the same features. Such a problem was investigated in detail, again via numerical analysis, in the studies by Barzykin *et al.* (1963, 1966, 1996).

It should be noted that apart from the kinetic function (4.4) for the monomolecular reaction of the order m other forms of kinetic function may be considered. Of special interest is, for example, the kinetic function corresponding to autocatalysis (reaction rate is chemically enhanced by the product of the same reaction) of the order m

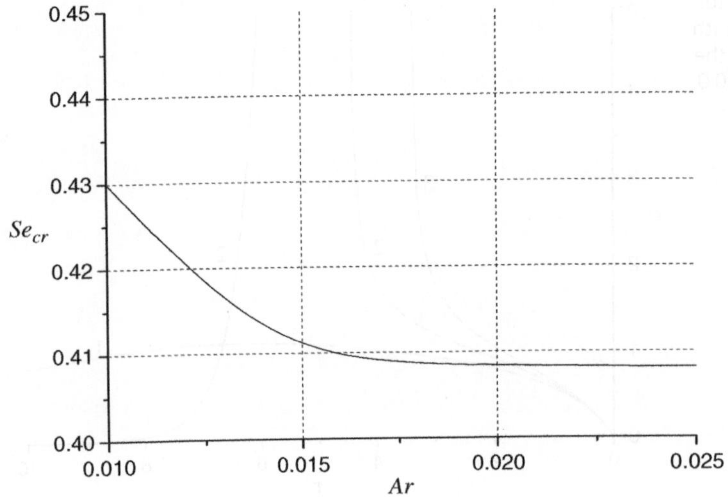

Figure 4.3 Dependence of the critical Semenov parameter on the Arrhenius number. Reaction of the first order $m = 1$.

$$\varphi(\eta) = (\eta + \eta_0)(1 - \eta)^m \tag{4.11}$$

where parameter η_0, the criterion of autocatalysis, expresses the ratio of the initial reaction rate to autocatalytic constant. The thermal explosion problem for the autocatalytic reaction of the first order was considered by Merzhanov and Dubovitskii (1960), Barzykin *et al.* (1996) and Novozhilov (2018).

Various aspects of the thermal explosion problem with reactant consumption have also been considered in a number of other studies. While, inevitably, most of results were obtained by numerical analysis, a limited number of revealing analytical asymptotic solutions have also been made available.

Thomas (1961) conducted a theoretical analysis of the problem considering a symmetrically heated uniform body undergoing the decomposition reaction of the order m. The major approximation involved was the estimation of the Arrhenius term in the conservation equation for reactants using the mean temperature. Therefore, spatial temperature variation was ignored, that is, the problem was essentially considered in the formulation similar to the Semenov one. This approximation, nevertheless, allowed the conservation equations for energy and reactant concentration to be decoupled. Therefore, a time-dependent solution for the concentration of reactant could be substituted into the energy equation. The latter one was investigated in the limit of large adiabatic temperature rise. This analysis led to asymptotic analytical expressions for the effect of reactant consumption on explosion critical conditions and its induction period. Conclusions of the study were validated against available numerical solutions.

Squire (1963) considered the problem for the reaction of the first order in a super-critical regime. This study also employed a number of simplifications assuming, for example, a linear relationship between temperature and reactant concentration. Critical conditions and the induction period were estimated numerically.

A study, similar in nature to those of Thomas (1961) and Squire (1963), was conducted by Adler and Enig (1964). These authors also used forms of conservation equations simplified with the use of average temperature and concentration. They have arrived at equations from which critical conditions may be found as a function of model parameters, assuming that the rise to the adiabatic combustion temperature separates sub- and super-critical regimes. These equations had to be solved numerically.

A more rigorous mathematical analysis has been developed by Kassoy and Linan (1978) and Babushok *et al.* (1990). Both these studies considered the Semenov (spatially uniform temperature) formulation, but with a differently defined nondimensional time, explicitly making the energy equation singularly perturbed. A critical value of the non-dimensional heat loss parameter was found in the form of an asymptotic formula. These asymptotic expressions agree reasonably well between the two papers. Babushok *et al.* (1990) used specific methods of non-linear and relaxation oscillations (Bogolyubov and Mitropolskii 1967; Mischenko and Rozov 1975) to separate dynamics in the phase space (θ, η) into slow, transient and fast (explosive) trajectories. There is a fundamental difficulty in establishing mathematically rigorous critical condition for the problem with reactant consumption. The authors point out that while rigorous separation between slow and slow-transient trajectories is possible, 'unfortunately, the boundary between the regions of transient and explosive reaction behaviours cannot be established rigorously ...'. In the view of this fundamental difficulty, the definition of critical condition adopted in the present book, that is temperature growth to the point of the maximum of the second derivative of the EGD (even without kinetic factor) function seems to be a reasonable approach.

Another attempt to conduct a comprehensive analysis of the problem was made by Shouman and El-Sayed (1992, 1997, 1998, 1999). It embraced both analytical and numerical approaches. They, in line with a number of previous studies, defined criticality as an inflection point of the time-dependent temperature solution. As follows from the explanation in Chapter 1, this definition leads to unrealistically high temperatures at critical conditions. (It is easy to show that the inflection point on the time-dependent temperature solution lies, for realistic values of the Semenov number, to the right from the inflection point of the EGD function and thus corresponds to even higher temperatures.) The results of this series of studies were applied by Shouman (1999) to analyse the problem of dust gas explosions.

While the studies reviewed above were mostly concerned with the mathematical side of the problem and employed simplest and highly idealized kinetic mechanisms, Prince and Williams (2012) made prediction of ignition times of practical propane and ethane mixtures considering detailed, deduced from a very complicated kinetic scheme (over 200 elementary reactions involving 40 species), kinetic mechanisms.

Of particular interest are also the dynamical regimes of thermal explosion with reactant consumption.

The problem with linear heating has been considered in detail by Barzykin (2004). Much more complicated behaviour, which has to be treated purely numerically, arises in the case of oscillating ambient conditions (background temperature). Analysis of such a problem, with spatially distributed temperature (Frank–Kamenetskii formulation with the three conventional cases of symmetry), has been performed by Novozhilov (2018).

The BVP that substitutes the case of the zero-order reaction (3.32) has the form:

$$
\begin{aligned}
&\frac{\partial \theta}{\partial \tau} = \frac{1}{\delta}\left(\frac{\partial^2 \theta}{\partial \xi^2} + \frac{m_s}{\xi}\frac{\partial \theta}{\partial \xi}\right) + \exp\left(\frac{\theta}{1 + \mathrm{Ar}\theta}\right)\varphi(\eta) \\
&\frac{d\eta}{d\tau} = \frac{1}{Td}\exp\left(\frac{\theta}{1 + \mathrm{Ar}\theta}\right)\varphi(\eta) \\
&\xi \in [0, 1]; \quad \tau \in [0, \infty] \\
&\frac{\partial \theta}{\partial \xi}(0, \tau) = 0; \quad \theta(1, \tau) = \mathrm{Zn} + A_0 \sin(\omega \tau)
\end{aligned}
\tag{4.12}
$$

or

$$\frac{\partial\theta}{\partial\xi}(0,\tau) = 0;\ \frac{\partial\theta}{\partial\xi}(1,\tau) = \mathrm{Zn} - \mathrm{Bi}(\theta - A_0\sin(\omega\tau))$$

Here, in order to consider a more general case, an extra nondimensional parameter, the Zeldovich number

$$Zn = \frac{\tilde{T}_a - \tilde{T}_0}{\left(\frac{\tilde{R}\tilde{T}_0^2}{\tilde{E}}\right)} \tag{4.13}$$

is introduced. This parameter measures the ratio of the difference between ambient and initial mixture temperatures to the characteristic temperature interval of the chemical reaction.

The parameters A_0 and ω are defined in exactly the same way as in Section 3.2, Chapter 3. Only the case of planar symmetry $m_s = 0$ is considered below.

In terms of chemical kinetics, reactions of the first and the second order, as well as the first-order autocatalysis are considered.

The behaviour of the system is controlled by the following sets of parameters. The most important is the Frank–Kamenetskii parameter δ, which describes critical conditions for the system evolution. In calculations, this parameter is being assigned a subcritical value (for the static, i.e. $A_0 = 0$) conditions. The positive difference between its critical value δ_{cr} at static ambient conditions and the actual value describes the degree of the subcriticality of the system.

Therefore, the following sets of parameters:

(δ, Ar, Td, Zn, A_0, ω) – for the boundary conditions of the first type, and,
(δ, Bi, Ar, Td, Zn, A_0, ω) – for the boundary conditions of the second type, control evolution of the system.

Critical conditions are described in terms of dependencies of nondimensional amplitude A_0 at the critical boundary on a particular parameter of interest, with all the other parameters being fixed. The latter dependence is determined by separating solutions growing up to the point $\theta = \theta_{F^{(2)}_{\max}}$ from bounded trajectories. In the present intrinsically non-linear problem, this can only be done numerically, to within a certain accuracy. The full description of a respective numerical procedure may be found in Novozhilov (2016, 2018).

In view of a relatively large number of parameters involved, it is essential to choose the most important dependencies that describe the system behaviour both qualitatively and quantitatively. To this extent, the most important parameters that affect the position of the critical boundary, as well as the ranges of their variation, are presented in Table 4.1.

Critical conditions in terms of dependencies $A_0(\omega)$ are presented in Figures 4.4–4.6. The dependency of critical conditions on different kinetic schemes is investigated in Figure 4.4. The

Table 4.1 Input data for the parametric analysis of critical conditions.

Parameter	Range of variation	Boundary conditions	Presenting figures
Oscillation frequency, ω	0.00 – 10.00	First type and second type	Figures 4.4–4.6
Biot number, Bi	5.00 – 20.00	Second type	Figures 4.5 and 4.6
Arrhenius number, Ar	0.00 – 0.30	Second type	Figure 4.7
Zeldovich number, Zn	−2.00 – 2.00	Second type	Figure 4.8

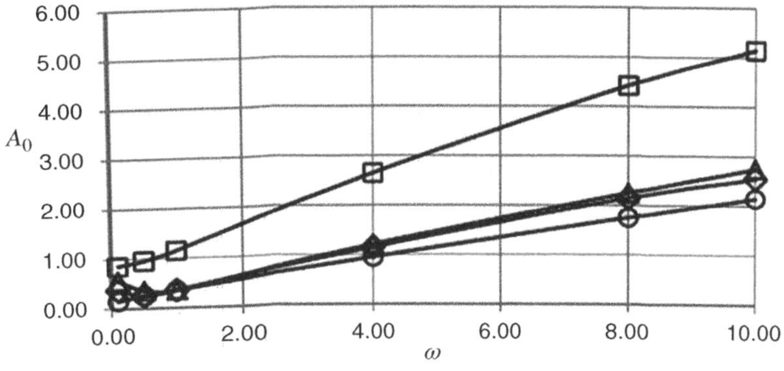

Figure 4.4 Critical dependencies $A_0(\omega)$. Boundary conditions of the first type. $\delta = 0.8$; Ar = 0.025; Td = 200.0; Zn = 0.0; $\eta_0 = 0.5$. ○ – zero-order (no kinetic effects); ◇ – first-order kinetics; Δ – second-order kinetics; □ – first-order autocatalysis. *Source:* Novozhilov (2018) / Springer Nature / CC BY 4.0.

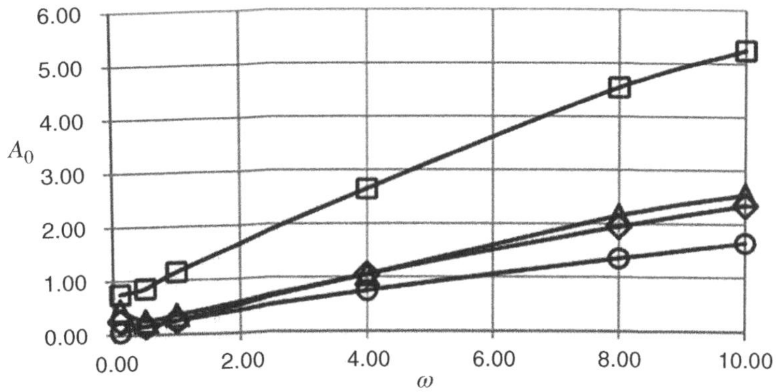

Figure 4.5 Critical dependencies $A_0(\omega)$. Boundary conditions of the second type. $\delta = 0.6$; Ar = 0.025; Td = 200.0; Zn = 0.0; Bi = 5.0; $\eta_0 = 0.5$. ○ – zero-order kinetics; ◇ – first-order kinetics; Δ – second-order kinetics; □ – first-order autocatalysis. *Source:* Novozhilov (2018) / Springer Nature / CC BY 4.0.

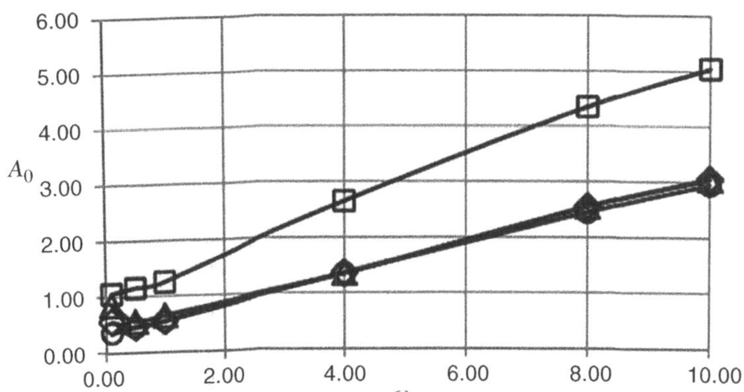

Figure 4.6 Critical dependencies $A_0(\omega)$. Boundary conditions of the second type. $\delta = 0.6$; Ar = 0.025; Td = 200.0; Zn = 0.0; Bi = 20.0; $\eta_0 = 0.5$. ○ – zero-order kinetics; ◇ – first-order kinetics; Δ – second-order kinetics; □ – first-order autocatalysis. *Source:* Novozhilov (2018) / Springer Nature / CC BY 4.0.

Frank–Kamenetskii parameter is fixed at $\delta = 0.8$, that is, below the critical one for the constant unperturbed ambient temperature case $A_0 = 0$, which is as well known (Table 2.1, Chapter 2) is $\delta_{cr} = 0.88$. As evident from Figure 4.4, the dependencies given by the first- and second-order kinetics are very close to the case of zero-order kinetics. This suggests that the effect of reactant consumption (close to critical conditions) is rather small. In contrast, autocatalysis kinetics exerts the major effect on critical conditions for thermal explosion. Respective values of the critical amplitude (at given frequency) are significantly higher than in the absence of any kinetic effects. This is due to the fact that the reaction in this case is slow initially (at small values of progress variable), before it grows to reach the maximum at $\eta = (1 - \eta_0)/2$ due to the accumulation of reaction products. As expected, the curve corresponding to the first-order kinetics lies between the curves for zero- and second-order reactions, as the increased degree of reactant consumption makes it more difficult for explosion to occur. The least favourable conditions for thermal explosion correspond to autocatalytic kinetics.

It is also evident from Figure 4.4 that, except for the range of low frequencies $0 < \omega \leq \approx 1$ for zero-, first- and second-order kinetics, all the critical dependencies of amplitude on frequency are very close to linear. A low-frequency range is, however, the most important for practical applications (Novozhilov 2016).

An important conclusion following immediately from Figure 4.4 is that, in line with somewhat earlier results by Novozhilov (2016) for the zero-order reaction, the system which is thermally stable under constant ambient conditions may develop instabilities (thermal explosion) if the latter conditions vary in the oscillating manner, even if the chemical reaction is being damped by reactant consumption.

Application of the boundary conditions of the second type results in similar trends. The corresponding results are presented in Figures 4.5 and 4.6 for the two different intensities of heat losses, that is, for different values of the parameter Bi.

The critical curve for the autocatalytic reaction is positioned significantly higher than the other curves, indicating much higher critical Frank–Kamenetskii parameters for thermal explosion. This is due to the fact that a significant pre-accumulation of reaction products is needed before the reaction reaches a considerable rate required for explosion to occur. Such pre-accumulation, in turn, requires a significant ambient temperature increase, that is, a significant amplitude of oscillations during the induction period.

A higher intensity of heat loss for the case presented in Figure 4.6 results in higher critical Frank–Kamenetskii parameters, compared to Figure 4.5, for all the kinetic mechanisms except autocatalysis. The critical boundary for the autocatalytic reaction is not affected by the change in Biot numbers. Again, this is due to the fundamental difference of this mechanism from the others. In the case of autocatalysis, the heat generation rate (controlled by the reaction rate) is smaller during the induction period than in the three other cases, and therefore not significantly affected by changed energy loss conditions at the boundary.

Figures 4.4–4.6 demonstrate essentially linear dependencies of critical ambient temperature oscillation amplitudes on the frequency of oscillations. This is explained as follows. The thermal stability of the reacting system is effectively determined by its behaviour at the temperatures close to the maximum (through oscillation cycle). If this maximum temperature is fixed, then the thermal explosion induction period (at this temperature) becomes larger than the period of oscillations as the frequency of the latter increases. Consequently, along the critical boundary, the oscillation amplitude must increase with the increase of frequency. Constant rate of increase indicates that along the critical boundary product of the maximum oscillation temperature and the time over which ambient temperature stays close to its maximum remains nearly constant. The latter time is inversely proportional to the frequency of oscillations.

It is instructive to interpret critical conditions presented in Figures 4.4–4.6 in terms of dimensional parameters. Using the definitions of non-dimensional amplitude and frequency (Section 3.2, Chapter 3), the conditions leading to thermal explosion may be expressed in the space of real physical parameters as

$$\tilde{A}_0 > \left(\frac{\tilde{R}\tilde{T}_0^2}{\tilde{E}}\right) f\left(\frac{\tilde{\rho}\tilde{c}_p\tilde{R}\tilde{T}_0^2}{\tilde{Q}\tilde{k}_0\tilde{E}}\tilde{\omega}\right) \tag{4.14}$$

where $f(\omega)$ represents any of the relevant critical curves from Figures 4.4 to 4.6. Then, the inequality (4.14) estimates the risk of thermal runaway in any system with particular thermodynamic and kinetic properties experiencing the oscillations of ambient temperature with given amplitude and frequency. The realistic values of the latter parameters have been estimated by Novozhilov (2016) for daily or seasonal temperature variations as $\tilde{A}_0 \le \approx 20\,°\mathrm{C}$ (overall ambient temperature variation within $\approx 40\,°\mathrm{C}$) and $6 \cdot 10^{-7}\mathrm{s}^{-1} \le \approx \tilde{\omega} \le \approx 6 \cdot 10^{-5}\mathrm{s}^{-1}$. Daily and seasonal variations are relevant for specific types of combustion systems (Novozhilov 2018), where thermal explosion may occur over prolonged time periods.

The effect of the Arrhenius number is illustrated in Figure 4.7. Critical dependencies $A_0(\mathrm{Ar})$ demonstrate a quite peculiar behaviour. The two competitive effects appear in governing Eqs. (4.3) and (4.9) when the Arrhenius number varies. The rate of the chemical reaction decreases with the increase in the Arrhenius number. Therefore, the generation of the energy sufficient for thermal runaway would require a larger critical oscillation amplitude. On the other hand, the consumption rate of the reactant drops along with the decrease in the reaction rate. The latter effect contributes to the decrease in the critical amplitude with rising Arrhenius number.

Competition between these two effects results in the existence of maximum on critical curves, observed in Figure 4.7. In the range of small Arrhenius numbers, an increase in the critical amplitude is dictated by the dependence of the heat generation rate on this parameter. As the Arrhenius number increases further, the kinetic effect (decrease in the reactant consumption rate) becomes dominant and critical amplitude gradually decreases. As the reactant consumption rate becomes small, it may be expected that the critical dependence $A_0(\mathrm{Ar})$ becomes eventually independent

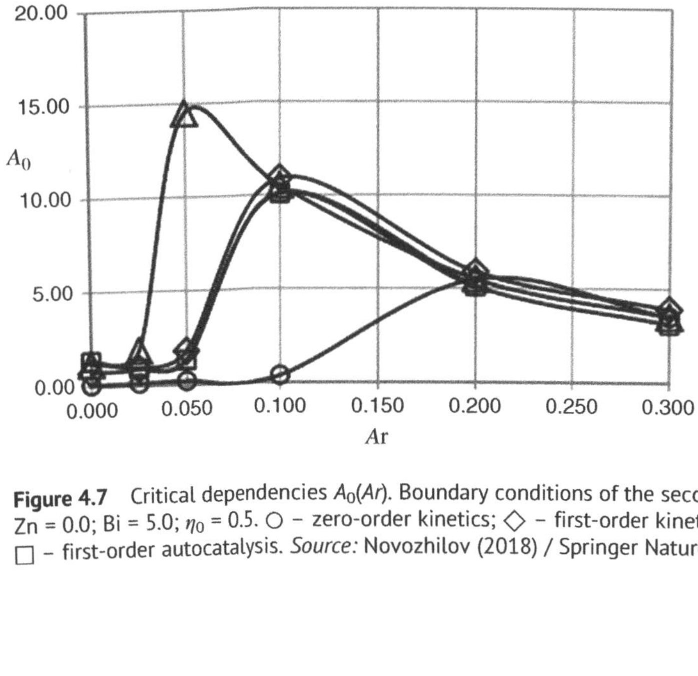

Figure 4.7 Critical dependencies $A_0(Ar)$. Boundary conditions of the second type. $\omega = 0.5$; $\delta = 0.6$; Td = 20.0; Zn = 0.0; Bi = 5.0; $\eta_0 = 0.5$. ○ – zero-order kinetics; ◇ – first-order kinetics; Δ – second-order kinetics; □ – first-order autocatalysis. *Source:* Novozhilov (2018) / Springer Nature / CC BY 4.0.

of a particular kinetic mechanism. This is exactly what is seen in Figure 4.7 where all the curves merge for Ar > ≈ 0.2.

It is also useful to consider the effects of other control parameters, for example, the Zeldovich number (Figure 4.8). Practically important range of Zeldovich numbers is estimated in a straightforward way (Novozhilov 2018) taking, for example, a typical value of Arrhenius number as Ar = 0.03. The characteristic temperature interval (at the ambient temperature of 300 K) would be of the order of 10 K. It is easy to see from here that the Zeldovich number value of the order of unity corresponds to approximately 10 °C difference between initial and ambient temperatures. Therefore, the range of Zeldovich numbers presented in Figure 4.8 ($-2 \leq \text{Zn} \leq 2$) covers the range of ambient temperature variations within approximately 40 °C. As explained above, this is a realistic range for seasonal variations. Daily variations are typically much smaller.

Positive values of the Zeldovich number refer to the cases where the additional energy supply from the surroundings to the system occurs, leading to a substantial decrease in the critical Frank–Kamenetskii parameters. In the opposite case of negative Zeldovich numbers, the system initially loses heat to the surroundings with lower temperature. This causes the amplitude of temperature variations, needed to drive the system toward thermal explosion, to increase. Hence, critical Frank–Kamenetskii parameters, at fixed amplitude, increase as well.

The above considerations explain the descending behaviour of the critical curves in Figure 4.8. The actual shape of these curves is perfectly linear. This indicates that the initial heat flux between the system and its surroundings, proportional to Zn, plays a decisive role in determining critical conditions. Indeed, the influence of ambient temperature oscillations is being felt by the system over a finite time scale, while the effect of the heat flux, controlled by the Zeldovich number, is being exerted on the system instantaneously.

Of particular significance is a discovery of an oscillatory mechanism of transition to thermal explosion, with temperature oscillations during an induction period (Novozhilov 2016). The model (4.12) allows implications of different kinetic mechanisms for this particular scenario to be investigated. It turns out that not only is this case of thermal instability development possible for any of the kinetics schemes considered, but actually represents an inevitable scenario upon the smooth transition of controlling parameters across the stability boundary.

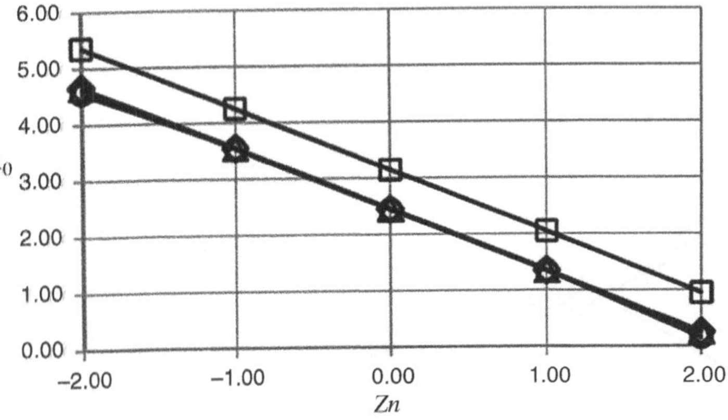

Figure 4.8 Critical dependencies $A_0(Zn)$. Boundary conditions of the second type. $\omega = 4.0$; $\delta = 0.1$; Ar = 0.025; Td = 200.0; Bi = 5.0; $\eta_0 = 0.5$. ○ – zero-order kinetics; ◇ – first-order kinetics; Δ – second-order kinetics; □ – first-order autocatalysis. *Source:* Novozhilov (2018) / Springer Nature / CC BY 4.0.

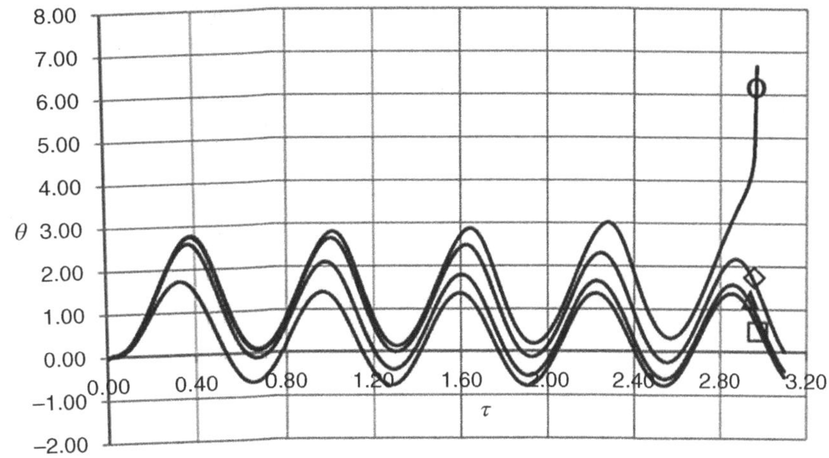

Figure 4.9 Degeneration of thermal explosion under kinetic conditions. Temperature time history at $\xi = 0$. Boundary conditions of the first type. $\delta = 0.6$; Ar = 0.0; Td = 200.0; Zn = 0.0; $\eta_0 = 0.5$; $\omega = 10.0$; $A_0 = 2.707$. ○ – zero-order kinetics; ◇ – first-order kinetics; Δ – second-order kinetics; □ – first-order autocatalysis. *Source:* Novozhilov (2018) / Springer Nature / CC BY 4.0.

Consider first kinetic effects in the case similar to the one investigated by Novozhilov (2016). This is shown in Figure 4.9 by the curve corresponding to the absence of kinetic effects (zero-order reaction). In this case, clearly, the reactant consumption transforms the system from super-critical to sub-critical conditions. The magnitude of the damping effect increases from the first-order to second-order kinetics, and further to the autocatalysis mechanism.

Such damping scenario is, in fact, not the only possibility. If controlling parameters are varied in such a way that critical boundaries (corresponding to respective cases with different kinetics) are crossed, instability may develop in two possible ways: the temperature either grows infinitely (for Ar = 0) in a monotonic fashion or through a number of oscillations. The series of Figures 4.10–4.12 demonstrate that both mechanisms of stability loss occur. Once the amplitude of oscillations is increased from $A_0 = 2.707$ (critical value for zero-order kinetics, Figure 4.9) to $A_0 = 4.32$, systems with first, second and autocatalysis kinetic mechanisms lose their stability one by one, all in an oscillatory fashion.

The fact that the oscillatory transition to the thermal runaway is not an exceptional, but rather a common scenario, is further evidenced by various cases presented in Figure 4.13 for the autocatalytic reaction. Here, various values of frequencies and amplitudes are considered. The autocatalytic kinetic mechanism exhibits especially wide spectrum of oscillatory transition scenarios due to the fact that the kinetic function in this case has a critical (maximum) point and therefore possesses some kind of intrinsic 'frequency'. This intrinsic frequency may resonate with the frequencies of externally imposed ambient temperature oscillations, giving rise to versatile oscillatory scenarios of transition to explosion.

In summary, for any type of considered kinetic mechanisms, reacting system may develop thermal instability if ambient conditions (temperature) vary in an oscillating manner with appropriate frequency and amplitude. This means that the depletion of the reactant (kinetic effects) does not have by itself a universal capacity to damp the development of thermal explosion.

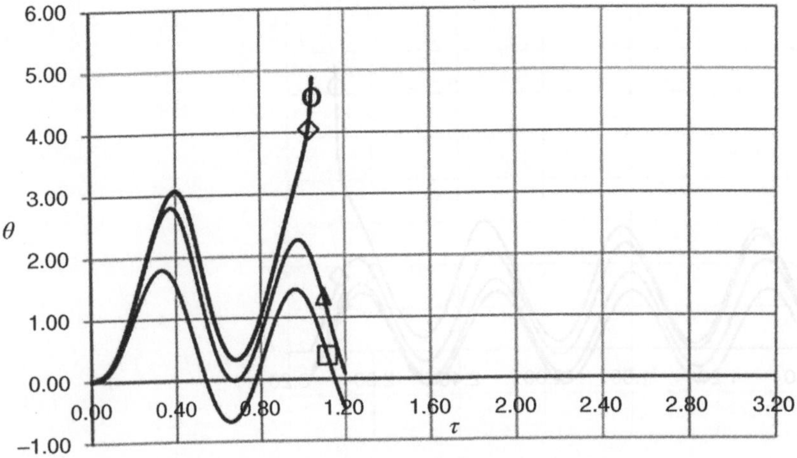

Figure 4.10 Thermal explosion development under different kinetic mechanisms. Temperature time history at $\xi = 0$. Boundary conditions of the first type. $\delta = 0.6$; Ar = 0.0; Td = 200.0; Zn = 0.0; $\eta_0 = 0.5$; $\omega = 10.0$; $A_0 = 2.80$. ○ – zero-order kinetics; ◇ – first-order kinetics; Δ – second-order kinetics; □ – first-order autocatalysis. *Source:* Novozhilov (2018) / Springer Nature / CC BY 4.0.

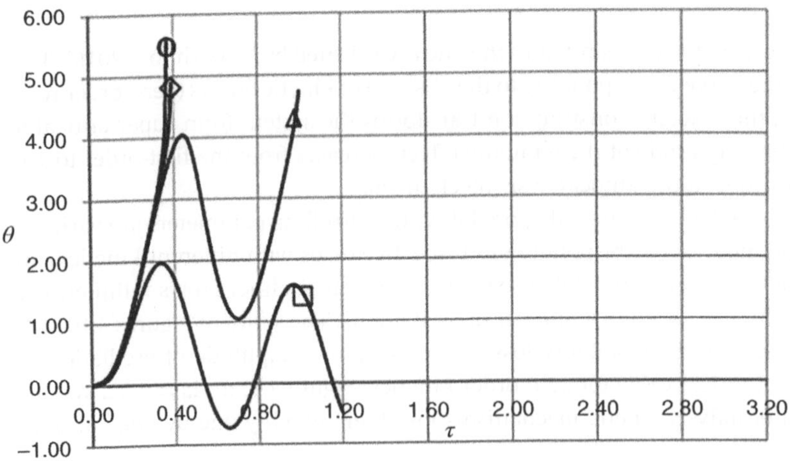

Figure 4.11 Thermal explosion development under different kinetic mechanisms. Temperature time history at $\xi = 0$. Boundary conditions of the first type. $\delta = 0.6$; Ar = 0.0; Td = 200.0; Zn = 0.0; $\eta_0 = 0.5$; $\omega = 10.0$; $A_0 = 3.02$. ○ – zero-order kinetics; ◇ – first-order kinetics; Δ – second-order kinetics; □ – first-order autocatalysis. *Source:* Novozhilov (2018) / Springer Nature / CC BY 4.0.

4.2 Conjugate Thermal Explosion

This section discusses the formulation of thermal explosion problem extended to two thermally interacting regions. Such problem formulation has been termed as the conjugate thermal explosion (Novozhilov 2008).

In contrast to classical formulations, the two quiescent reacting regions with different thermal and kinetic properties, having a common boundary (Figure 4.14), are considered. The problem

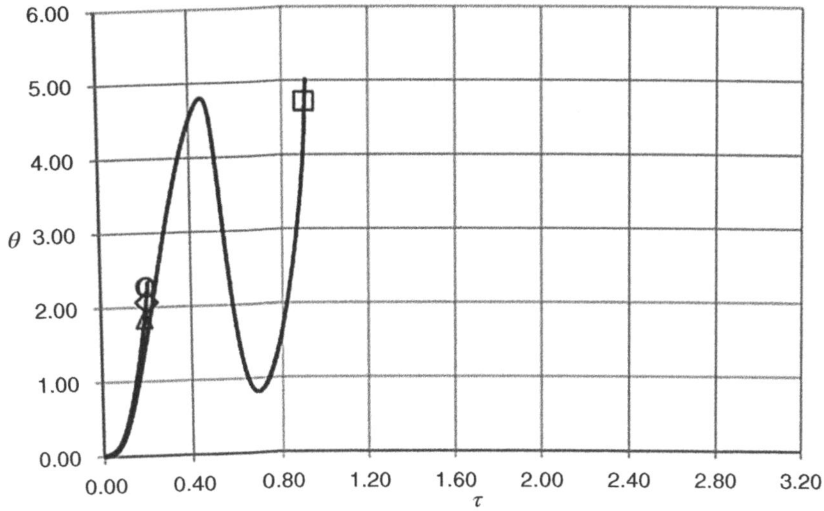

Figure 4.12 Thermal explosion development under different kinetic mechanisms. Temperature time history at $\xi = 0$. Boundary conditions of the first type. $\delta = 0.6$; Ar = 0.0; Td = 200.0; Zn = 0.0; $\eta_0 = 0.5$; $\omega = 10.0$; $A_0 = 4.32$. ○ – zero-order kinetics; ◇ – first-order kinetics; Δ – second-order kinetics; □ – first-order autocatalysis. *Source:* Novozhilov (2018) / Springer Nature / CC BY 4.0.

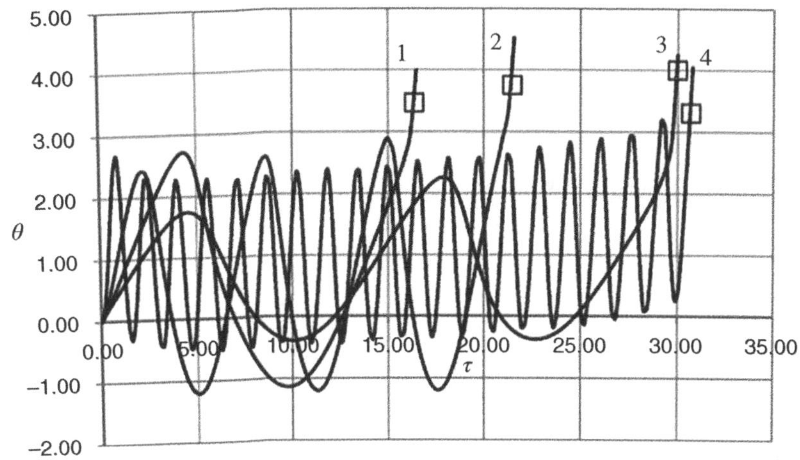

Figure 4.13 Oscillatory explosion development for the autocatalytic reaction. Temperature time history at $\xi = 0$. □ – first-order autocatalysis. Boundary conditions of the second type. Ar = 0.025; Td = 200.0; Zn = 0.0. 1 – $\delta = 0.6$, Bi = 20.0, $\omega = 0.5$, $A_0 = 1.2$; 2 – $\delta = 0.6$, Bi = 20.0, $\omega = 1.0$, $A_0 = 1.3$; 3 – $\delta = 0.8$, Bi = 5.0, $\omega = 0.5$, $A_0 = 0.6$; 4 – $\delta = 0.8$, Bi = 5.0, $\omega = 4.0$, $A_0 = 2.4$. *Source:* Novozhilov (2018) / Springer Nature / CC BY 4.0.

is considered in the three conventional symmetrical formulations (planar, cylindrical and spherical). For the sake of convenience of referencing, the regions are notionally termed 'gas' (the 'inner' region, adjacent to the symmetry boundary) and 'solid' (the 'outer' region having the contact boundary with surroundings).

The steady-state theory is considered first.

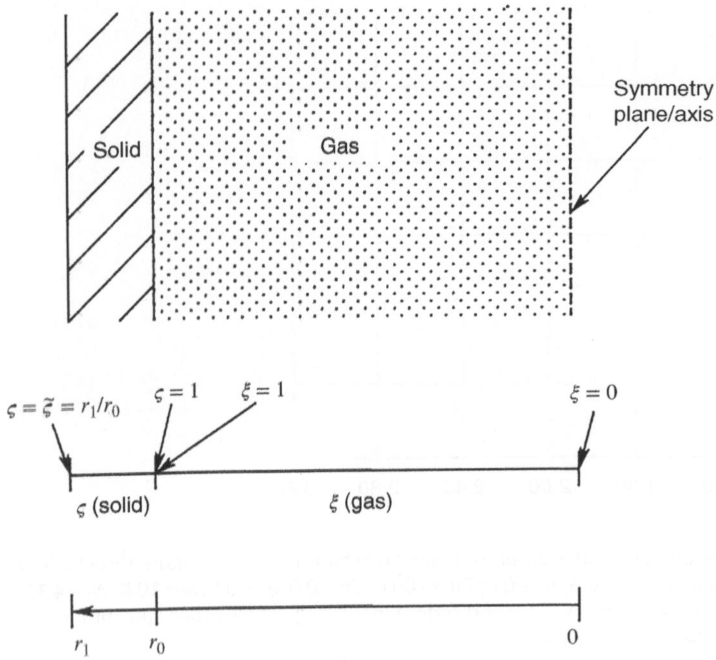

Figure 4.14 Sketch of the problem formulation. *Source:* Reproduced with permission from Novozhilov (2008) / Taylor & Francis.

The set of equations and boundary conditions for the two regions is written as follows:

Gas

$$\frac{d^2\theta_g}{d\xi^2} + \frac{m}{\xi}\frac{d\theta_g}{d\xi} = -\delta_g \exp(\theta_g), \quad 0 \le \xi \le 1 \tag{4.15}$$

$$\frac{d\theta_g}{d\xi} = 0, \quad \xi = 0 \tag{4.16}$$

Solid

$$\frac{d^2\theta_s}{d\varsigma^2} + \frac{m}{\varsigma}\frac{d\theta_s}{d\varsigma} = -\delta_s \exp(\theta_s), \quad 1 \le \varsigma \le \tilde{\varsigma} \tag{4.17}$$

$$\frac{d\theta_s}{d\varsigma} = -\mathrm{Bi} \cdot \theta_s, \quad \varsigma = \tilde{\varsigma} \tag{4.18}$$

Boundary condition of the second type are set at the interface between the outer region and surroundings.

Coupling conditions for heat transfer in the conduction mode at the interface between the two regions are set in a natural way

$$\theta_g = \theta_s, \quad \xi = \varsigma = 1 \tag{4.19}$$

$$\frac{d\theta_g}{d\xi} = \lambda_{sg}\frac{d\theta_s}{d\varsigma}, \quad \xi = \varsigma = 1 \tag{4.20}$$

where $\lambda_{sg} = \frac{\lambda_s}{\lambda_g}$ is the ratio between the thermal conductivities of the two regions.

Considering for the concreteness the cylindrical symmetry (the cases of planar and spherical symmetries are reviewed at the end of the Section) with the general solution of the form (2.35, 2.36), Eqs. (4.15) and (4.17) are transformed into

$$\theta_g = \ln\left(\frac{8-b}{\delta_g}\right) - 2\ln\left(a\xi^{k_1} + \frac{1}{a}\xi^{k_2}\right);\quad k_{1,2}^2 - 2k_{1,2} + \frac{b}{8} = 0 \tag{4.21}$$

$$\theta_s = \ln\left(\frac{8-c}{\delta_s}\right) - 2\ln\left(d\varsigma^{l_1} + \frac{1}{d}\varsigma^{l_2}\right);\quad l_{1,2}^2 - 2l_{1,2} + \frac{c}{8} = 0 \tag{4.22}$$

The symmetry condition, combined with the second of Eq. (4.21), leads to

$$b = 0,\quad k_1 = 2,\quad k_2 = 0 \tag{4.23}$$

The solution for the inner region is, therefore, simplified to

$$\theta_g = \ln\left(\frac{8}{\delta_g}\right) - 2\ln\left(a\xi^2 + \frac{1}{a}\right) \tag{4.24}$$

For the outer region, the boundary condition (4.18), being applied to solution (4.22), results in

$$2\frac{dl_1\varsigma^{l_1-1} + \left(\frac{1}{d}\right)l_2\varsigma^{l_2-1}}{d\varsigma^{l_1} + \left(\frac{1}{d}\right)\varsigma^{l_2}} = \mathrm{Bi}\cdot\left[\ln\left(\frac{8-c}{\delta_s}\right) - 2\ln\left(d\varsigma^{l_1} + \frac{1}{d}\varsigma^{l_2}\right)\right] \tag{4.25}$$

Coupling conditions (4.19) and (4.20) at $\xi = \varsigma = 1$ can be transformed into

$$\frac{\delta_g}{8}\left(a + \frac{1}{a}\right)^2 = \frac{\delta_s}{8-c}\left(d + \frac{1}{d}\right)^2 \tag{4.26}$$

and

$$\frac{2a}{\left(a + \frac{1}{a}\right)} = \lambda_{sg}\frac{\left[dl_1 + \frac{1}{d}l_2\right]}{\left(d + \frac{1}{d}\right)} \tag{4.27}$$

respectively.

The latter two equations may be combined to give

$$\frac{8-c}{\delta_s} = \frac{2}{\delta_g}(\lambda_{sg})^2 \times \left[\left(2(\lambda_{sg})^{-1} - l_1\right)\cdot d + \left(2(\lambda_{sg})^{-1} - l_2\right)\cdot\frac{1}{d}\right]\left(dl_1 + \frac{1}{d}l_2\right) \tag{4.28}$$

which, being substituted into (4.25), results in

$$2\frac{dl_1\varsigma^{l_1-1} + \left(\frac{1}{d}\right)l_2\varsigma^{l_2-1}}{d\varsigma^{l_1} + \left(\frac{1}{d}\right)\varsigma^{l_2}} = \mathrm{Bi}\cdot\ln\left[\frac{2}{\delta_g}(\lambda_{sg})^2\frac{\left[\left(2(\lambda_{sg})^{-1} - l_1\right)\cdot d + \left(2(\lambda_{sg})^{-1} - l_2\right)\cdot\frac{1}{d}\right]\left(dl_1 + \frac{1}{d}l_2\right)}{\left(d\varsigma^{l_1} + \frac{1}{d}\varsigma^{l_2}\right)^2}\right] \tag{4.29}$$

Equation (4.29), with respect to parameter d, is the central result for cylindrical geometry. It determines critical conditions for explosion as a relation between the Biot number, Frank–Kamenetskii parameters δ_g, δ_s and conductive properties of the media.

Equation (4.27) gives relation between c, l_1 and l_2. Since l_1 and l_2 must also satisfy the second of Eq. (4.22), the possible pairs (l_1, l_2) can be found as follows:

$$l_1^1 = \frac{-\beta + \sqrt{\beta^2 - 4\alpha\gamma}}{2\alpha};\ l_2^1 = \frac{-\beta - \sqrt{\beta^2 - 4\alpha\gamma}}{2\alpha} \tag{4.30}$$

or

$$l_1^2 = \frac{-\beta - \sqrt{\beta^2 - 4\alpha\gamma}}{2\alpha};\ l_2^2 = \frac{-\beta + \sqrt{\beta^2 - 4\alpha\gamma}}{2\alpha} \tag{4.31}$$

where

$$\alpha = 1 + \frac{\delta_s}{4\delta_g}(\lambda_{sg})^2\left(d - \frac{1}{d}\right)^2 \tag{4.32}$$

$$\beta = -\frac{\delta_s}{4\delta_g}(\lambda_{sg})^2\left(2(\lambda_{sg})^{-1}\left(d^2 - \frac{1}{d^2}\right) + 4\left(\frac{1}{d^2} - 1\right)\right) + 2 \tag{4.33}$$

$$\gamma = \left(1 - \frac{\delta_s}{\delta_g}(\lambda_{sg})^2\left[(\lambda_{sg})^{-1}\left(1 + \frac{1}{d^2}\right) - \frac{1}{d^2}\right]\right) \tag{4.34}$$

Critical condition is equivalent to the disappearance of the solution of Eq. (4.29) with respect to d, for either pair l_i^j.

Since Eq. (4.29) is invariant with respect to the transformations $(d \mapsto -d)$ and $\left(d \mapsto \frac{1}{d};\right.$ $\left. l_1 \mapsto l_2;\ l_2 \mapsto l_1\right)$, it is sufficient to investigate the existence of a solution for any one pair (l_1, l_2), that is, (4.30) or (4.31).

The typical behaviour of the LHS and the RHS of Eq. (4.29) is shown in Figure 4.15. The critical condition is found as the condition of merging of two roots of Eq. (4.29) into one (Figure 4.15), with the LHS and RHS curves having the same tangent.

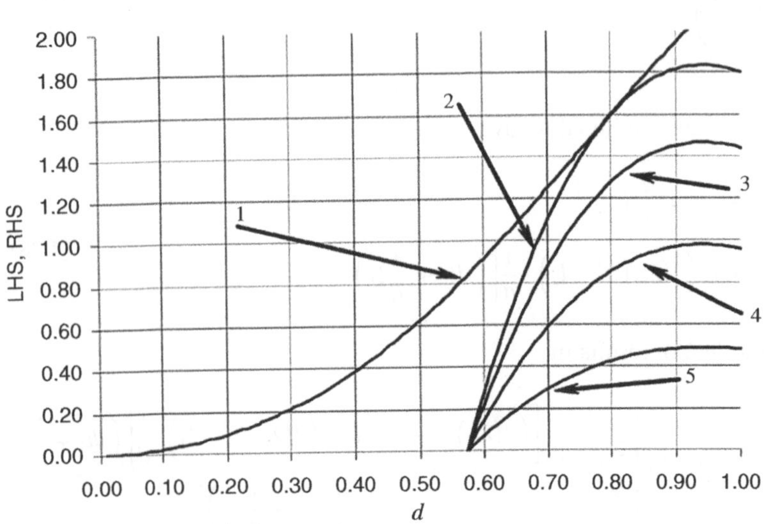

Figure 4.15 Illustration of the solution of Eq. (4.29). $\delta_g = 1.0$; $\delta_s = 2.5$; $\xi = 1.1$; $\lambda_{sg} = 1.0$. 1 – LHS of Eq. (4.29); 2 – RHS of Eq. (4.29), Bi = 3.5; 3 – RHS of Eq. (4.29), Bi = 3.0; 4 – RHS of Eq. (4.29), Bi = 2.0; 5 – RHS of Eq. (4.29), Bi = 1.0. *Source:* Reproduced with permission from Novozhilov (2008) / Taylor & Francis.

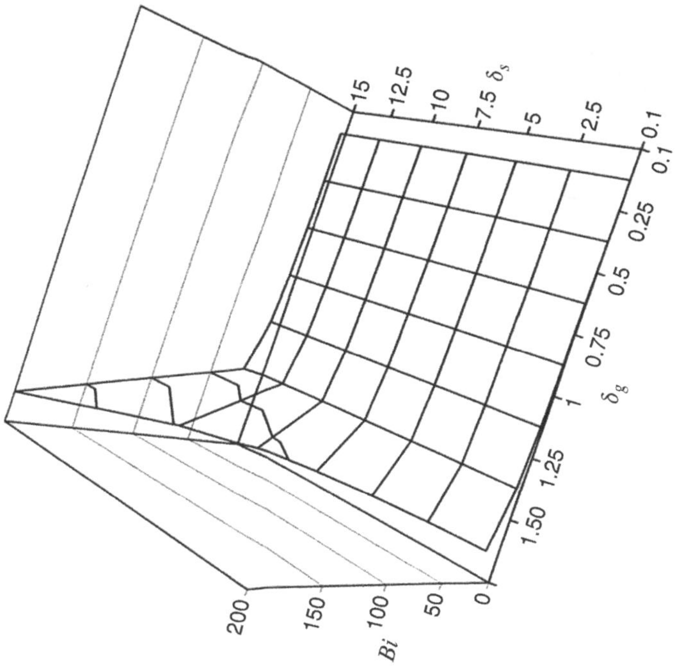

Figure 4.16 Critical surface Bi = Bi(δ_g, δ_s) for the case of cylindrical symmetry. λ_{sg} = 1.0; $\tilde{\varsigma}$ = 1.1.
Source: Reproduced with permission from Novozhilov (2008) / Taylor & Francis.

Critical condition can be plotted, therefore, as a surface $Bi = Bi\left(\delta_g, \delta_s, \lambda_{sg}, \tilde{\varsigma}\right)$. This surface is presented in Figure 4.16 for particular values of thermal conductivities λ_g, λ_s and dimension $\tilde{\varsigma}$. The super-critical region is below the surface.

The question of interest is, of course, how the latter critical conditions compare to those given by the classical theory for a single region. The critical curve for a single region (curve 1) is plotted in Figure 4.17 along with the several (for different values of δ_g, sub-critical for a single region problem with $Bi = \infty$) cross-sections of the critical surface (Figure 4.16). Thermal explosion conditions correspond to the sets of parameters below the particular curve.

Figure 4.17 demonstrates the difference in critical conditions between double-region and single-region (classical) cases. This difference is qualitatively determined by whether heat generation in a gas or solid region dominates (i.e. by relative magnitudes of Frank–Kamenetskii parameters δ_g and δ_s). For δ_s much smaller than δ_g, energy generation in the gas region must be balanced by energy losses. Therefore, the Biot number is finite as $\delta_s \to 0$, in contrast to the single-region asymptotic, where Bi $\to 0$ as $\delta_s \to 0$. This explains the behaviour of curves 2, 3 and 4 in Figure 4.17 to the left of intersections with curve 1. This difference in the asymptotic curves behaviour reflects the fundamental fact that due to the double-layer heat transfer configuration thermal explosion for a given value of Bi may occur for values of δ_s, which are below critical for a single-region problem.

In the opposite case, when δ_s is large compared to δ_g, the behaviour of the system is essentially dominated by energy generation and dissipation from the solid region. However, heat generated in the solid region must also heat up the gas region in order for thermal explosion to occur. For this reason, critical conditions exist at large values of δ_s (curves 2, 3 and 4, Figure 4.17), while for a single-region problem such values of δ_s are super-critical (in the latter case $\delta \to \delta_{cr} = 2.0$ as the Biot-number approaches infinity asymptotically).

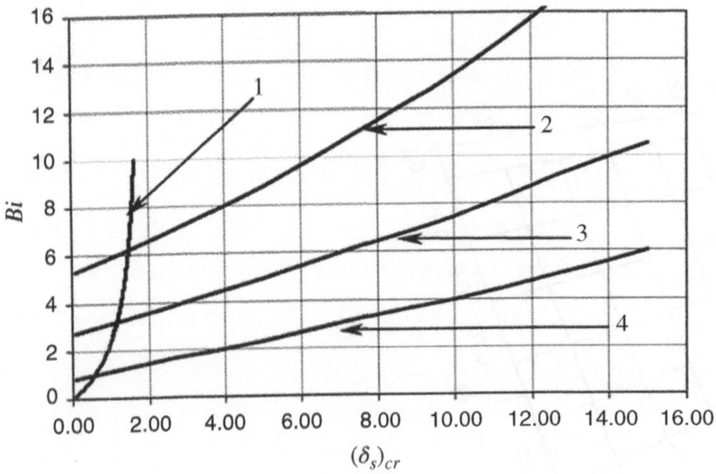

Figure 4.17 Comparison between the critical conditions for conjugate thermal explosion and single-region thermal explosion. Cylindrical symmetry. 1 – critical condition dependence $\delta_{cr} = \delta_{cr}(\mathrm{Bi})$ for a single region; 2 – dependence $(\delta_s)_{cr} = (\delta_s)_{cr}(\mathrm{Bi})$ for $\delta_g = 1.25$; 3 – dependence $(\delta_s)_{cr} = (\delta_s)_{cr}(\mathrm{Bi})$ for $\delta_g = 1.0$; 4 – dependence $(\delta_s)_{cr} = (\delta_s)_{cr}(\mathrm{Bi})$ for $\delta_g = 0.5$. *Source:* Reproduced with permission from Novozhilov (2008) / Taylor & Francis.

The temporal development of the conjugate thermal explosion may be followed by solving numerically non-steady equations similar to (2.63), subject to coupling conditions, for each of the regions.

As the time scales (2.2) are different between the regions, the uniform time scale for composite sample may be introduced as $\sqrt{\tilde{t}_g \tilde{t}_s}$, where $\tilde{t}_g$ and $\tilde{t}_s$ are time scales for the gas and the solid regions, respectively. Note that time scales and the Frank–Kamenetskii parameters may generally vary independently.

Taking the above into account, it is clear that the set of equations describing unsteady conjugate thermal explosion involves the ratio of the two time scales $\tau_{sg} = \sqrt{\frac{\tilde{t}_s}{\tilde{t}_g}}$. This set is written in the following form (Novozhilov 2008):

$$\frac{\partial \theta_g}{\partial \tau} = (\tau_{sg})^{1/2}\left[\frac{1}{\delta_g}\left(\frac{\partial^2 \theta_g}{\partial \xi^2} + \frac{1}{\xi}\frac{\partial \theta_g}{\partial \xi}\right) + \exp(\theta_g)\right], \quad 0 \le \xi \le 1 \tag{4.35}$$

$$\frac{\partial \theta_s}{\partial \tau} = (\tau_{sg})^{-1/2}\left[\frac{1}{\delta_s}\left(\frac{\partial^2 \theta_s}{\partial \varsigma^2} + \frac{1}{\varsigma}\frac{\partial \theta_s}{\partial \varsigma}\right) + \exp(\theta_s)\right], \quad 1 \le \varsigma \le \tilde{\varsigma} \tag{4.36}$$

$$\frac{d\theta_g}{d\xi} = 0, \quad \xi = 0 \tag{4.37}$$

$$\theta_g = \theta_s, \quad \xi = \varsigma = 1 \tag{4.38}$$

$$\frac{d\theta_g}{d\xi} = \lambda_{sg}\frac{d\theta_s}{d\varsigma}, \quad \xi = \varsigma = 1 \tag{4.39}$$

$$\frac{d\theta_s}{d\varsigma} = -\mathrm{Bi}\cdot\theta_s, \quad \varsigma = \tilde{\varsigma} \tag{4.40}$$

Figure 4.18 presents the case where the parameters δ_g and δ_s are fixed below the critical value for a single region (at given Bi). The latter choice is demonstrated in Figure 4.19, where the set

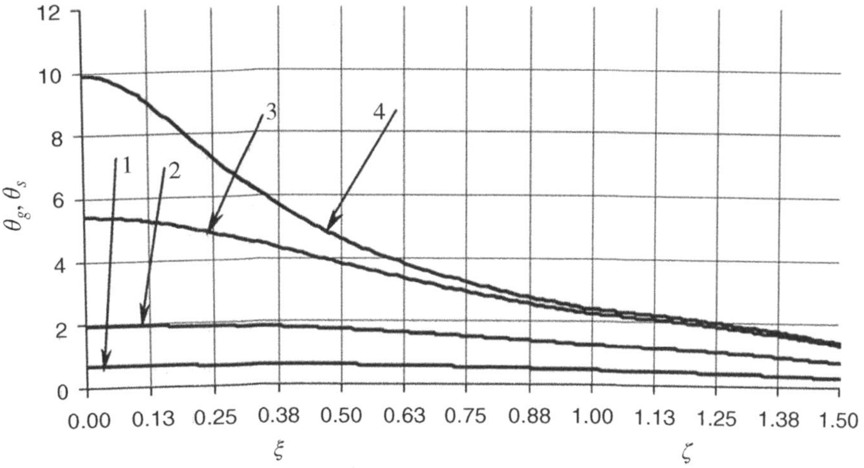

Figure 4.18 Unsteady simulation of conjugate thermal explosion development for cylindrical symmetry. $\delta_s = 0.2$; $\delta_g = 0.8$; Bi = 2.0; $\tilde{\varsigma} = 1.5$; $\lambda_{sg} = 2.2$; $\tau_{sg} = 1.0$. 1 – $\tau = 1.00$; 2 – $\tau = 32.50$; 3 – $\tau = 33.30$; 4 – $\tau = 33.32$. Symmetry axis on the left. *Source:* Reproduced with permission from Novozhilov (2008) / Taylor & Francis.

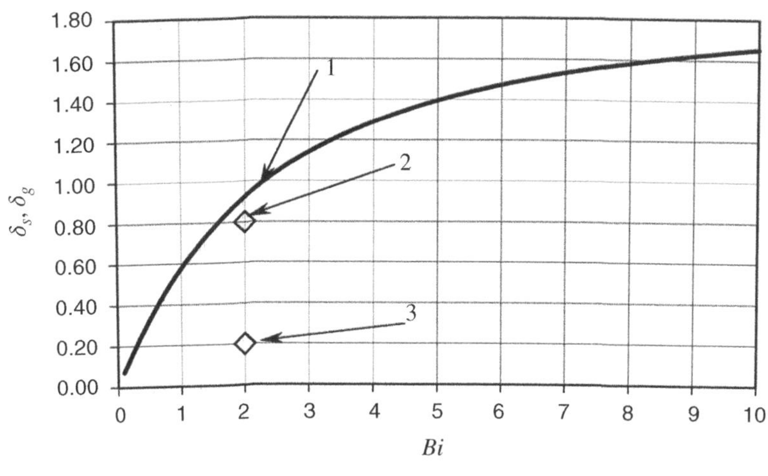

Figure 4.19 Comparison of the simulation parameters with the critical conditions for a single-media region. Cylindrical symmetry. 1 – critical condition dependence $\delta_{cr} = \delta_{cr}(\text{Bi})$ for a single region; 2 – values (δ_g, Bi) used in numerical simulations, Figure 4.18; 3 – values (δ_s, Bi) used in numerical simulations, Figure 4.18. *Source:* Reproduced with permission from Novozhilov (2008) / Taylor & Francis.

of numerical parameters is shown in comparison with the critical region for a single-medium systems. It is clear that the both parameters (δ_g, δ_s) are below their respective values for a single region case.

The results presented in Figure 4.18 demonstrate that thermal explosion in double-medium systems is possible even if each of the regions, considered as a separate system, is at subcritical condition. The thermal runaway occurs initially in the gas (inner) region since energy loss is made more difficult by the presence of the outer layer. The latter provides an additional thermal resistance and is also chemically active. Later on, the thermal runaway in the inner region triggers thermal instability in the solid (outer) region. This explosion manifests itself as an unlimited temperature rise at the inner boundary of the solid region (at $\varsigma = 1$).

The case illustrated in Figure 4.20 features reversed, compared to the case of Figure 4.18, magnitudes of thermal conductivities (i.e. the thermal conductivity of the inner region is much higher than that of the outer region). The latter effect, that is, temperature peak at $\varsigma = 1$, is seen in a more pronounced way in Figure 4.20 due to poor heat conduction through the solid region, and correspondingly larger influence of its chemical reaction term near the vicinity of the contact surface $\xi = \varsigma = 1$.

Therefore, thermal explosion in the outer region may be classified as the ignition of this portion of the reacting mixture at the boundary, heated by the adjacent (gas) region.

Ignition may generally be seen as the limiting case of thermal explosion (Zeldovich *et al.* 1980; Barzykin 2004), where the point of maximum temperature during runaway is located at the boundary of the system. Such a condition clearly requires energy supply to the system through that boundary otherwise, due to energy loss, the point of maximum temperature will inevitably be located inside the domain. Therefore, ignition may only be realized as a dynamical regime of thermal explosion. In the present example, energy generated within the inner region and transferred through the contact surface to the outer region drives the ignition of the latter at the boundary $\varsigma = 1$.

It is essential to emphasize that explosion development illustrated in Figures 4.18 and 4.20 is only possible due to the presence of neighbouring regions and thermal interaction between them. Otherwise, the thermal runaway cannot occur in either region as their respective values of Frank–Kamenetskii parameters (Figures 4.18–4.20) are below critical.

Position of maximum temperature during the thermal runaway does not have to be necessarily at the centre of domain as in Figures 4.18 and 4.20. It may also occur at the interface between the two regions, or inside the solid region. These cases are illustrated in Figure 4.21 and are associated with high values of the Frank–Kamenetskii parameter in the latter region, compared to that in the gas.

Finally, Figure 4.22 demonstrates the system behaviour for conditions below critical. Temperature profiles approach the steady-state solution asymptotically at $\tau \to \infty$.

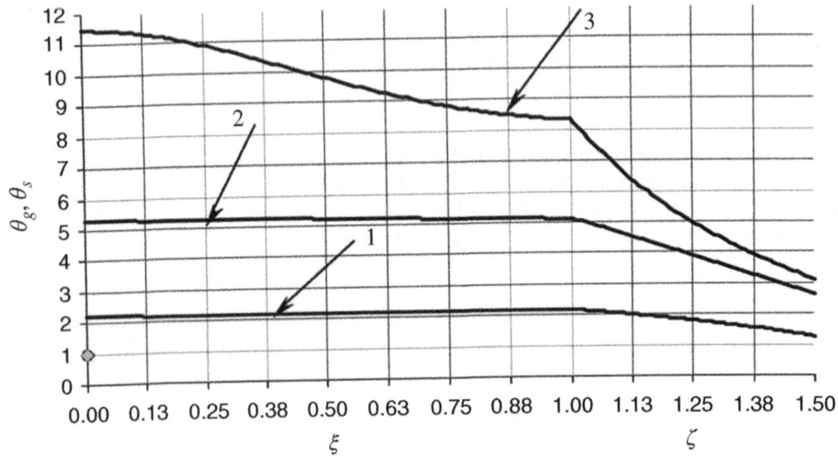

Figure 4.20 Unsteady simulation of conjugate thermal explosion development for cylindrical symmetry. $\delta_s = 0.2$; $\delta_g = 0.8$; Bi = 2.0; $\tilde{\xi} = 1.5$; $\lambda_{sg} = 0.0265$; $\tau_{sg} = 1.0$. 1 – $\tau = 0.900$; 2 – $\tau = 1.010$; 3 – $\tau = 1.017$. Symmetry axis on the left. *Source:* Reproduced with permission from Novozhilov (2008) / Taylor & Francis.

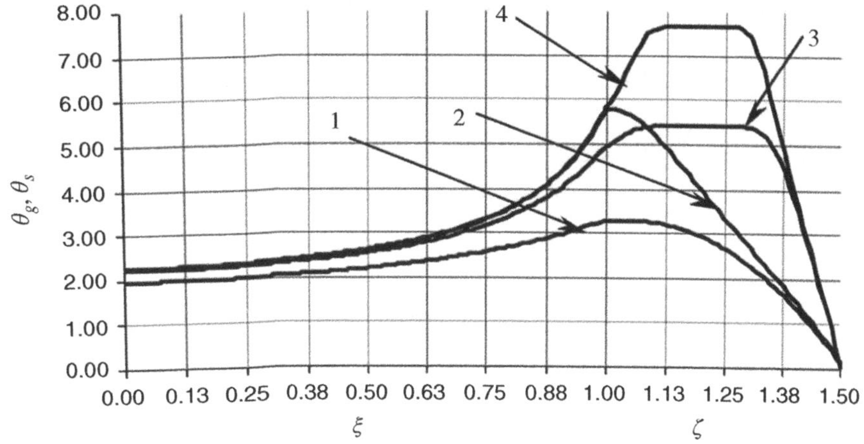

Figure 4.21 Illustration of thermal explosion development at the interface (curves 1 and 2) and in the solid region (curves 3 and 4). Cylindrical symmetry. Bi = 2.0; $\tilde{\xi} = 1.5$. 1 – $\tau = 0.164$, $\delta_s = 12.48$, $\delta_g = 0.17$, $\lambda_{sg} = 26.50$, $\tau_{sg} = 0.50$; 2 – $\tau = 0.170$, $\delta_s = 12.48$, $\delta_g = 0.17$, $\lambda_{sg} = 26.50$, $\tau_{sg} = 0.50$; 3 – $\tau = 0.169$, $\delta_s = 124.80$, $\delta_g = 0.17$, $\lambda_{sg} = 2.65$, $\tau_{sg} = 0.50$; 4 – $\tau = 0.170$, $\delta_s = 124.80$, $\delta_g = 0.17$, $\lambda_{sg} = 2.65$, $\tau_{sg} = 0.50$. Symmetry axis on the left. *Source:* Reproduced with permission from Novozhilov (2008) / Taylor & Francis.

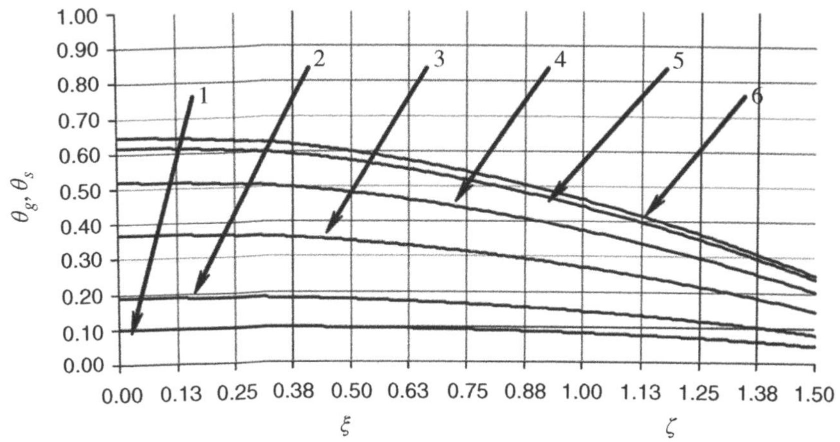

Figure 4.22 System behaviour at subcritical conditions. Cylindrical symmetry. $\delta_s = 0.2$; $\delta_g = 0.2$; Bi = 2.0; $\tilde{\xi} = 1.5$; $\lambda_{sg} = 1.0$; $\tau_{sg} = 1.0$. 1 – $\tau = 0.1$; 2 – $\tau = 0.2$; 3 – $\tau = 0.5$; 4 – $\tau = 1.0$; 5 – $\tau = 2.0$; 6 – $\tau = 5.0$. Symmetry axis on the left. *Source:* Reproduced with permission from Novozhilov (2008) / Taylor & Francis.

The analysis provided above (for the case of cylindrical symmetry) demonstrates that critical conditions for the double-medium reactive system (conjugate thermal explosion) are significantly different from critical conditions in the classical thermal explosion theory. In particular, the thermal runaway may occur even when both regions are in sub-critical states, considered as a single medium. On the other hand, under certain super-critical conditions for a single-medium system, the conjugate thermal explosion may be prevented by sufficiently intensive energy exchange with the surroundings.

The other two cases of symmetry (planar and spherical) are reviewed below.

For the planar case ($m = 0$), solutions for the gas and solid regions have the form (2.29)

$$\theta_g = -\ln\left(\frac{\delta_g}{2C_1^2} ch^2(C_1\xi + C_2)\right), \quad 0 \le \xi \le 1 \tag{4.41}$$

$$\theta_s = -\ln\left(\frac{\delta_s}{2K_1^2} ch^2(K_1\varsigma + K_2)\right), \quad 1 \le \varsigma \le \tilde{\varsigma} \tag{4.42}$$

with the undetermined constants C_1, C_2, K_1, K_2.

Boundary conditions (4.16), (4.18), (4.19) and (4.20) can be easily shown to lead to the following relations:

$$-2K_1 \cdot th(K_1\tilde{\varsigma} + K_2) = \mathrm{Bi} \cdot \ln\left(\frac{\delta_s}{2K_1^2} ch^2(K_1\tilde{\varsigma} + K_2)\right) \tag{4.43}$$

$$\frac{\delta_g}{C_1^2} ch^2(C_1) = \frac{\delta_s}{K_1^2} ch^2(K_1 + K_2) \tag{4.44}$$

$$C_1 \cdot th(C_1) = \lambda_{sg} K_1 \cdot th(K_1 + K_2) \tag{4.45}$$

$$C_2 = 0 \tag{4.46}$$

The solutions (4.41) and (4.42) for θ_g, θ_s are invariant with respect to transformations

$$C_1 \mapsto -C_1; \quad C_2 \mapsto -C_2 \tag{4.47}$$

$$K_1 \mapsto -K_1; \quad K_2 \mapsto -K_2 \tag{4.48}$$

Therefore, one needs only look for solutions satisfying C_i, $K_i \ge 0$.

Taking this into account, the problem is reducible from Eqs. (4.43–4.46) to finding non-negative solutions (K_1, K_2) of the following set of two equations:

$$-2K_1 \cdot th(K_1\tilde{\varsigma} + K_2) = \mathrm{Bi} \cdot \ln\left(\frac{\delta_s}{2K_1^2} ch^2(K_1\tilde{\varsigma} + K_2)\right) \tag{4.49}$$

$$\frac{\delta_g}{\mathrm{Arsh}^2\left(\sqrt{\frac{\delta_s}{\delta_g}}\lambda_{sg} sh(K_1 + K_2)\right)} \cdot \left[1 + \frac{\delta_s}{\delta_g}(\lambda_{sg})^2 sh^2(K_1 + K_2)\right] = \frac{\delta_s}{K_1^2} ch^2(K_1 + K_2) \tag{4.50}$$

and investigating conditions for the existence of such solutions.

It should be noted that Eq. (4.50) may be re-written taking into account the identity

$$\mathrm{Arsh}(x) = \ln\left(x + \sqrt{x^2 + 1}\right) \tag{4.51}$$

As for the case of the spherical symmetry ($m = 2$), it has already been noted in Section 2.3, Chapter 2, that exact solutions of Eqs. (4.15) and (4.17) are not known.

For the single-region case with the conditions of the second type at the external boundary, the procedure to find critical conditions has been explained in Section 3.1, Chapter 3.

A similar procedure may be extended to the case of two adjacent regions. This is illustrated in Figure 4.23 in the variables (x_g, y_g), (x_s, y_s) defined, using the change of variables (3.11) for the gas and the solid regions, respectively. In the new variables, the original Frank–Kamenetskii equations are transformed to equations of the form (3.12) in each of the regions.

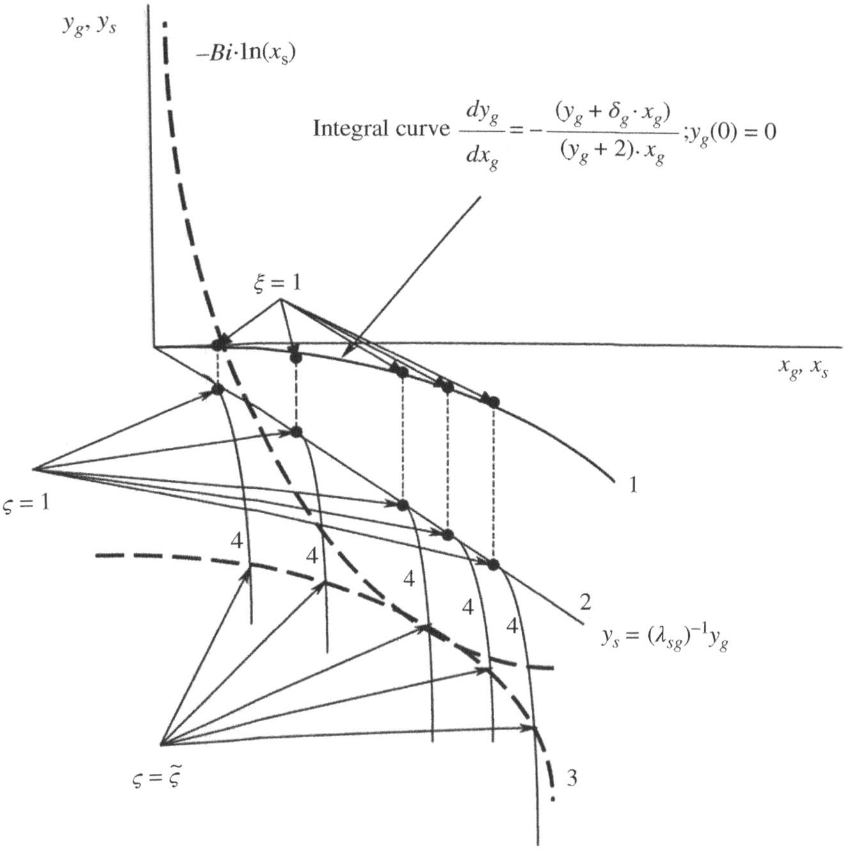

Figure 4.23 Determination of critical conditions for conjugate thermal explosion in the case of spherical symmetry. Construction of curves 1, 3 and 4 and line 2 is described in the text. *Source:* Reproduced with permission from Novozhilov (2008) / Taylor & Francis.

Curve 1 in Figure 4.23 is an integral curve $y_g = \tilde{y}_g(x_g)$ for the gas region, obtained exactly in the same way as in Section 3.1.

Line 2, $y_s = (\lambda_{sg})^{-1} \cdot y_g$, is an image of the interface condition (4.39).

Curve 3 is generated by the following procedure. Each point on line 2 uniquely determines a pair (x_s^0, y_s^0), from which the integral curve $y_s = \tilde{y}_s(x_s)$ of equation identical to (3.12) can be originated. (Note that according to the interface condition (4.20) (x_s^0, y_s^0) is an image, under vertical projection, of the point (x_g, y_g), corresponding to $\xi = 1$ at curve 1, down to line 2.)

Note that the starting point (x_s^0, y_s^0) on line 2 is generally a regular point of the equation:

$$\frac{dy_s}{dx_s} = -\frac{(y_s + \delta_s x_s)}{(y_s + 2)x_s} \tag{4.52}$$

The family of integral curves of Eq. (4.52) originating on line 2 is shown in Figure 4.23 as curve 4. Each of those determines, in view of formulae (3.11), the ODE

$$\frac{d\theta_s}{d\varsigma} = \Phi(\varsigma, \theta_s), \quad 1 \le \varsigma \le \tilde{\varsigma} \tag{4.53}$$

The initial value $\theta_s(1)$ of its solution is determined from the condition (following from (3.11) at $\varsigma = 1$).

$$x_s^0 = \exp(\theta_s(1)) \tag{4.54}$$

Curve 3 in Figure 4.23 is the locus of points on the family of curve 4, determined by the condition that such points correspond to the integration of the respective Eq. (4.53) from $\varsigma = 1$ to $\varsigma = \tilde{\varsigma}$.

Criticality is given by a condition of a single, equal-tangent intersection between curve 3 and curve $y_s = -\text{Bi}\ln(x_s)$.

The numerical analysis, performed on the cases possessing planar and spherical symmetries, suggests runaway development features qualitatively similar to those for the cylindrical symmetry. However, there are also anticipated quantitative differences. For example, thermal explosion in the planar symmetry case may occur at much larger values of the conductivity ratio λ_{sg}, compared to the cylindrical symmetry (with all the other parameters being identical).

4.3 Diffusion Thermal Explosion

Consideration of kinetic effects has been performed in Section 4.1 on the assumption of uniform reactant distribution in the reactive system volume. In general, in the course of thermal explosion development, reactant concentrations may not be uniform. Figure 4.24 shows schematically an example of a non-uniform reactant concentration distribution in a closed vessel. Accounting for diffusion effects presents an additional challenge to the analysis of the thermal explosion problem complicating, for example, the kinetic (i.e. species conservation) equation.

In the most extreme case, reactants (for the reactions of the order one or higher) are completely unmixed initially. Such a problem will be referred to in this book as diffusion thermal explosion.

Consideration of such a problem is a natural extension of the concept of conjugate thermal explosion. Indeed, if reactants are fully separated initially, then the regions of their location would interact by exchanging diffusion fluxes, analogously to thermal fluxes exchange in the case of conjugate thermal explosion.

At the same time, the concept of the diffusion thermal explosion, that is explosion with non-premixed, initially, reactants is a natural and practically important extension of the classical theory.

Figure 4.24 Schematic example of the non-uniform reactant concentration distribution in a closed vessel.

The theoretical analysis of the problem of diffusion thermal explosion may be performed using an approximate method developed by Maksyutov *et al.* (1982). They apply this method to the reaction of the second order involving two different species.

The problem, in principle, can be considered in both closed vessel, or open space. Maksyutov *et al.* (1982) consider the case of an open space, with the problem possessing the plane symmetry (one-dimensional problem, infinitely wide case of reacting layer shown in Figure 2.5). In such a configuration, initially, temperature is uniform, reactants with the concentrations $\tilde{C}_1(\tilde{x},\tilde{t})$ and $\tilde{C}_2(\tilde{x},\tilde{t})$ are separated by the plane $\tilde{x} = 0$

$$\begin{aligned} &\tilde{T}(\tilde{x},0) = \tilde{T}_0; \quad -\infty < \tilde{x} < \infty \\ &\tilde{C}_1(\tilde{x},0) = \tilde{C}_1^0; \quad \tilde{C}_2(\tilde{x},0) = 0; \quad \tilde{x} \le 0 \\ &\tilde{C}_1(\tilde{x},0) = 0, \quad \tilde{C}_2(\tilde{x},0) = \tilde{C}_2^0, \quad \tilde{x} > 0 \end{aligned} \tag{4.55}$$

where $\tilde{C}_{1,2}^0 \equiv \tilde{C}_0$ are the initial values of respective concentrations.

Introducing non-dimensional variables, adiabatic time scale $\tilde{t}_{ad}$, defined by Eq. (4.2) (with $m = 2$ for the reaction of the second order), is used for time scaling. Since the problem is considered in open space, a natural way to introduce spatial scaling is the use of diffusion coefficient (which is assumed to be equal for both species) $\tilde{D}$.

$$\xi = \frac{\tilde{x}}{\sqrt{\tilde{D}\tilde{t}_{ad}}} \tag{4.56}$$

Non-dimensional concentrations may be defined as

$$\begin{aligned} C_1(\xi,\tau) &= \frac{\tilde{C}_1(\xi,\tau)}{\left[\tilde{C}_1^0 + \tilde{C}_2^0\right]} = \frac{1}{2}\frac{\tilde{C}_1(\xi,\tau)}{\tilde{C}_0} \\ C_2(\xi,\tau) &= \frac{\tilde{C}_2(\xi,\tau)}{\left[\tilde{C}_1^0 + \tilde{C}_2^0\right]} = \frac{1}{2}\frac{\tilde{C}_1(\xi,\tau)}{\tilde{C}_0} \end{aligned} \tag{4.57}$$

Thus defined C_1 and C_2 have a meaning of molar fractions.

Progress variables are defined as follows:

$$\eta_1(\xi,\tau) = 1 - C_1(\xi,\tau), \quad \eta_2(\xi,\tau) = 1 - C_2(\xi,\tau) \tag{4.58}$$

The following trivial relationships hold:

$$\eta_1 + \eta_2 = C_1 + C_2 = 1 \tag{4.59}$$

Development of the steady-state theory for the open space is not possible. Therefore, the non-steady formulation is considered.

The diffusion thermal explosion problem, for the reaction of the second order and planar symmetry, is described by the following set of PDEs:

$$\begin{aligned} \frac{\partial\theta}{\partial\tau} &= \frac{\partial^2\theta}{\partial\xi^2} + \varphi(\eta_1)\exp\left(\frac{\theta}{1+\mathrm{Ar}\theta}\right) \\ \frac{\partial\eta_1}{\partial\tau} &= \frac{\partial^2\eta_1}{\partial\xi^2} + \vartheta\varphi(\eta_1)\exp\left(\frac{\theta}{1+\mathrm{Ar}\theta}\right) \\ \eta_2 &= (1-\eta_1) \end{aligned} \tag{4.60}$$

It is assumed here that reactants have equal thermal conductivities and heat capacities, Lewis number $Le = \tilde{\lambda}/\left(\tilde{\rho}\tilde{c}_p\tilde{D}\right)$ is equal to unity; $\varphi(\eta_1)$ is kinetic function, and ϑ non-dimensional parameter

$$\varphi(\eta_1) = (1-\eta_1)\eta_1; \quad \vartheta = \frac{\tilde{\rho}\tilde{c}_p\tilde{R}\tilde{T}_0^2}{\tilde{Q}\tilde{E}\tilde{C}_0} \tag{4.61}$$

Initial and boundary conditions, written in terms of progress variables, are

$$\begin{aligned} &\theta(\xi,0) = 0; \quad -\infty < \xi < \infty \\ &\eta_1(\xi,0) = 0, \quad \eta_2(\xi,0) = 1, \quad \xi \le 0 \\ &\eta_1(\xi,0) = 1, \quad \eta_2(\xi,0) = 0, \quad \xi > 0 \\ &\frac{\partial\eta_1}{\partial\xi} = \frac{\partial\eta_2}{\partial\xi} = 0; \quad \xi \to \pm\infty \\ &\frac{\partial\theta}{\partial\xi} = 0; \quad \xi \to \pm\infty \end{aligned} \tag{4.62}$$

Assuming the pronounced self-acceleration of the chemical reaction, thermal explosion development may be approximated by the two consecutive stages.

Mixing of reactants is a necessary condition for thermal explosion and is considered as a first stage. Chemical reaction terms are neglected during this stage. Mixing occurs in the layer positioned around the plane $\xi = 0$; from general considerations of the diffusion process, it is clear that the width of the mixing zone grows with time as $\sim \sqrt{\tilde{D}\tilde{t}} = \sqrt{\tilde{D}\tilde{t}_{ad}}\sqrt{\tau}$. As soon as reactants are mixed, the chemical reaction starts.

Half-width (at half-maximum) of the profile $C_1(\xi,\tau)C_2(\xi,\tau) = \varphi(\eta_1)$ (at $\tau > 0$) may serve as a measure of half-width of the mixing zone. At the same time, this is the zone where a noticeable reaction rate is expected to develop at a later stage.

At $\tau > 0$, the profile $\varphi(\eta_1) = (1-\eta_1)\eta_1$ always has the maximum at $\xi = 0$, equal to $(C_1C_2)_{\max} = [(1-\eta_1)\eta_1]_{\max} = \frac{1}{4}$. Therefore, the coordinate $b = \xi_*(\tau)$ at which $C_1C_2(\xi_*(\tau),\tau) = \frac{1}{2}(C_1C_2)_{\max} = \frac{1}{8}$ serves as an estimation of the half-width of the mixing (reaction) zone.

Maksyutov *et al.* (1982) postulate that the diffusion stage ends once the thickness $\tilde{b}$ becomes equal to the critical value of the half-width of the reacting layer (Figure 2.5) $\tilde{r}_{cr}$ (i.e. such $\tilde{r} = \tilde{r}_{cr}$ that delivers the critical value δ_{cr} of the Frank–Kamenetskii parameter). This condition allows the duration of diffusion stage, called the physical induction time τ^p_{ind}, to be determined quantitatively.

The end of the diffusion stage marks the point where the rate of energy release by the chemical reaction starts to exceed the rate of heat dissipation from the mixing zone. At the second, the chemical reaction stage, energy generation prevails over energy dissipation from reaction zone, and the process is controlled by the self-acceleration of the exothermic reaction. In the adopted approximation, diffusion terms may be neglected, and thermal explosion is considered as occurring in the adiabatic mode. This consideration provides quantitative estimation of the chemical induction time τ^c_{ind}.

Total explosion induction time is estimated as a sum of induction times of both stages.

Mathematically, for the diffusion stage, the problem degenerates into

$$\frac{\partial \eta_1}{\partial \tau} = \frac{\partial^2 \eta_1}{\partial \xi^2} \tag{4.63}$$

with the relevant (for the progress variable η_1) initial and boundary conditions (4.62).

The solution of the latter problem is

$$\eta_1(\xi,\tau) = 1 - C_1(\xi,\tau) = \frac{1}{\sqrt{\pi}}\int_{-\infty}^{\frac{\xi}{2\sqrt{\tau}}} \exp(-s^2)\,ds = 1 - \frac{1}{2}\mathit{erfc}\left(\frac{\xi}{2\sqrt{\tau}}\right), \quad \tau > 0 \tag{4.64}$$

Further (for $\tau > 0$)

$$\begin{aligned} C_1(\xi,\tau) &= 1 - \eta_1(\xi,\tau) = \frac{1}{2}\mathit{erfc}\left(\frac{\xi}{2\sqrt{\tau}}\right) \\ C_2(\xi,\tau) &= 1 - \frac{1}{2}\mathit{erfc}\left(\frac{\xi}{2\sqrt{\tau}}\right) \\ C_1C_2 &= (1-\eta_1)\eta_1 = \frac{1}{2}\mathit{erfc}\left(\frac{\xi}{2\sqrt{\tau}}\right)\left[1 - \frac{1}{2}\mathit{erfc}\left(\frac{\xi}{2\sqrt{\tau}}\right)\right] \end{aligned} \tag{4.65}$$

Recalling that the reaction of the second order ($m = 2$) is considered and taking into account that $\tilde{b} = \sqrt{\tilde{D}\tilde{t}_{ad}}b = \sqrt{\tilde{D}\frac{\tilde{\rho}\tilde{c}_p\tilde{R}\tilde{T}_0^2}{\tilde{Q}\tilde{k}_0\tilde{C}_0^2\tilde{E}}b}$, $\delta = \frac{\tilde{Q}\tilde{k}_0\tilde{C}_0^2\tilde{E}\tilde{r}^2}{\tilde{\lambda}\tilde{R}\tilde{T}_0^2}$, the condition defining the end of the diffusion stage is written simply ($Le = 1$) as

$$b^2\left(\tau_{ind}^p\right) = \delta_{cr} \tag{4.66}$$

Since the critical value of the Frank–Kamenetskii parameter for the planar symmetry case is known, $\delta_{cr} \approx 0.88$ (Table 2.1), Eq. (4.66) fully determines the physical induction time, based on solution (4.65) and definition of the half-width b.

For the mathematical treatment of the reaction stage, it may be assumed that $(1-\eta_1)\eta_1 = [(1-\eta_1)\eta_1]_{\max} = \frac{1}{4}$.

Equation (4.60) degenerates then into the adiabatic thermal explosion problem

$$\frac{d\theta}{d\tau} = [(1-\eta_1)\eta_1]_{\max}\exp\left(\frac{\theta}{1+\text{Ar}\theta}\right) \tag{4.67}$$

with the solution

$$\tau = \frac{1}{[(1-\eta_1)\eta_1]_{\max}}\int_0^{\theta}\exp\left(\frac{s}{1+\text{Ar}\cdot s}\right)ds \tag{4.68}$$

from which the chemical induction time τ_c may be estimated as

$$\tau_{ind}^c = \frac{1}{[(1-\eta_1)\eta_1]_{\max}}\int_0^{\theta_{F_{\max}^{(2)}}}\exp\left(\frac{s}{1+\text{Ar}\cdot s}\right)ds \tag{4.69}$$

Here, the upper limit of integration is taken as a maximum of the second derivative of the EGD function, in line with the thermal explosion criterion adopted in the book.

The total induction time of diffusion thermal explosion is estimated as $\tau_{ind} = \tau_{ind}^p + \tau_{ind}^c$.

It should be emphasized that in view of a number of assumptions discussed above, the theoretical analysis of Maksyutov *et al.* (1982) is approximate and requires experimental confirmation.

As a more accurate approach to the analysis of the diffusion thermal explosion problem, Maksyutov *et al.* (1982) performed a numerical analysis of the set of Eqs. (4.60)–(4.62) and obtained various spatial and temporal characteristics of explosion development.

4.4 Spotted Thermal Explosion

Spotted thermal explosion is another problem with substantially non-uniform initial conditions. Described informally, the problem refers to the presence of the initial hot spot in the reactive mixture (Figure 4.25). The chemical reaction rate in this region is significantly higher than that in the surrounding cold mixture. Released energy is being transferred away from the spot, involving new layers of the surrounding mixture into the

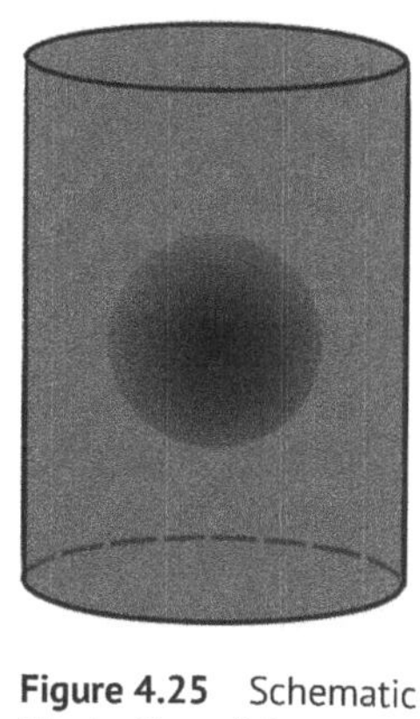

Figure 4.25 Schematic illustration of the initial hot spot in the reacting mixture.

reaction. This thermal wave progression may eventually lead (under certain conditions such as the critical minimum size of the spot) to thermal explosion.

From an applied point of view, such a problem has been traditionally considered in relation to the susceptibility of certain energetic materials (e.g. condensed phase explosives) to the formation of the localized regions of high temperatures under external impact. The latter may take a form of light, neutrons, shock wave, etc.

The present chapter neglects the effects of natural convection. Therefore, conduction is considered as the only mechanism of heat transfer. Detailed consideration of the full problem, involving natural convection, is presented in Chapter 5.

Theoretical analysis of the problem presents considerable mathematical difficulties. A number of approximate analytical methods, involving the idealizations of various nature, have been developed by Zinn (1962) Friedman (1963a, b), Boddington (1963) and Thomas (1965). These methods have been critically discussed by Merzhanov (1966) and found to produce results that deviate significantly from those obtained from numerical simulations. This outcome is not surprising, taking into account the complexity of the problem in both physical and mathematical sense (not even mentioning further complications, due to detailed chemical kinetics effects, omitted in the above studies). A more advanced approximate method has been proposed by Seplyarskii and Afanasiev (1989).

At present, the most accurate analysis, apart from direct physical experiments, relies on the numerical integration of governing equations (Merzhanov *et al.* 1963; Merzhanov 1966)..

In the simplest form, considering Π-shaped spot in the infinite space and neglecting kinetic effects, the problem is described by the energy balance equation, identical to (2.10)

$$\frac{\partial \theta}{\partial \tau} = \frac{1}{\delta}\left(\frac{\partial^2 \theta}{\partial \xi^2} + \frac{m}{\xi}\frac{\partial \theta}{\partial \xi}\right) + \exp\left(\frac{\theta}{1 + \text{Ar}\theta}\right) \quad \tau \geq 0; \ \xi \geq 0 \tag{4.70}$$

where the Frank–Kamenetskii parameter is constructed using the half-width of the spot profile, and the initial and boundary conditions have the form

$$\begin{aligned} &\theta(0,\xi) = 0, \ \ 0 \leq \xi \leq 1 \\ &\theta(0,\xi) = -\text{Zn}, \ \ \xi > 1 \\ &\frac{\partial \theta}{\partial \xi}(0,\tau) = 0; \ \ \frac{\partial \theta}{\partial \xi}(\infty,\tau) = 0 \end{aligned} \tag{4.71}$$

Temperature histories at the vessel's centre, for spherical spot, are presented in Figure 4.26. It is apparent that for small values of the Frank–Kamenetskii parameter initial energy stored in the hot spot quickly dissipates (curves 1 and 2). Self-heating of the spot becomes evident (points of maximum of curves 3 and 4) as the Frank–Kamenetskii parameter approaches its certain critical value. Explosion develops at the values of δ exceeding critical (curves 5 and 6). As the value of the Frank–Kamenetskii parameter grows, the temperature curve resembles more and more the one of adiabatic thermal explosion since the surface of the spot does not have sufficient time to cool down.

Critical conditions for spotted thermal explosion are controlled essentially by the Zeldovich number (i.e. initial amount of energy in the spot). It turns out that the dependence of critical conditions on this parameter, for different symmetries, may be approximated by the following formula:

$$\delta_{cr}^{(m)} = \alpha_m(\ln|\text{Zn}|)^{\beta_m} \tag{4.72}$$

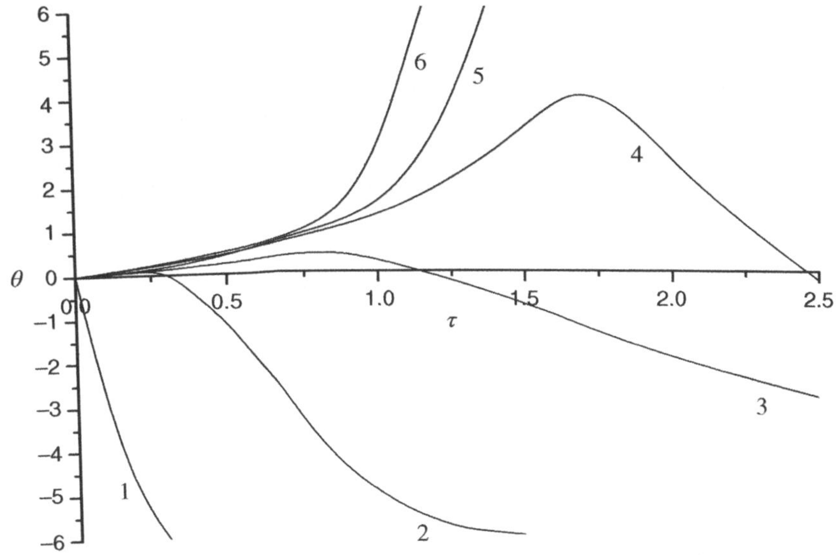

Figure 4.26 Temporal development of spotted thermal explosion. Single Π-shaped spot, reaction of the zeroth order. Spherical symmetry (m = 2) Temperature at the centre of the vessel. Ar = 0.02; Zn = − 10.0. 1 − δ = 0.1; 2 − δ = 5.5; 3 − δ = 15.0; 4 − δ = 20.0; 5 − δ = 20.5; 6 − δ = 25.0.

Table 4.2 Parameters of the formula (4.72).

Symmetry index, m	α_m	β_m
0	2.66	1.30
1	7.39	0.83
2	12.10	0.60

where the numerical parameters α_m and β_m, obtained from numerical simulations, are presented in Table 4.2.

There are two aspects in the development of spot thermal explosion: the explosion of the spot itself and the initiation of the wave-like heating process of the rest of the mixture. In order to understand the role of the chemical reaction induced beyond the spot, Merzhanov *et al.* (1963) considered the cases where such the latter reaction was artificially suppressed. It was found that the reaction in the mixture surrounding the spot does not have, in fact, any noticeable effect on critical conditions. Moreover, even replacing the spot with the vessel having wall temperatures equal to the surrounding temperature, one can observe only small changes in $\delta_{cr}^{(m)}$ and characteristic features of explosion development. Thus, there is no any specifics of spotted thermal explosion related to the presence of the surrounding mixture.

The latter portion of the mixture is relevant only for the consideration of characteristics of thermal/reaction wave developing in this part of the mixture itself. According to the results of numerical simulations, the initial speed of such a wave is well approximated by the equation

$$\frac{d\xi}{d\tau} = \frac{\gamma}{\delta} \tag{4.73}$$

where the parameter γ is only weakly dependent on Zn and δ and is in the range $(1.1 - 1.4) \times 10^3$. Feng *et al.* (1992) focused attention exclusively on the region surrounding hot spot by replacing the latter by a constant power source at the centre of the spherical reacting volume. They were able to develop an analytical approach capable of predicting critical (i.e. minimum) power supply required for the ignition of the mixture.

Apart from the simplest problem described above, a number of studies considered extended problem formulations. Thermal explosion developing in the system of several spots has been analysed by Zaturska (1974, 1975). The same problem, taking into account the effects of natural convection, is considered in Chapter 5 of the present book. Knyazeva *et al.* (1988) investigated the effect of different initial temperature distributions on explosion development, while Burkina and Knyazeva (1991) analysed the influence of autocatalysis on critical conditions.

4.5 Experimental Validation of the Theory of Thermal Explosion

As any physical theory, the theory of thermal explosion requires experimental validation. A significant amount of experimental research, conducted throughout the years of the theory development, has been compared to theoretical predictions.

Early validations (Frank–Kamenetskii 1939) focused on the few reactions with known kinetics whose explosive behaviour was believed to be of thermal nature. These are decompositions of azomethane, ethylazide and methylnitrate, as well as hydrogen sulphide oxidation. For the first three of these reactions, which have simple monomolecular kinetics, predicted critical runaway conditions conformed very well with the theory predictions. The fourth reaction has rather complicated autocatalytic kinetics, and agreement was somewhat less favourable.

Further, the theory was applied to the prediction of critical runaway conditions for such reactions for which it has not yet been observed experimentally. In this way, critical conditions for the autoignition of nitrous oxide, never observed before, were derived. Later experiments by Zeldovich and Yakovlev (1938) not only confirmed the existence of such phenomenon but also the quantitative validity of the theory predictions.

The theory also correctly predicts the location of the so-called third explosion limit of $H_2 - O_2$ mixtures, on the assumption that this limit is of thermal nature (Frank–Kamenetskii 2016). Comparisons were made with the experiments by Chirkov (1937) and Ziskin (1942).

Another substance, which demonstrated quite good agreement between theory and experiments (under the assumption of purely thermal nature of runaway), was diethyl peroxide (Harris 1940).

A very convenient method for the investigation of thermal explosion is thermogravimetric analysis (TGA). Heating of the reactive sample in a cylindrical vessel under well-controlled conditions (typically the linear law of the change of ambient temperature, spatially uniform sample temperature, boundary conditions of either the first of the third type) induces, at certain critical temperature, thermal explosion. If the kinetics of chemical transformation is known and can be sufficiently simplified, measured critical temperature may be compared with that predicted by the classical thermal explosion theory.

Table 4.3 Comparison between experiments and Semenov thermal explosion theory. Critical thermal explosion temperatures and pre-explosionspiepr 132 heat-ups.

Vessel geometry		Critical temperature $\tilde{T}_{cr}$, K		Pre-explosion heat-up $\Delta\tilde{T}_{cr}$, K	
Diameter $\tilde{d}$, mm	**Aspect ratio l/d**	**Experiment**	**Semenov theory**	**Experiment**	**Semenov theory**
(a) Dinitroxydiethylnitramine					
6.2	2	431	430	11	8.30
8.5	2	425.5	425	8	7.45
10.3	2	418.5	418	9	7.70
10.3	1	421.5	420.5	9	7.80
12.7	1	420.5	419	6	7.75
32.0	2	435	433	9	8.25
36.5	2	432	430	10	8.15
37.5	2	431	429	8	8.10
49.0	2	426	427	8	8.05
62.0	1	425	425	9	7.95
(b) tetryl					
6.1	2	424.5	428	8	9.5
8.4	2	421.5	422	12	9.4
8.7	2	412.5	413	10	8.9
11.7	2	412.5	412	9	8.9
37.5	3	426	425	15	10
45.0	3	424	424.5	24	10
49.0	3	425	423.0	9	10

Source: Developed based on data by Barzykin (2004).

Table 4.3 presents the data obtained for the two substances, dinitroxydiethylnitramine and tetryl. The table lists critical temperatures, as well as the pre-explosion heat-up of the substance, that is, the difference between sample and ambient temperatures at critical condition. A very good agreement between the theory and the experiment is apparent.

5

Thermal Explosion in Dynamic Mixtures

5.1 Thermal Explosion in Flow Reactor

Chapters 2, 3 and 4 considered thermal explosions in a quiescent medium. Although reasonable (and strictly valid for the case of solid fuels) in many cases, this assumption is still an idealization of real physical conditions. Motion within reacting media (i.e. dynamical conditions) may exert a major influence on the critical conditions of thermal explosion and its development in time.

This difference may be predicted already considering the comparison between the Semenov and the Frank-Kamenetskii theories of thermal explosion, presented in Chapter 2. It has been demonstrated that the existence of heat conduction fluxes, even in the absence of motion within the medium, is sufficient to cause significant discrepancies between the two formulations. Dynamic conditions would further complicate physico-chemical environment, leading to not only conductive but also convective fluxes, two ways coupled with non-uniformities in reaction rate and temperature distributions.

Note that Semenov model is strictly valid, in essence, under conditions idealized by perfectly stirred reactor (PSR) assumptions. These are assumptions of perfect mixing (homogeneity) of reacting mixture.

The present section considers thermal explosion in PSR system in more detail, in particular taking into account effects of reactants consumption.

Possibilities of thermal explosion development under natural and forced (however, less restrictive than PSR) convection conditions are considered in Sections 5.2 and 5.3, respectively.

Most fundamental analysis of thermal explosion regimes in PSR (which is also called homogeneous chemical reactor) is due, primarily, to early works of Zeldovich and co-authors (Zeldovich 1941; Zeldovich and Zysin 1941).

PSR is shown schematically in Figure 5.1, as the simplest case of a wide pipe with zero hydrodynamical resistance. Different geometrical arrangements are possible, for example, mixture may be injected at the periphery of the cylinder and removed at the centre, along the cylinder axis (Zeldovich *et al.* 1980). Similar to Chapter 4, all the reactants are assumed to have equal concentrations, and all the stoichiometric coefficients are assumed being equal to unity. Reactant enters the system with certain initial concentration $\tilde{C}_0$ and temperature $\tilde{T}_{in}$. Within the reactor, by the very virtue of PSR assumptions, spatially uniform reaction transforms the mixture to another state described by values $\tilde{C}_{ex}$ and $\tilde{T}_{ex}$. The latter values are constant over the volume of the system, and the stream leaving the reactor has the same exit properties $(\tilde{C}_{ex}, \tilde{T}_{ex})$.

Thermal Explosion: Theory and Application, First Edition. Vasily B. Novozhilov.

Figure 5.1 Schematic of perfectly stirred (homogeneous) chemical reactor. Black arrows indicate intensive forced mixing.

Consider now the basic, most fundamental, mathematical formulation of the process in a homogeneous chemical reactor. Inflow temperature of supplied reactants is considered here, for simplicity but without loss of generality, as equal to ambient temperature. This non-restrictive assumption is consistent with the same approach adopted in Chapter 2. Therefore, the definition of non-dimensional temperature remains, as usual, given by Eq. (2.2). In the case of inflow temperature different from the ambient, the definition may be conveniently modified by making use of effective initial temperature, that is the steady-state temperature of the system in the absence of chemical reaction (Merzhanov and Abramov 1977; Barzykin 2004).

Governing equations may be written quite easily by modifying formulations already considered in Chapter 4.

The Semenov energy balance equation retains its form (4.3), with the only exception that the heat transfer coefficient $\tilde{h}$ in the Semenov parameter expression (4.2) must be replaced by the effective value $\tilde{h}_{eff}$, which takes into account combined effect of conductive heat exchange through the walls and convective exchange due to flow through the reactor.

The kinetic equation (4.9) must be modified to take into account the removal of reactants by the exit stream. In dimensional form, such a removal is described by the term $-\frac{\tilde{\dot{\mho}}}{\tilde{V}}(\tilde{C}-\tilde{C}_0)$, where $\tilde{\dot{\mho}}$ is the volumetric flow rate through the reactor. It is easy to see that the dimension of this term is consistent with the dimension of the reaction rate.

The problem, therefore, is described by the following set of equations:

$$\begin{aligned}\frac{d\theta}{d\tau} &= \exp\left(\frac{\theta}{1+Ar\theta}\right)\varphi(\eta) - \frac{1}{Se}\theta \\ \frac{d\eta}{d\tau} &= \frac{1}{Td}\exp\left(\frac{\theta}{1+Ar\theta}\right)\varphi(\eta) - \frac{1}{Da}\eta\end{aligned} \tag{5.1}$$

with the initial conditions

$$\theta(0) = 0; \eta(0) = 0$$

Here

$$Se = \frac{\tilde{Q}\tilde{k}_0\tilde{C}_0^m\tilde{V}\tilde{E}}{\tilde{h}_{eff}\tilde{S}\tilde{R}\tilde{T}_0^2}; \quad Td = \frac{\tilde{Q}\tilde{E}\tilde{C}_0}{\tilde{\rho}\tilde{c}_p\tilde{R}\tilde{T}_0^2} \tag{5.2}$$

and an additional emerging non-dimensional parameter is the Damköhler number

$$Da = \frac{\tilde{V}\tilde{Q}\tilde{k}_0\tilde{C}_0^m\tilde{E}}{\tilde{\dot{\mho}}\tilde{\rho}\tilde{c}_p\tilde{R}\tilde{T}_0^2} \tag{5.3}$$

This is a ratio of the flow time scale $\frac{\tilde{V}}{\dot{\tilde{\mho}}}$ (which may be interpreted as a flow residence time in the reactor) to the chemical reaction time scale, which is given in the present case by the adiabatic time scale $\frac{\tilde{\rho}\tilde{c}_p\tilde{R}\tilde{T}_0^2}{\tilde{Q}\tilde{k}_0\tilde{C}_0^m\tilde{E}}$.

Consider first the possible stationary states of the reactor. At such a state, the degree of reactant heating must be proportional to the degree of chemical transformation, that is, $\theta \propto \eta$. This fact also obviously follows formally from the set of Eq. (5.1) upon setting time derivatives on the left-hand side (LHS) of both equations to zero

$$\theta = \mathrm{N}\nu \cdot \eta \tag{5.4}$$

where non-dimensional parameter, the Novozhilov number

$$\mathrm{N}\nu = \frac{Td\, Se}{Da} = \frac{\tilde{Q}\tilde{E}\tilde{C}_0\dot{\tilde{\mho}}}{\tilde{h}_{eff}\tilde{S}\tilde{R}\tilde{T}_0^2} \tag{5.5}$$

is the reactant molar supply rate at which energy generated by chemical reaction is balanced by energy removal at flow temperature equal to characteristic temperature interval.

Due to this time-independent relation between the unknown variables, the set of Eq. (5.1) reduces to the single ODE

$$\frac{d\theta}{d\tau} = F(\theta) \tag{5.6}$$

formally similar to the Semenov model (2.5), but with the different EGD function

$$F(\theta) = \exp\left(\frac{\theta}{1 + Ar\theta}\right)\left(1 - \frac{1}{\mathrm{N}\nu}\theta\right)^m - \frac{1}{Se}\theta \tag{5.7}$$

Obviously, the model may also be written equivalently, using the relation (5.4), in terms of the single ODE for the progress variable η.

The principal difference of the EGD function (5.7) from the EGD function (2.6) in the basic Semenov model is that the function (5.7) takes into account the effect of reactant consumption, while the function (2.6) is the dependence of the temperature change rate on temperature at the initial value of reactant concentration. This difference results in the fact that the EGD function of the flow reactor model contains the two parameters, *Se* and *Nν*, in contrast to the classical Semenov model with the only parameter, the Semenov number *Se*. The former two parameters may, however, be varied independently, and their ratio is proportional to the volumetric flow rate $\frac{N\nu}{Se} \propto \dot{\tilde{\mho}}$. The effect of changes in the energy removal rate on thermal explosion possibility, which is of primary interest, may be investigated by fixing the Novozhilov number *Nν* and varying the Semenov number *Se*. Due to the proportionality, discussed just above, higher values of the parameter *Se* would correspond to smaller flow rates $\dot{\tilde{\mho}}$ and therefore slower energy removal rates and the other way round.

Qualitatively, therefore, the steady-state theory of the homogeneous chemical reactor is very similar to the classical Semenov steady-state theory.

In contrast to the Semenov model with chemical kinetics (Eqs. (4.3), (4.9)), where the only possible stationary state is trivial (full reactants consumption and system temperature equal to ambient), in the PSR model different stationary states may exist.

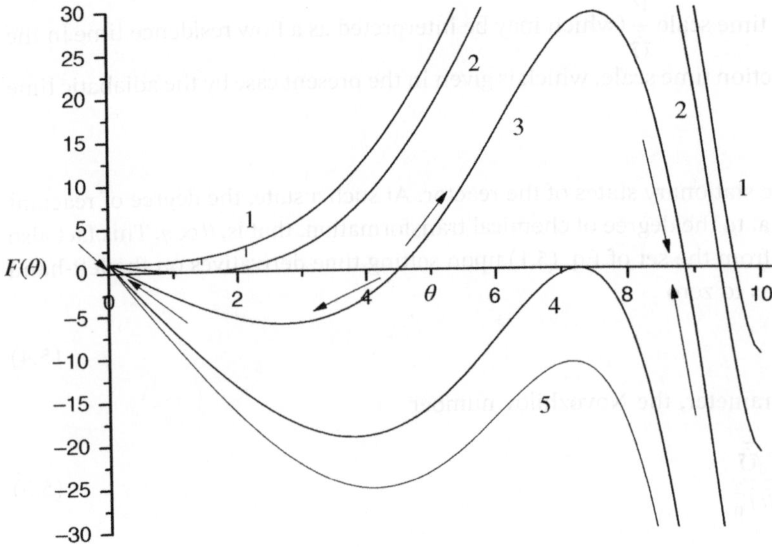

Figure 5.2 EGD function of the PSR model with different values of the Semenov parameter. $Ar = 0.01$. $N_v = 10.0$. 1 – $Se = 1.0$; 2 – $Se = Se_{cr} \approx 0.47$; 3 – $Se = 0.2$; 4 – $Se = 0.11$; 5 – $Se = 0.096$. Arrows – directions of the system evolution in different regions of phase space for the case with three roots (curve 3).

Depending on the value of the Semenov parameter, the function $F(\theta)$ may have one, two or three roots in the half-space $\theta \geq 0$. These qualitatively different cases are shown in Figure 5.2.

The cases with one root correspond to either a rather high or a rather low rate of energy removal from the system.

The case of curve 5, Figure 5.2, corresponds to a high rate of energy removal, that is to a high velocity of flow in the reactor. This root corresponds to the stable stationary state of the system with small flow heat-up in the reactor. Reaction proceeds at a slow rate, and consumption of reactants (degree of chemical transformation) is small.

In the opposite case (curve 1, Figure 5.2), the energy removal rate is low, and the residence time of the flow within the reactor is larger than the induction time of the chemical reaction. These conditions lead to a stationary state with reaction at high temperature and full consumption of reactants. Essentially, this scenario describes a thermal explosion. In contrast to a similar solution in the classical Semenov theory (Chapter 2), this stationary state describes a regime with realistic combustion temperature.

Critical condition $Se = Se_{cr}$ of transition from a slow reaction regime to thermal explosion (autoignition) is given, exactly as in the classical Semenov theory, by the requirement that the curves $F(\theta)$ and $\theta = 0$ are tangent (curve 2, Figure 5.2). Therefore, exactly as in the Semenov theory, it is given by the Legendre transform requirement (2.15) for the convex part of the first term on the RHS of Eq. (5.7).

The set of equations analogous to (2.16, 2.17) reads

$$\frac{1}{(1+Ar\theta_{cr})^2}\exp\left(\frac{\theta_{cr}}{1+Ar\theta_{cr}}\right)\left(1-\frac{1}{Nv}\theta_{cr}\right)^m - \frac{m}{Nv}\exp\left(\frac{\theta_{cr}}{1+Ar\theta_{cr}}\right)\left(1-\frac{1}{Nv}\theta_{cr}\right)^{m-1} - \frac{1}{Se_{cr}} = 0 \tag{5.8}$$

$$\exp\left(\frac{\theta_{cr}}{1+Ar\theta_{cr}}\right)\left(1-\frac{1}{Nv}\theta_{cr}\right)^m - \frac{1}{Se_{cr}}\theta_{cr} = 0$$

Considering, as in Chapter 2, the limit $Ar \to 0$ would imply the critical conditions given by the set of equations

$$\begin{aligned} &\exp(\theta_{cr})\left(1-\frac{1}{Nv}\theta_{cr}\right)^{m} \\ &-\frac{m}{Nv}\exp(\theta_{cr})\left(1-\frac{1}{Nv}\theta_{cr}\right)^{m-1}-\frac{1}{Se_{cr}}=0 \\ &\exp(\theta_{cr})\left(1-\frac{1}{Nv}\theta_{cr}\right)^{m}-\frac{1}{Se_{cr}}\theta_{cr}=0 \end{aligned} \tag{5.9}$$

which results in the single quadratic equation

$$\theta_{cr}^2 + (m - Nv - 1)\theta_{cr} + Nv = 0 \tag{5.10}$$

It is clear that the second of Eq. (5.9) cannot have non-positive roots; therefore, both the roots

$$\begin{aligned} \theta_{cr}^{(1)} &= \frac{-(m-Nv-1)-\sqrt{(m-Nv-1)^2-4Nv}}{2} \\ \theta_{cr}^{(2)} &= \frac{-(m-Nv-1)+\sqrt{(m-Nv-1)^2-4Nv}}{2} \end{aligned} \tag{5.11}$$

are positive.

The value of the Semenov parameter, corresponding to the smaller of the two roots,

$$Se_{cr} = \theta_{cr}^{(1)}\exp\left(-\theta_{cr}^{(1)}\right)\left(1-\frac{1}{Nv}\theta_{cr}\right)^{-m} \tag{5.12}$$

describes critical condition of thermal explosion, manifested by disappearance of the root $\theta_{cr}^{(1)}$ and transition to a stationary state at high (combustion) temperature close to $\theta_{cr}^{(2)}$. This transition corresponds to the transition from the curve of type 3 to the curve of type 1 in Figure 5.2.

The larger root $\theta_{cr}^{(2)}$ and the corresponding Semenov number $Se^{(2)}$ describe one of the possible states, where thermal explosion has occurred (specifically, the state where the EGD function is tangent to the θ-axis.

The values of the Semenov parameters Se_{cr} and $Se^{(2)}$ correspond to EGD curves with two roots (curves 2 and 4, Figure 5.2, respectively).

In the range of the Semenov parameter values $Se^{(2)} < Se < Se_{cr}$, the EGD function $F(\theta)$ has three roots. An example is given by curve 3 in Figure 5.2. Exactly as is the case in the classical Semenov theory, the middle of these three roots is unstable, while the other two roots are stable (Figure 5.2). The root with the least value corresponds to the slow reaction regime, while the root with the largest value to the combustion regime resulting from autoignition of the mixture stream.

It should be noted that the state corresponding to the middle root is unstable under an assumption, most common in the stability analysis, that external parameters (e.g. flow rate) controlling the system, remain constant.

In contrast to the systems with constant mass and volume, considered in the basic Semenov theory, a flow reactor can be dynamically controlled. In the most straightforward way, such control is achieved by varying flow rate through the reactor. In dynamically controlled conditions, stability property of a particular state may change. Thus, the state corresponding to the middle root of the EGD function shown as curve 3 in Figure 5.2 may become stable under some, specifically chosen control strategy. Operation of the reactor in the thus stabilized state corresponding to the middle root may sometimes have practical advantages.

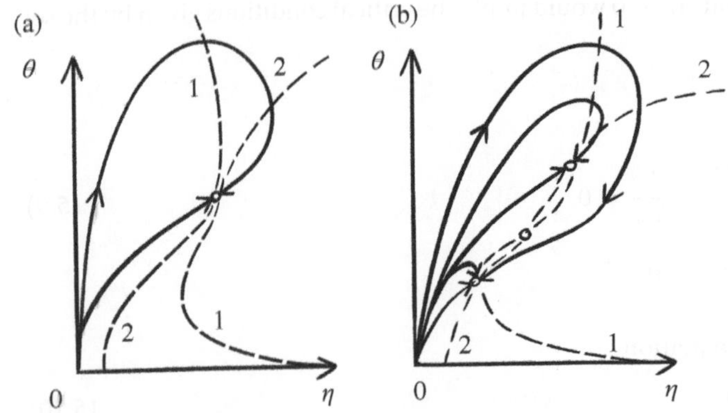

Figure 5.3 Examples of phase trajectories (qualitative) of perfectly stirred (homogeneous) chemical reactor with one (a) and three (b) stationary states. Dashed curves: 1 – locus of the points $\frac{d\theta}{d\eta} = 0$; 2 – locus of the points $\frac{d\eta}{d\theta} = 0$.

Non-steady evolution of processes in PSR is governed by the set of Eq. (5.1). Such non-steady behaviour may be quite complicated. For example, Figure 5.3 presents, qualitatively, examples of phase trajectories of chemical reactor system with one or three stationary states.

Figure 5.3 (a) qualitatively demonstrates phase trajectories corresponding to thermal explosion. These trajectories approach a stationary state with high-temperature and substantially consumed reactants.

The system shown in Figure 5.3 (b) has three stationary states, two stable and one unstable, in the middle. The lower stable state represents chemical equilibrium with slow reaction, while the higher one develops as a result of autoignition.

Apart from attraction to stable stationary states (with phase trajectories similar to Figure 5.3), there may also exist attractors in the form of limit cycles, encircling either one or three stationary states. These limit cycles correspond to oscillatory regimes. In total, 35 different phase portraits may exist for a chemical reactor system with a basic non-autocatalytic reaction (Vaganov *et al.* 1978). Their study describes boundaries, in the parameter space, dividing the regions with different phase portraits. Besides, transitions between these regions have been studied. An important generalization is presented by the case of autocatalytic reaction. Different regimes of such reactions were studied by Abramov *et al.* (1978).

Extensive literature is dedicated to various problems associated with design, operation and control of chemical reactors.

Problems of stability and control of various operation regimes of chemical reactors, including more specific regimes, such as auto-oscillations, are considered by Frank-Kamenetskii (2016).

Different aspects of the design and operation of chemical reactors are also discussed, for example, in the monographs by Aris (1961) and, recently, by Conesa (2020).

5.2 Thermal Explosion Under Natural Convection Conditions

Natural convection, inevitably occurring (except for very specific cases) in media with temperature gradients, affects the dynamics of heat transfer processes in reacting mixtures. Generally, it intensifies the process of heat transfer between hot and cold layers of the flow and therefore increases the rate of cooling of the mixture. Due to this reason, the development of natural convection must also affect the critical conditions of thermal explosion. Given excessive cooling of the mixture, it should be expected that critical parameters of thermal explosion, describing the contribution of chemical reaction to energy generation, increase as natural convection becomes more and more intensive. Although the basic, the Semenov and the Frank-Kamenetskii models ignore this effect, the necessity of its consideration was evident from the very beginning of the development of the thermal explosion theory. This was discussed first by Frank-Kamenetskii (1939).

The effect of natural convection on critical conditions of thermal explosion was investigated, using unsteady numerical simulations, by Shtessel *et al.* (1971). Merzhanov and Shtessel (1971) investigated this effect experimentally. It should be noted that the latter results (based on unsteady simulations or experiments) used uniform initial temperature distribution in the mixture. As has already being mentioned in Chapter 2, in the non-steady thermal explosion theory, the result (that is, occurrence or non-occurrence of explosion) depends, generally, on initial distribution.

Results of the above studies may be summarized as dependence of critical conditions on the Rayleigh number

$$Ra = \frac{\tilde{g}\tilde{\beta}\tilde{\rho}\left(\tilde{T}_0 - \tilde{T}_a\right)\tilde{r}^3}{\tilde{\mu}\tilde{\kappa}} \tag{5.13}$$

Here, the Rayleigh number is defined based on the characteristic temperature difference; $\tilde{r}$ is the characteristic length scale of the vessel.

Specifically, the critical boundary has the form

$$f(Ra) = 1 + 0.062Ra^{\frac{1}{3}}; \quad Ra < 10^8 \tag{5.14}$$

This dependence, shown in Figure 5.4, has been established for the flat (i.e. containing the mixture between parallel walls) and cylindrical vessels.

There are also shown (qualitatively) here the regions where the convection is negligible (regions 1 and 4), and the regions where convection accompanies either non-explosive or explosive mixture evolution (regions 2 and 3).

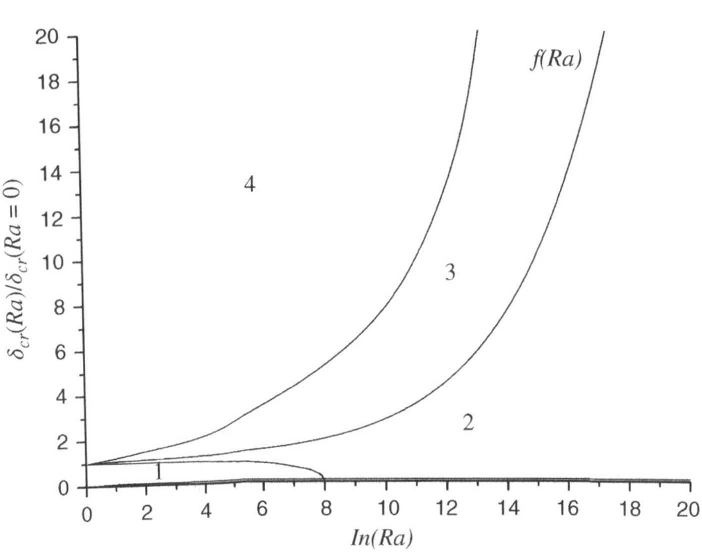

Figure 5.4 Dependence (5.14) of critical Frank-Kamenetskii parameter (normalized by its value in the absence of natural convection) on Rayleigh number. Uniform initial temperature distribution. Regions with different controlling mechanisms. 1 – non-explosive regime without convection; 2 – non-explosive regime with convection; 3 – explosive regime with convection and 4 – explosive regime. Characteristic time of convection development is larger than the explosion induction time.

In those regions where convection develops (regions 2 and 3), it affects not only the critical condition but also the duration of the induction period (although too much lesser extent).

In the cases considered by Merzhanov and Shtessel (1971), their theoretical analysis and experimental observations show that convection develops in the upper part of the vessel only, while heat conduction is the dominant mechanism of heat transfer in the bottom part. They also investigated the effect of external heat exchange under natural convection conditions. At $Bi \to \infty$, the influence of natural convection on explosion critical conditions is maximal (critical Frank-Kamenetskii parameter depends strongly on Rayleigh number). As the Biot number decreases, the influence on natural convection on critical conditions of the explosion weakens. It disappears almost completely at $Bi \to 0$.

More complicated is the case with non-uniform initial temperature distribution, considered by Novozhilov and Byrne (2013a, b) and Novozhilov (2014). As only the induction period of thermal explosion is considered, deviations of pressure, density and temperature from their values corresponding to the state of mechanical and thermal equilibrium of the mixture (hydrostatic pressure $\tilde{p}_0(\tilde{z}) = \tilde{\rho}_0 \tilde{g} \tilde{z} + const$ and constant values of density $\tilde{\rho}_0$ and temperature $\tilde{T}_0$) are considered to be small

$$\begin{aligned} &\tilde{p} = \tilde{p}_0 + \tilde{p}'; \quad \tilde{\rho} = \tilde{\rho}_0 + \tilde{\rho}'; \quad \tilde{T} = \tilde{T}_0 + \tilde{T}' \\ &\left|\frac{\tilde{p}'}{\tilde{p}_0}\right| \ll 1; \quad \left|\frac{\tilde{\rho}'}{\tilde{\rho}_0}\right| \ll 1; \quad \left|\frac{\tilde{T}'}{\tilde{T}_0}\right| \ll 1 \end{aligned} \tag{5.15}$$

As the Mach number is low, a weakly compressible approximation is also being used. This means that density changes only in response to changes in temperature and not the pressure. This is enforced by using a constant (averaged over the height) background pressure in the equation of state

$$\langle \tilde{p}_0 \rangle = \tilde{\rho} \tilde{R} \tilde{T} \tag{5.16}$$

As a result, small variations in density and pressure are related as

$$\tilde{\rho}' = -\tilde{\rho}_0 \tilde{\beta} \tilde{T}'; \text{ or } \tilde{\rho} = \left(1 - \tilde{\beta} \tilde{T}'\right) \tilde{\rho}_0; \ \tilde{\beta} = \frac{1}{\tilde{T}_0} \tag{5.17}$$

Following the conventional procedure of derivation of governing equations for natural convection flow of reacting heat-conducting gas (Landau and Lifshitz 1987), the set of equations

$$\begin{aligned} &\frac{\partial \tilde{\rho}}{\partial \tilde{t}} + \tilde{\nabla}\left(\tilde{\rho} \vec{\tilde{u}}\right) = 0 \\ &\frac{\partial \vec{\tilde{u}}}{\partial \tilde{t}} + \left(\vec{\tilde{u}} \cdot \tilde{\nabla}\right) \vec{\tilde{u}} = -\frac{\tilde{\nabla} \tilde{p}'}{\tilde{\rho}_0} + \frac{\tilde{\mu}}{\tilde{\rho}_0} \tilde{\Delta} \vec{\tilde{u}} - \vec{\tilde{g}} \tilde{\beta} \tilde{T}' \\ &\frac{\partial \tilde{T}'}{\partial \tilde{t}} + \vec{\tilde{u}} \cdot \tilde{\nabla} \tilde{T}' = \tilde{\kappa} \tilde{\Delta} \tilde{T}' + \frac{\tilde{Q}}{\tilde{\rho}_0 \tilde{c}_p} \tilde{w} \\ &\tilde{\rho} = \left(1 - \tilde{\beta} \tilde{T}'\right) \tilde{\rho}_0 \end{aligned} \tag{5.18}$$

is obtained. Here, the coefficient of thermal expansion

$$\tilde{\beta} = -\frac{1}{\tilde{\rho}} \frac{\partial \tilde{\rho}}{\partial \tilde{T}}\bigg|_{\tilde{T} = \tilde{T}_0} \tag{5.19}$$

is given by the formula (5.17).

The set of Eq. (5.18) is slightly different from the conventional set for natural convection flow in the Boussinesq approximation (Landau and Lifshitz 1987) since density variation is also being accounted for in the continuity equation and not just in the buoyancy term.

Further, the reaction of the zero order (no reactant consumption) is considered.

In order to make Eq. (5.18), non-dimensional, a particular special scale $\tilde{L}$ (for example, the size of the flow domain) needs to be introduced. In a natural convection problem, the natural time and velocity scales are $\sqrt{\frac{\tilde{L}}{\tilde{g}}}$ and $\sqrt{\tilde{L}\tilde{g}}$, respectively. The density scale $\tilde{\rho}_0$ and the pressure scale $\tilde{L}\tilde{g}\tilde{\rho}_0$ are used. The following non-dimensional variables are introduced (non-dimensional temperature θ is defined in the same conventional way as in Chapter 2)

$$\tau = \frac{\tilde{t}}{\sqrt{\tilde{L}/\tilde{g}}}; \rho = \frac{\tilde{\rho}}{\tilde{\rho}_0}; p' = \frac{\tilde{p}'}{\tilde{L}\tilde{g}\tilde{\rho}_0}; \vec{u} = \frac{\tilde{\vec{u}}}{\sqrt{\tilde{L}\tilde{g}}}$$

$$Pe = \frac{\tilde{L}\sqrt{\tilde{L}\tilde{g}}}{\tilde{\kappa}}; \; Re = \frac{\tilde{L}\sqrt{\tilde{L}\tilde{g}\tilde{\rho}_0}}{\tilde{\mu}} \tag{5.20}$$

$$\delta = \sqrt{\frac{\tilde{L}}{\tilde{g}}} \frac{\tilde{A}_0 \tilde{Q}}{\tilde{\rho}_0 \tilde{c}_p \tilde{T}_0^{(1-\alpha)}} \frac{1}{Ar} \exp\left(-\frac{1}{Ar}\right)$$

The definition of Frank-Kamenetskii parameter is different here from the one in Chapter 2.

With the help of the above scales and non-dimensional parameters, the set of Eq. (5.18) may be transformed to the non-dimensional form

$$\begin{aligned}
&\frac{\partial \rho}{\partial \tau} + \nabla\left(\rho \vec{u}\right) = 0 \\
&\frac{\partial \vec{u}}{\partial \tau} + \left(\vec{u} \cdot \nabla\right)\vec{u} = -\nabla \tilde{p}' + \frac{1}{\text{Re}} \tilde{\Delta} \tilde{\vec{u}} - Ar\theta \vec{e}_g \\
&\frac{\partial \theta}{\partial \tau} + \vec{u} \cdot \nabla \theta = \frac{1}{Pe} \Delta \theta + \delta (1 + Ar\theta)^{\alpha} \exp\left(\frac{\theta}{1 + Ar\theta}\right) \\
&\rho = 1 - Ar\theta
\end{aligned} \tag{5.21}$$

Here $\vec{e}_g = \frac{\tilde{\vec{g}}}{\left|\tilde{\vec{g}}\right|}$; also the formulae (5.17), (5.19) and the form of reaction rate term considered in Chapter 1 have been used.

The set of Eq. (5.21) has to be solved numerically. Novozhilov and Byrne (2013a, b) and Novozhilov (2014) used the Lattice Boltzmann (LB) computational method for this purpose.

A thermal (i.e. hot spherical region, or ball) is used as a basic form of non-uniform initial temperature distribution. Several thermal explosion scenarios are discussed below for either a single thermal or a pair of thermals localized in the computational domain, i.e. for the initial distributions of the form

$$\begin{aligned}
&\theta = \theta_{\max} \exp\left(-b\left|\vec{x} - \vec{x}_0\right|\right) \text{ (single thermal)} \\
&\theta = \theta_{\max} \exp\left(-b\left|\vec{x} - \vec{x}_1\right| - b\left|\vec{x} - \vec{x}_2\right|\right) \text{ (double-thermal system)}
\end{aligned} \tag{5.22}$$

where the parameter $b > 0$ controls the width of the distribution; $\vec{x}_0$, $\vec{x}_1$ and $\vec{x}_2$ are fixed points inside the flow domain.

The maximum temperature $\theta_{\max}$ allows the Rayleigh number to be defined as

$$Ra = \mathrm{Re}^2 \mathrm{Pr} \theta_{\max} \tag{5.23}$$

Using this definition, the dependence of the critical Frank-Kamenetskii parameter on Rayleigh number can be established.

In the following examples, a cubic domain with no-slip wall boundary conditions and fixed temperature $\theta = 0$ at the walls is considered.

Appropriate variations of the length scale $\tilde{L}$ and the maximum temperature $\theta_{\max}$ allow the Rayleigh number range from 10^3 to 10^8 to be considered.

The other parameters (with the properties of air being used) are $\mathrm{Pr} = 0.7$, $Ar = 0.1$ and $\alpha = 0.5$. Further details of numerical method implementation may be found in Novozhilov and Byrne (2013a).

Figures 5.5–5.41 illustrate five particular, different scenarios of thermal explosion development under natural convection conditions. The values of Rayleigh numbers in these examples are at the

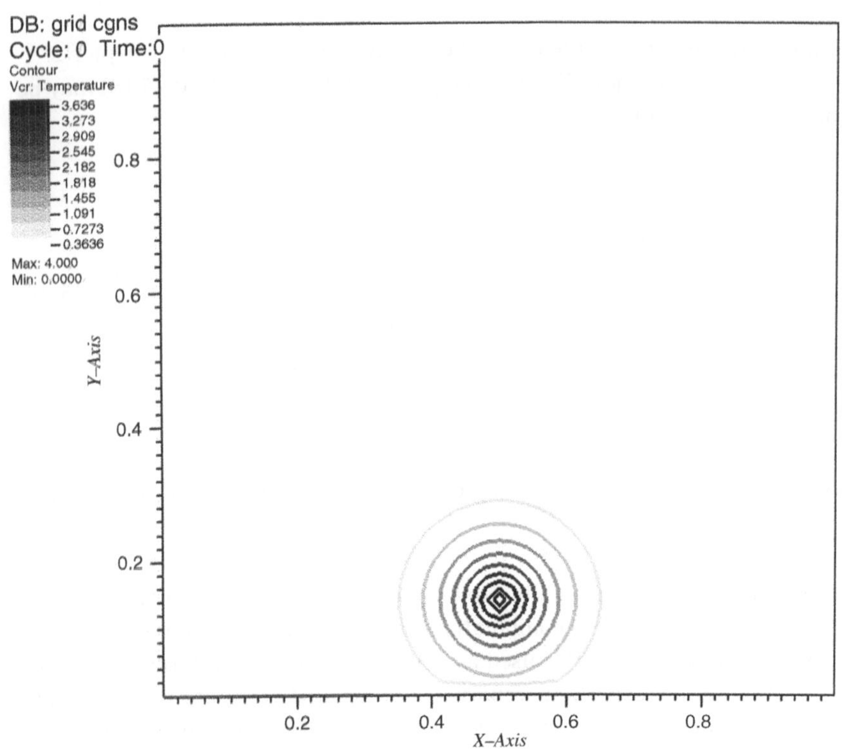

Figure 5.5 Unstable thermal (Frank-Kamenetskii parameter δ above critical). $Ra = 4 \cdot 10^3$; $\theta_{\max} = 4.0$. Isolines of non-dimensional temperature. $\tau = 0$. Isoline values are shown on the scale panel on the left-hand side of the figure.

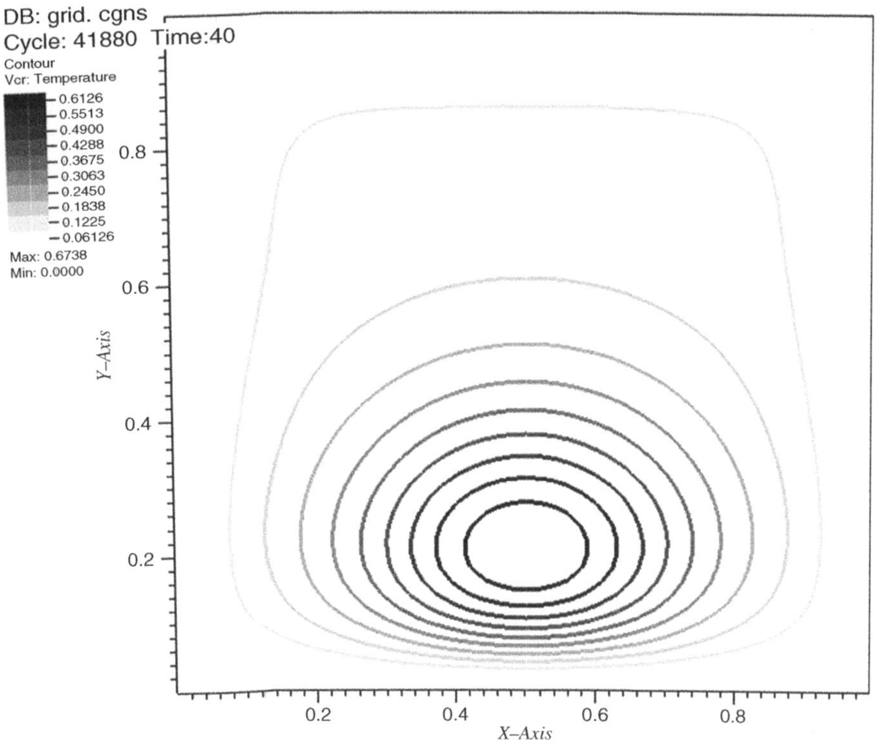

Figure 5.6 Unstable thermal (Frank-Kamenetskii parameter δ above critical). $Ra = 4 \cdot 10^3$; $\theta_{max} = 4.0$. Isolines of non-dimensional temperature. $\tau = 40$. Isoline values are shown on the scale panel on the left-hand side of the figure.

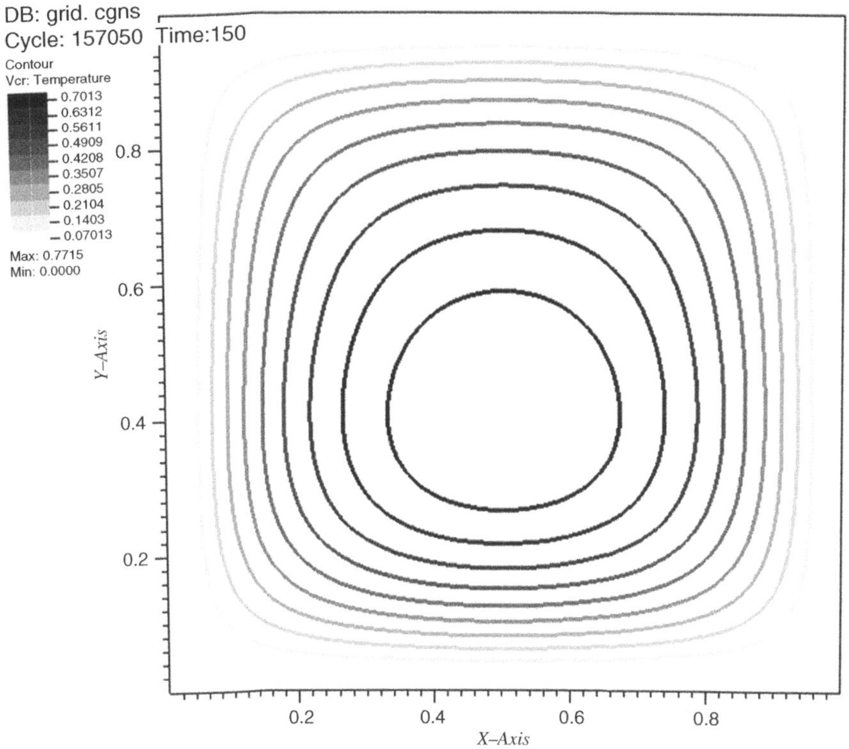

Figure 5.7 Unstable thermal (Frank-Kamenetskii parameter δ above critical). $Ra = 4 \cdot 10^3$; $\theta_{max} = 4.0$. Isolines of non-dimensional temperature. $\tau = 150$. Isoline values are shown on the scale panel on the left-hand side of the Figure.

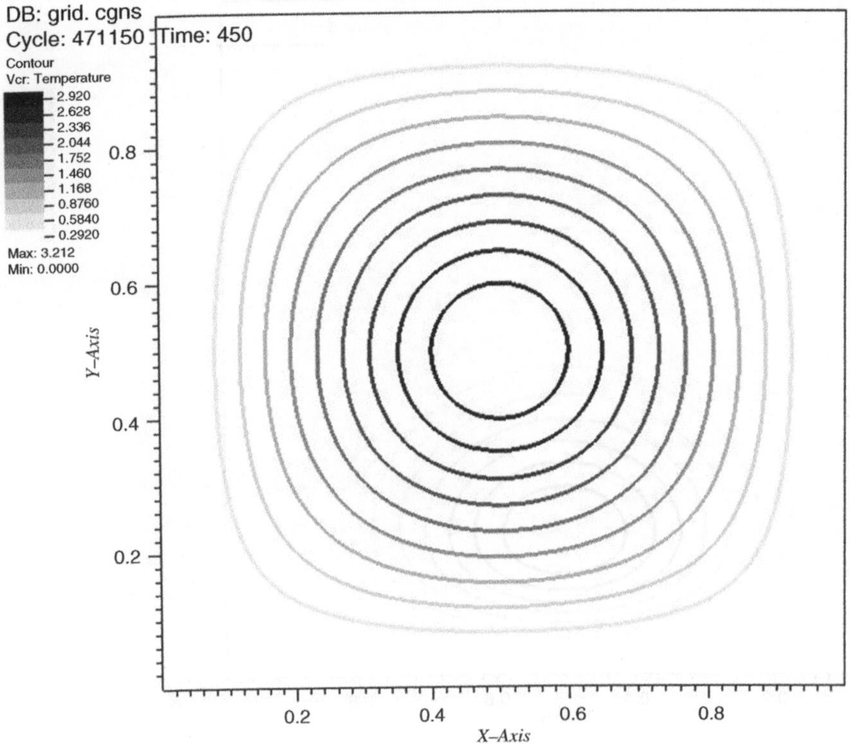

Figure 5.8 Unstable thermal (Frank-Kamenetskii parameter δ above critical). $Ra = 4 \cdot 10^3$; $\theta_{max} = 4.0$. Isolines of non-dimensional temperature. $\tau = 450$. Isoline values are shown on the scale panel on the left-hand side of the figure.

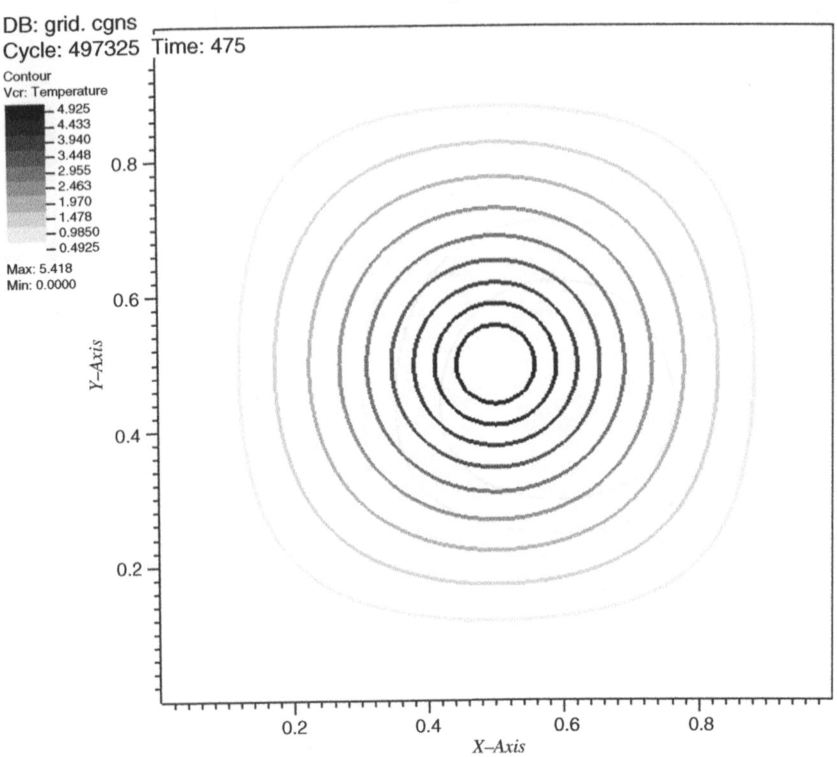

Figure 5.9 Unstable thermal (Frank-Kamenetskii parameter δ above critical). $Ra = 4 \cdot 10^3$; $\theta_{max} = 4.0$. Isolines of non-dimensional temperature. $\tau = 475$. Isoline values are shown on the scale panel on the left-hand side of the figure.

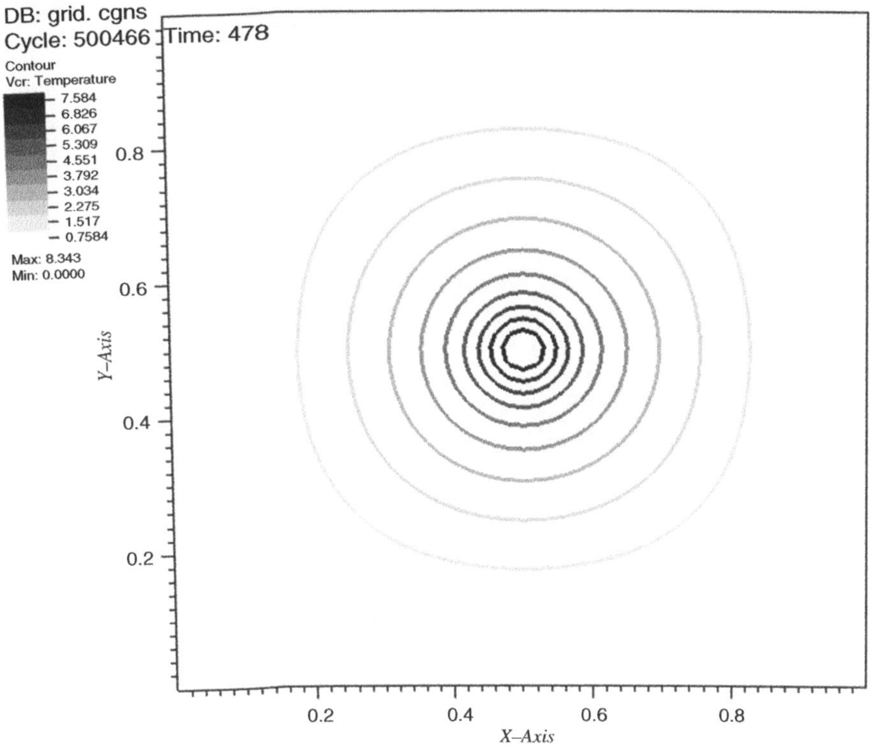

Figure 5.10 Unstable thermal (Frank-Kamenetskii parameter δ above critical). $Ra = 4 \cdot 10^3$; $\theta_{max} = 4.0$. Isolines of non-dimensional temperature. $\tau = 478$. Isoline values are shown on the scale panel on the left-hand side of the figure.

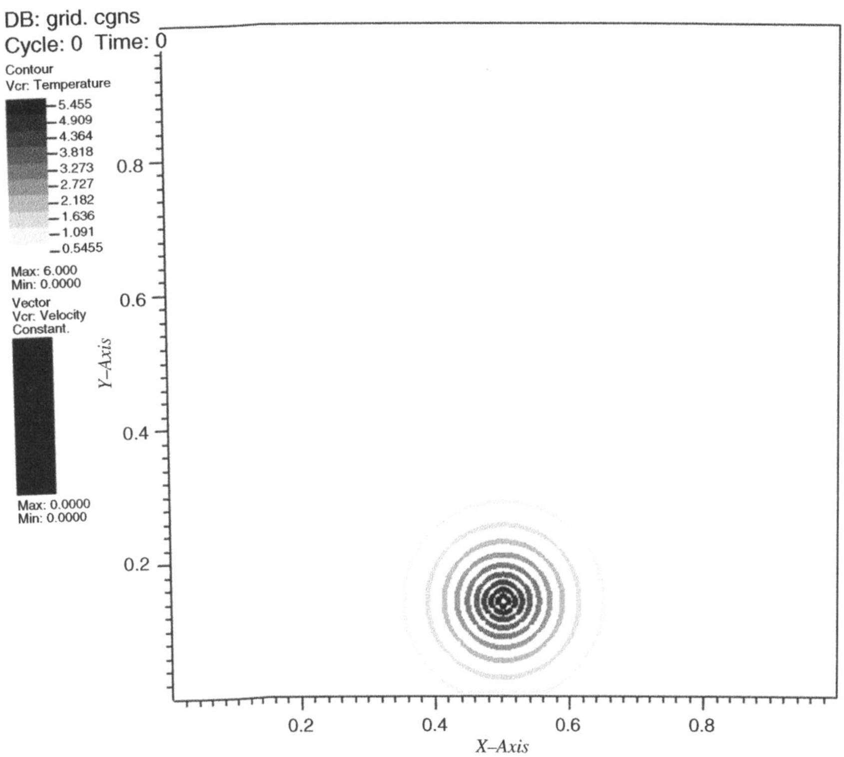

Figure 5.11 Stable thermal (Frank-Kamenetskii parameter δ below critical). $Ra = 6 \cdot 10^7$; $\theta_{max} = 6.0$. Isolines of non-dimensional temperature and non-dimensional velocity field. $\tau = 0$. Isoline values and velocity magnitude are shown on the scale panel on the left-hand side of the figure.

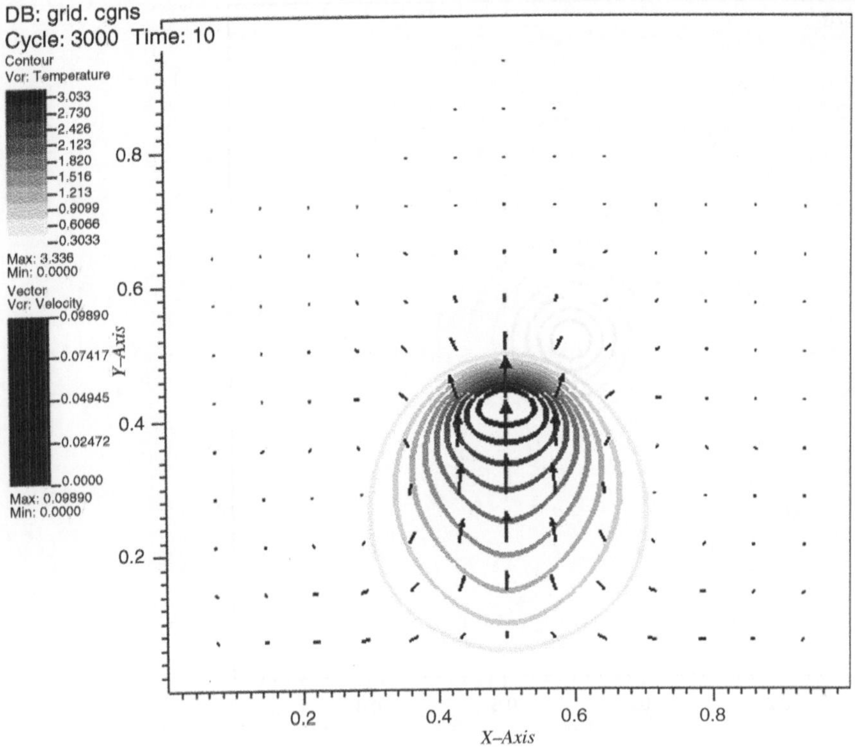

Figure 5.12 Stable thermal (Frank-Kamenetskii parameter δ below critical). $Ra = 6 \cdot 10^7$; $\theta_{\max} = 6.0$. Isolines of non-dimensional temperature and non-dimensional velocity field. $\tau = 10$. Isoline values and velocity magnitude are shown on the scale panel on the left-hand side of the figure.

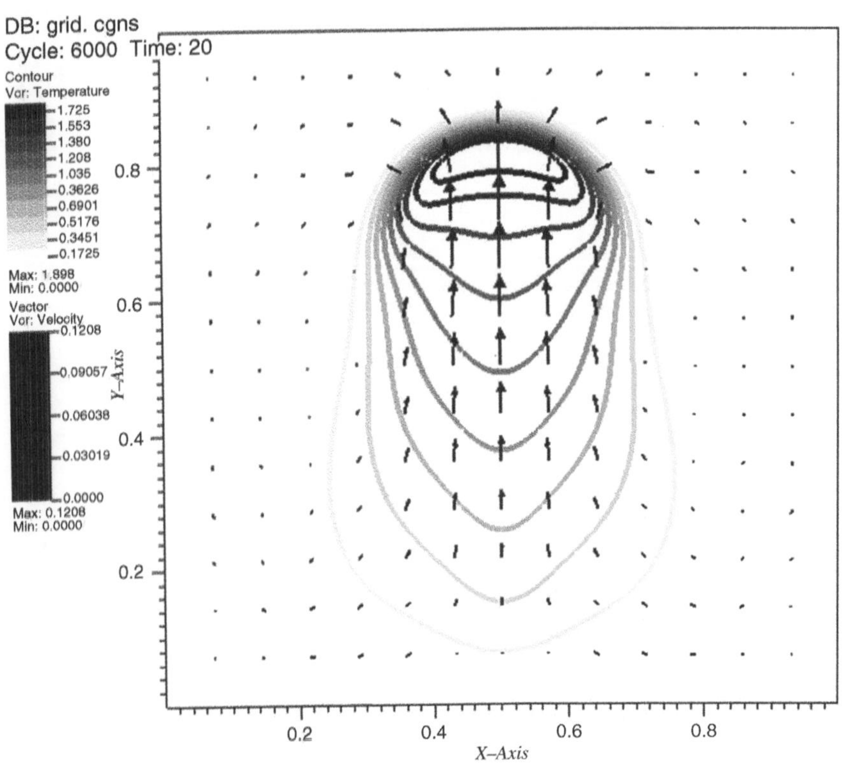

Figure 5.13 Stable thermal (Frank-Kamenetskii parameter δ below critical). $Ra = 6 \cdot 10^7$; $\theta_{\max} = 6.0$. Isolines of non-dimensional temperature and non-dimensional velocity field. $\tau = 20$. Isoline values and velocity magnitude are shown on the scale panel on the left-hand side of the figure.

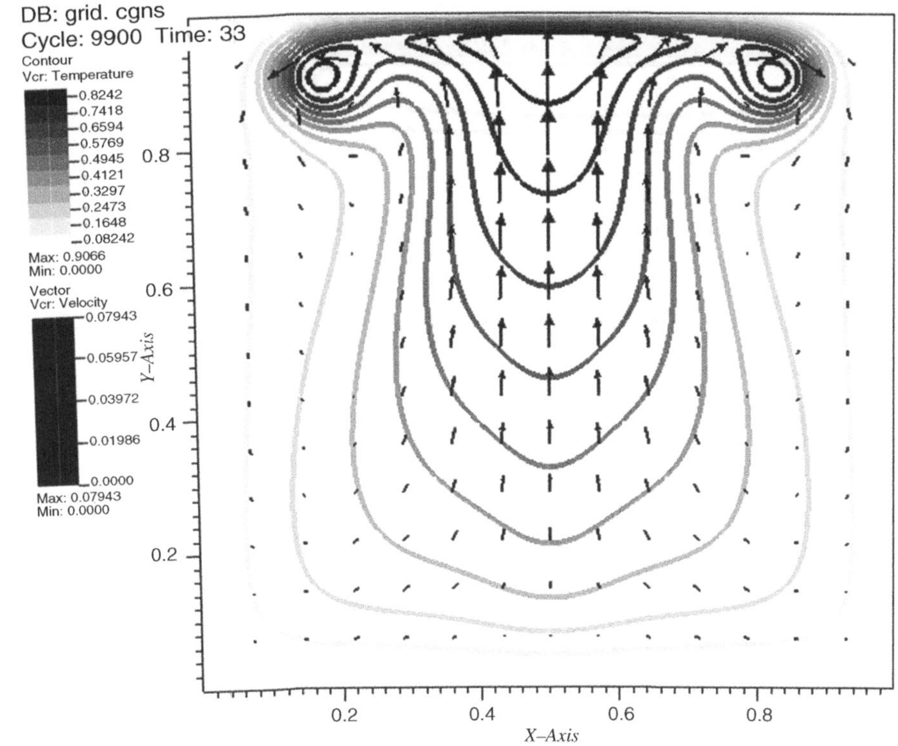

Figure 5.14 Stable thermal (Frank-Kamenetskii parameter δ below critical). $Ra = 6 \cdot 10^7$; $\theta_{max} = 6.0$. Isolines of non-dimensional temperature and non-dimensional velocity field. τ = 33. Isoline values and velocity magnitude are shown on the scale panel on the left-hand side of the figure.

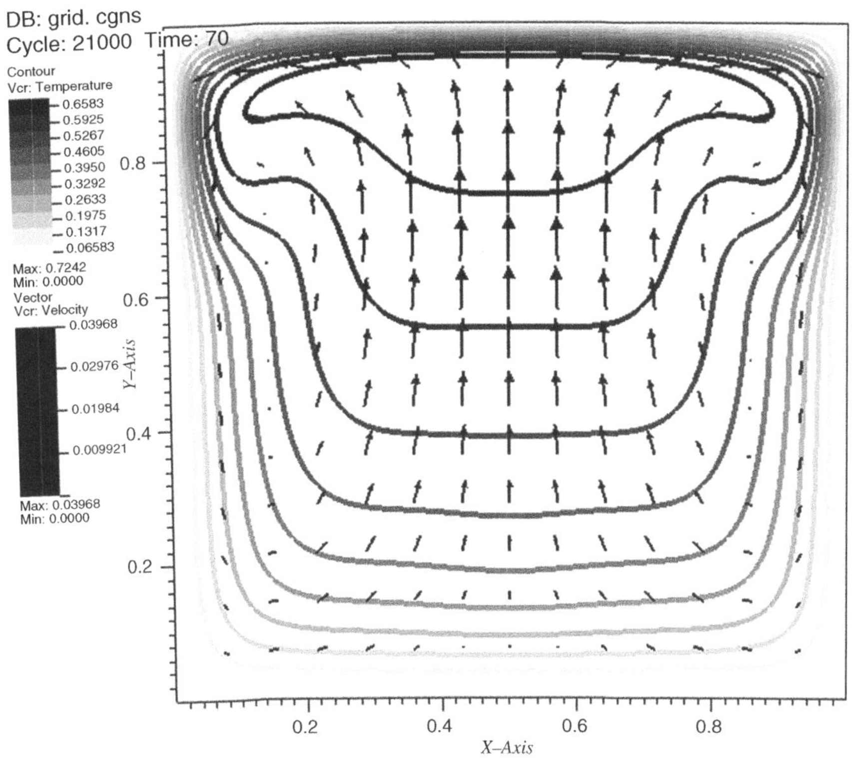

Figure 5.15 Stable thermal (Frank-Kamenetskii parameter δ below critical). $Ra = 6 \cdot 10^7$; $\theta_{max} = 6.0$. Isolines of non-dimensional temperature and non-dimensional velocity field. τ = 70. Isoline values and velocity magnitude are shown on the scale panel on the left-hand side of the figure.

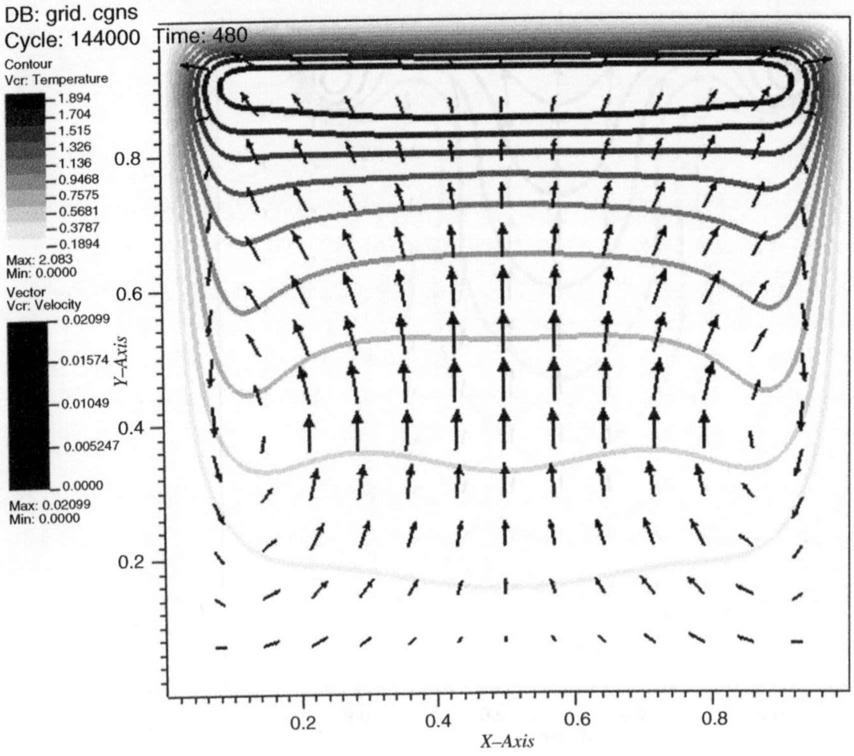

Figure 5.16 Stable thermal (Frank-Kamenetskii parameter δ below critical). $Ra = 6 \cdot 10^7$; $\theta_{max} = 6.0$. Isolines of non-dimensional temperature and non-dimensional velocity field. $\tau = 480$. Isoline values and velocity magnitude are shown on the scale panel on the left-hand side of the figure.

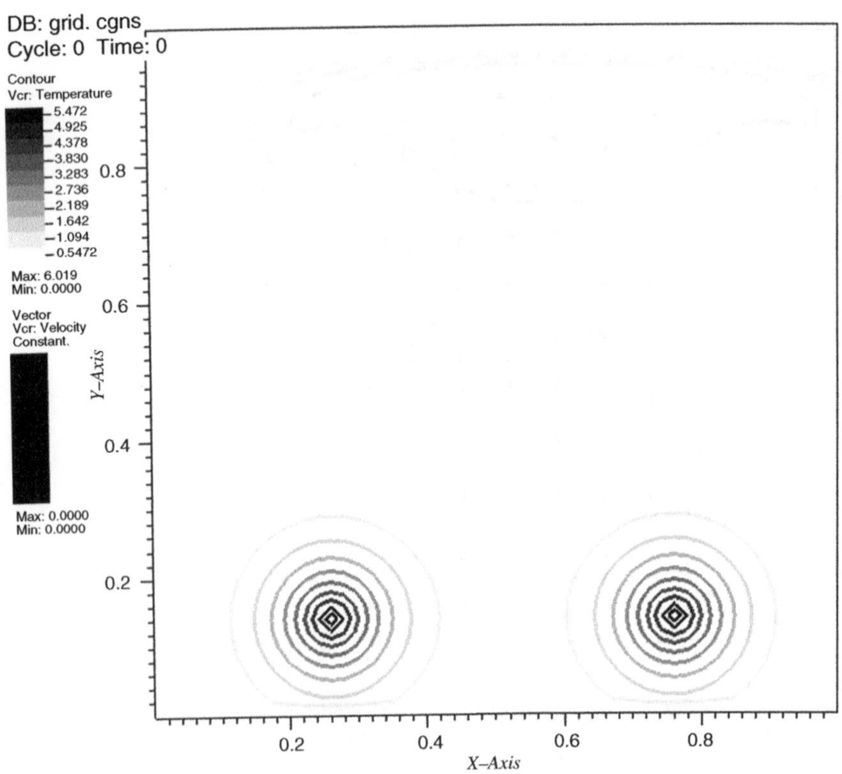

Figure 5.17 Stable double-thermal system (Frank-Kamenetskii parameter δ below critical). $Ra = 6 \cdot 10^7$; $\theta_{max} = 6.0$. Isolines of non-dimensional temperature and non-dimensional velocity field. $\tau = 0$. Isoline values and velocity magnitude are shown on the scale panel on the left-hand side of the figure.

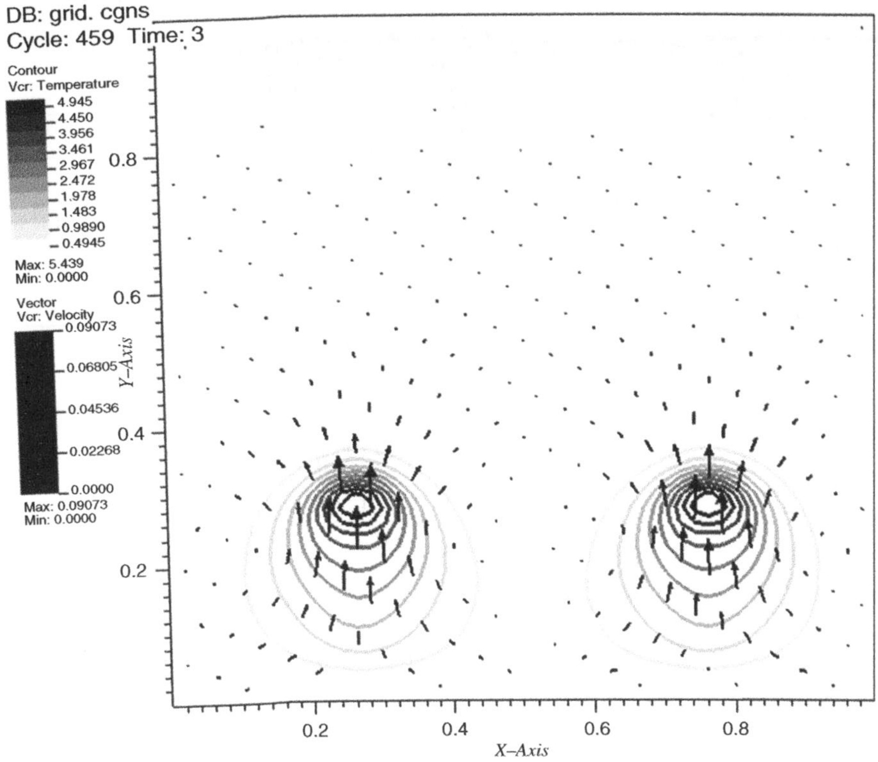

Figure 5.18 Stable double-thermal system (Frank-Kamenetskii parameter δ below critical). $Ra = 6 \cdot 10^7$; $\theta_{\max} = 6.0$. Isolines of non-dimensional temperature and non-dimensional velocity field. $\tau = 3$. Isoline values and velocity magnitude are shown on the scale panel on the left-hand side of the figure.

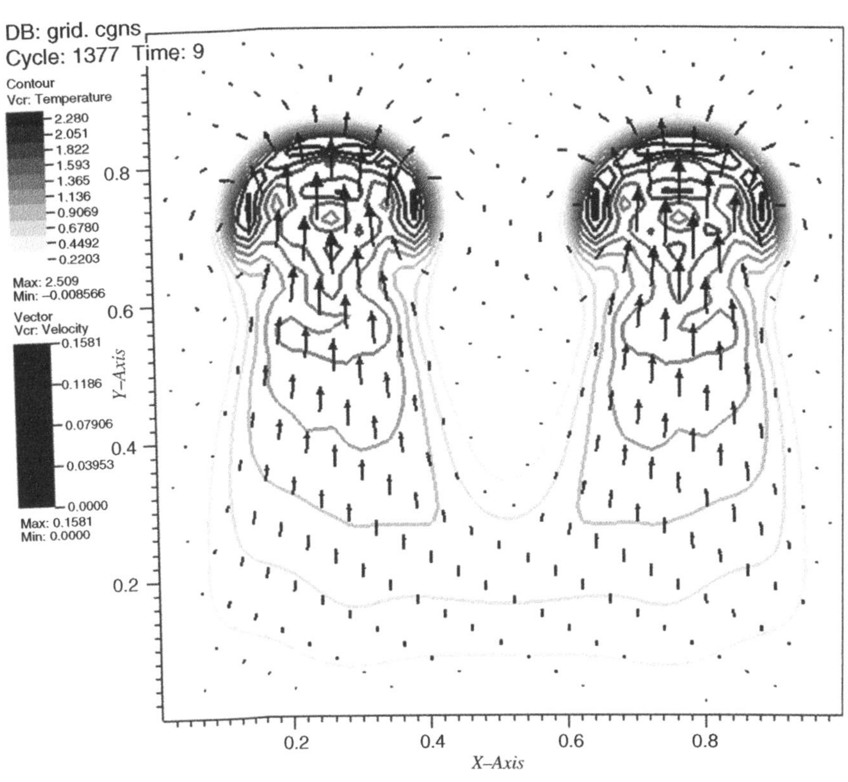

Figure 5.19 Stable double-thermal system (Frank-Kamenetskii parameter δ below critical). $Ra = 6 \cdot 10^7$; $\theta_{\max} = 6.0$. Isolines of non-dimensional temperature and non-dimensional velocity field. $\tau = 9$. Isoline values and velocity magnitude are shown on the scale panel on the left-hand side of the figure.

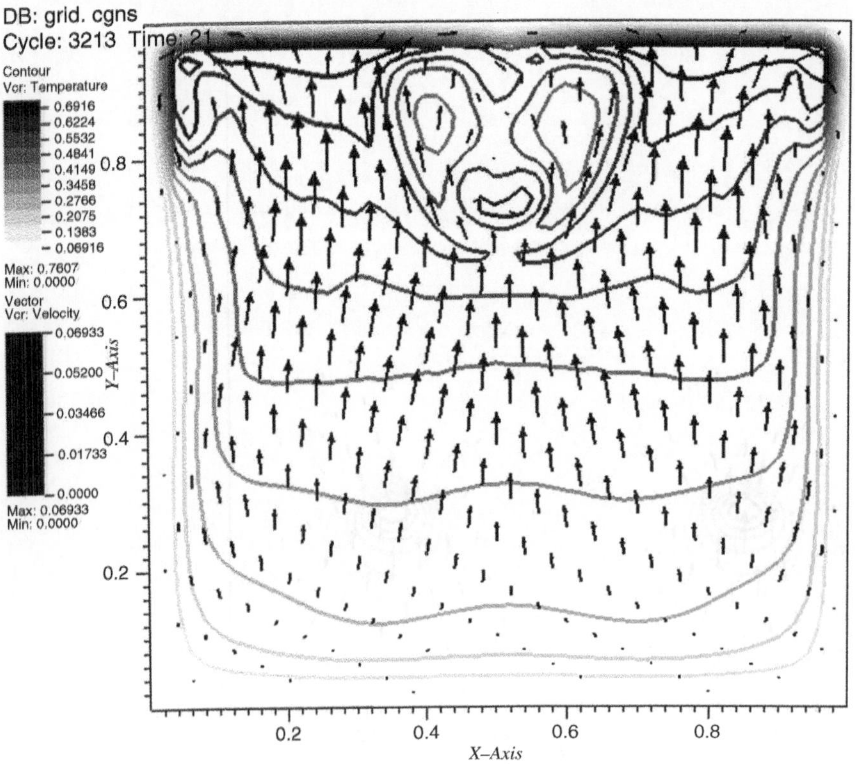

Figure 5.20 Stable double-thermal system (Frank-Kamenetskii parameter δ below critical). $Ra = 6 \cdot 10^7$; $\theta_{max} = 6.0$. Isolines of non-dimensional temperature and non-dimensional velocity field. $\tau = 21$. Isoline values and velocity magnitude are shown on the scale panel on the left-hand side of the figure.

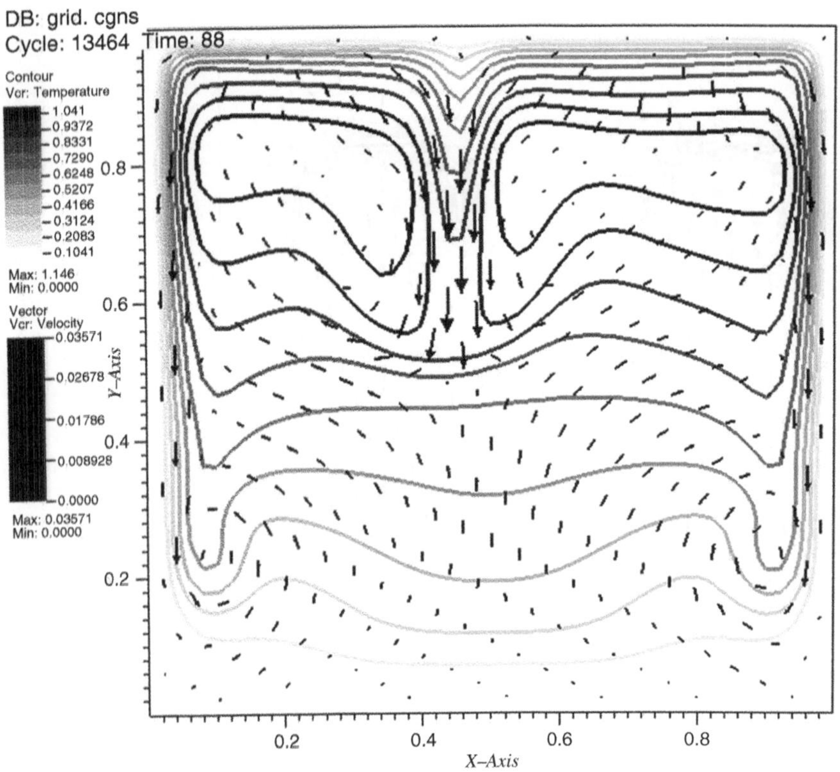

Figure 5.21 Stable double-thermal system (Frank-Kamenetskii parameter δ below critical). $Ra = 6 \cdot 10^7$; $\theta_{max} = 6.0$. Isolines of non-dimensional temperature and non-dimensional velocity field. $\tau = 88$. Isoline values and velocity magnitude are shown on the scale panel on the left-hand side of the figure.

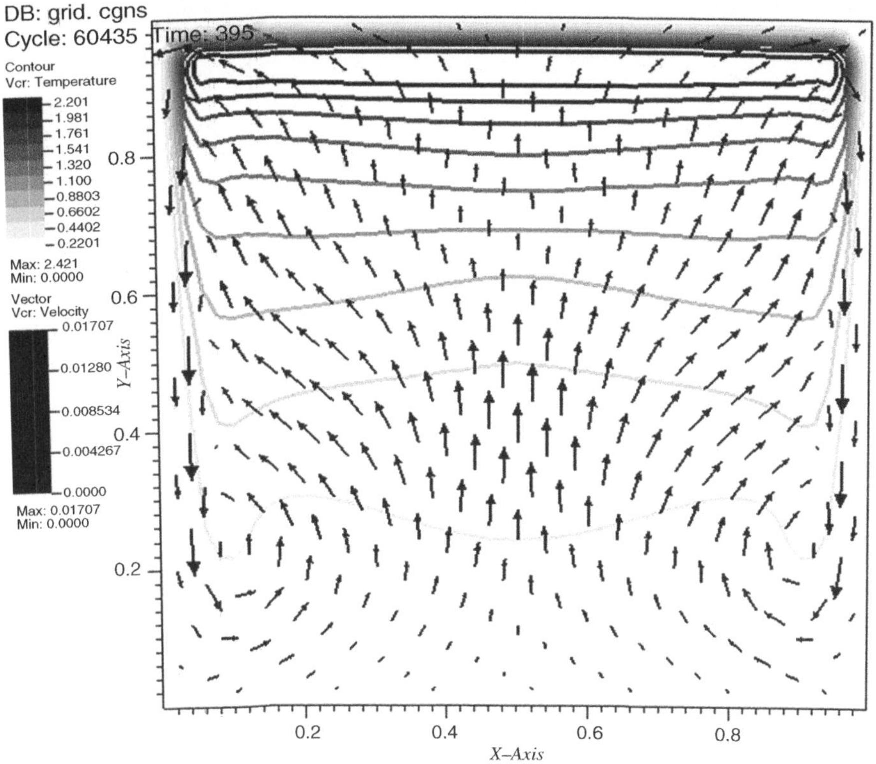

Figure 5.22 Stable double-thermal system (Frank-Kamenetskii parameter δ below critical). $Ra = 6 \cdot 10^7$; $\theta_{max} = 6.0$. Isolines of non-dimensional temperature and non-dimensional velocity field. $\tau = 395$. Isoline values and velocity magnitude are shown on the scale panel on the left-hand side of the figure.

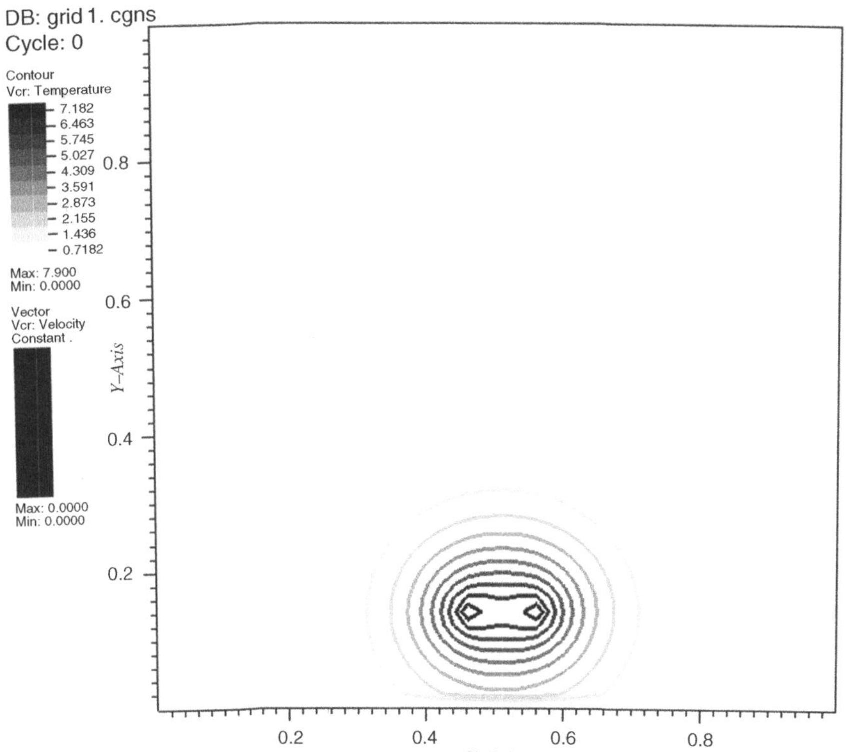

Figure 5.23 Unstable double-thermal system (Frank-Kamenetskii parameter δ above critical). $Ra = 8 \cdot 10^7$; $\theta_{max} = 8.0$. Isolines of non-dimensional temperature and non-dimensional velocity field. $\tau = 0$. Isoline values and velocity magnitude are shown on the scale panel on the left-hand side of the figure.

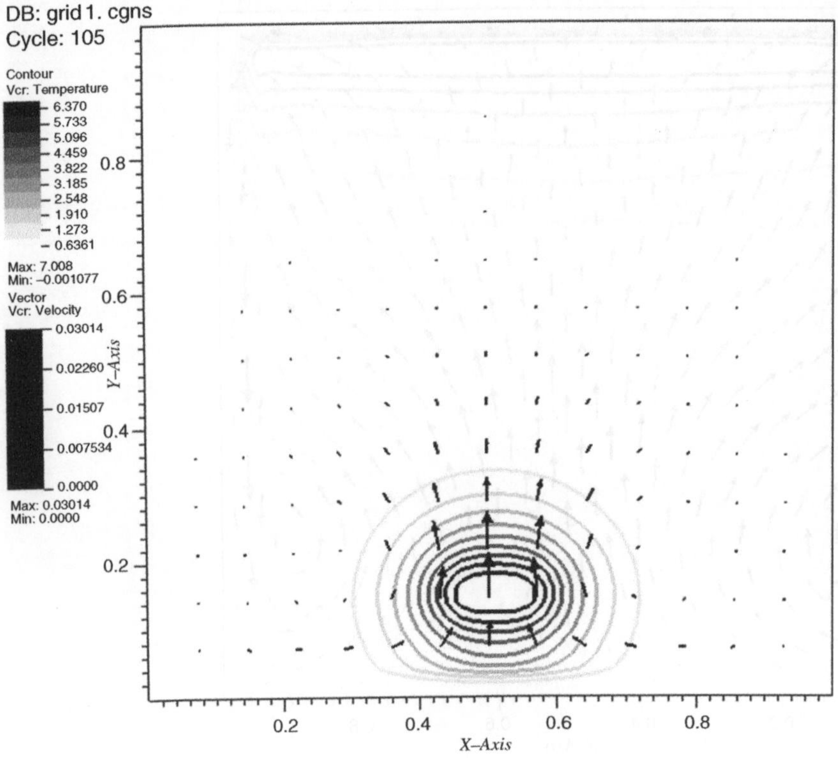

Figure 5.24 Unstable double-thermal system (Frank-Kamenetskii parameter δ above critical). $Ra = 8 \cdot 10^7$; $\theta_{max} = 8.0$. Isolines of non-dimensional temperature and non-dimensional velocity field. $\tau = 0.90$. Isoline values and velocity magnitude are shown on the scale panel on the left-hand side of the figure.

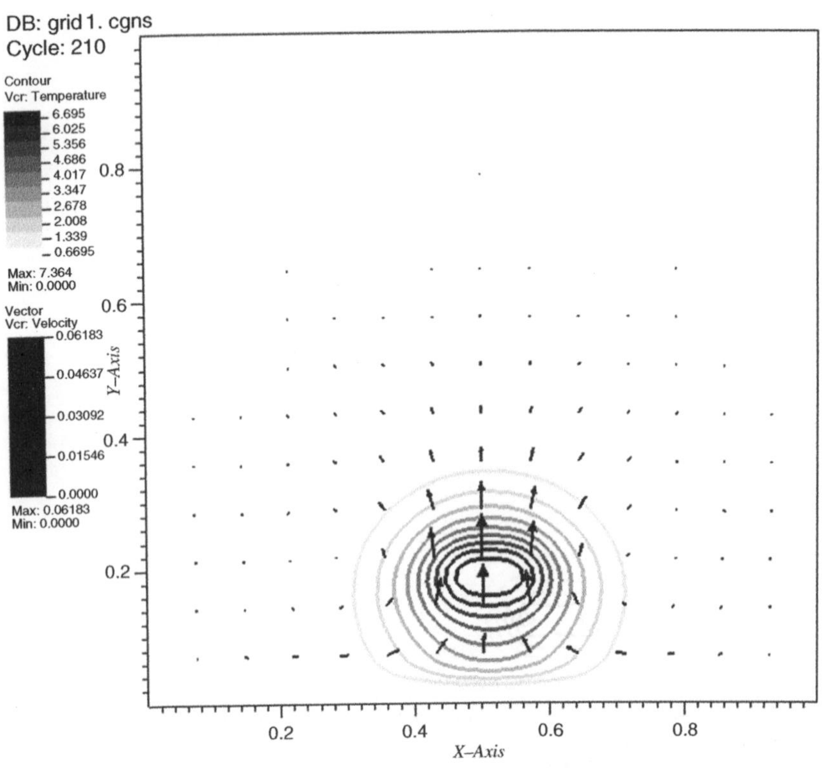

Figure 5.25 Unstable double-thermal system (Frank-Kamenetskii parameter δ above critical). $Ra = 8 \cdot 10^7$; $\theta_{max} = 8.0$. Isolines of non-dimensional temperature and non-dimensional velocity field. $\tau = 1.80$. Isoline values and velocity magnitude are shown on the scale panel on the left-hand side of the figure.

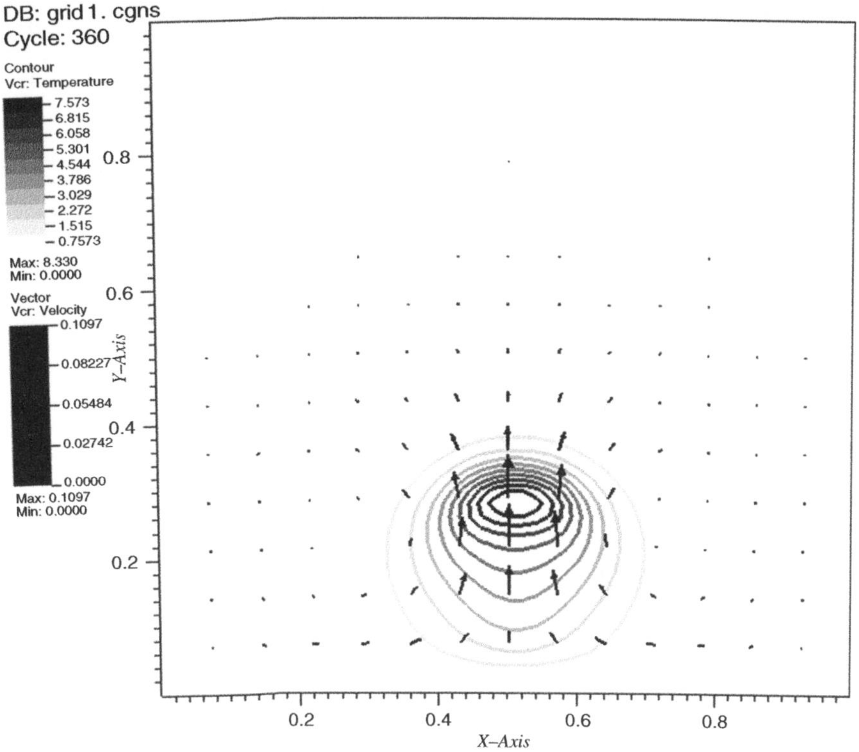

Figure 5.26 Unstable double-thermal system (Frank-Kamenetskii parameter δ above critical). $Ra = 8 \cdot 10^7$; $\theta_{\max} = 8.0$. Isolines of non-dimensional temperature and non-dimensional velocity field. $\tau = 3.09$. Isoline values and velocity magnitude are shown on the scale panel on the left-hand side of the figure.

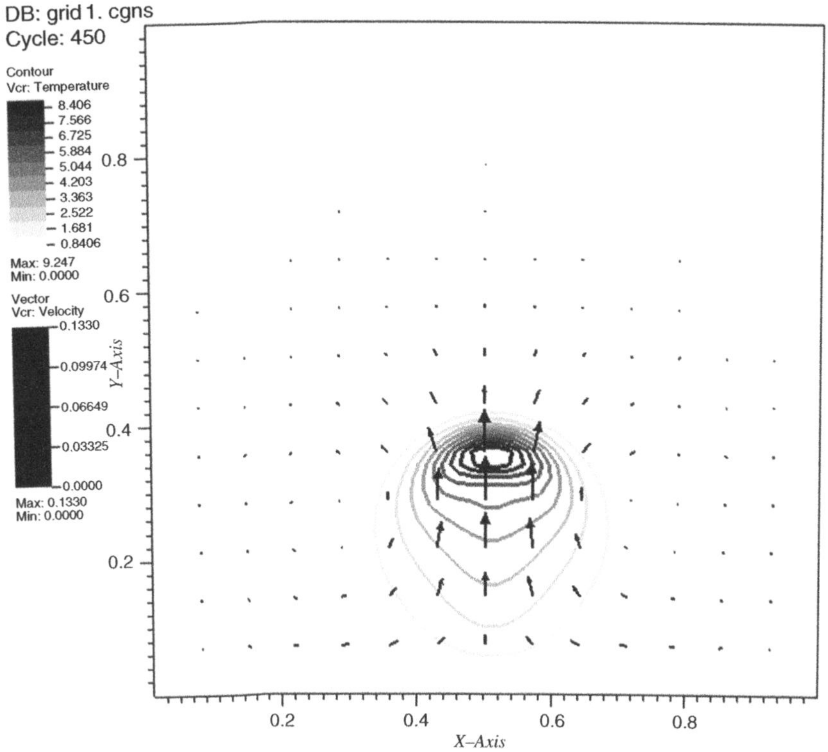

Figure 5.27 Unstable double-thermal system (Frank-Kamenetskii parameter δ above critical). $Ra = 8 \cdot 10^7$; $\theta_{\max} = 8.0$. Isolines of non-dimensional temperature and non-dimensional velocity field. $\tau = 3.86$. Isoline values and velocity magnitude are shown on the scale panel on the left-hand side of the figure.

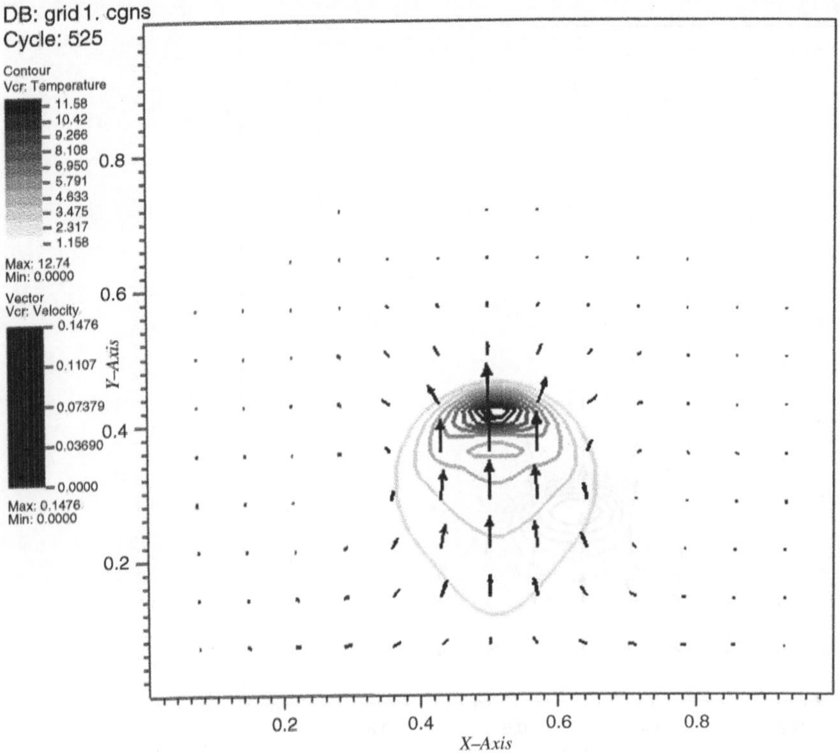

Figure 5.28 Unstable double-thermal system (Frank-Kamenetskii parameter δ above critical). $Ra = 8 \cdot 10^7$; $\theta_{max} = 8.0$. Isolines of non-dimensional temperature and non-dimensional velocity field. $\tau = 4.5$. Isoline values and velocity magnitude are shown on the scale panel on the left-hand side of the figure.

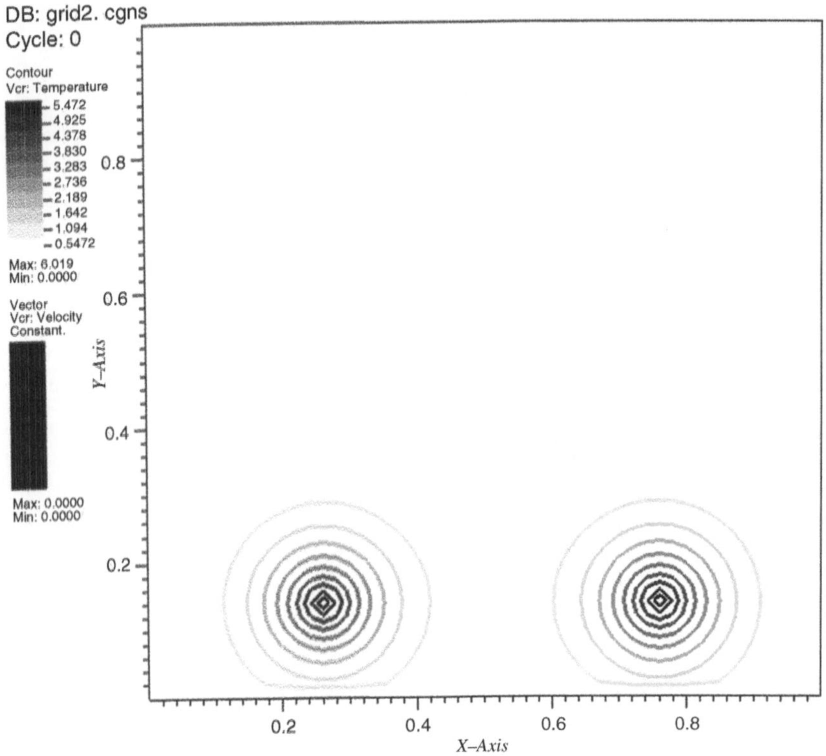

Figure 5.29 Unstable double-thermal system (Frank-Kamenetskii parameter δ above critical). $Ra = 6 \cdot 10^7$; $\theta_{max} = 6.0$. Isolines of non-dimensional temperature and non-dimensional velocity field. $\tau = 0$. Isoline values and velocity magnitude are shown on the scale panel on the left-hand side of the figure.

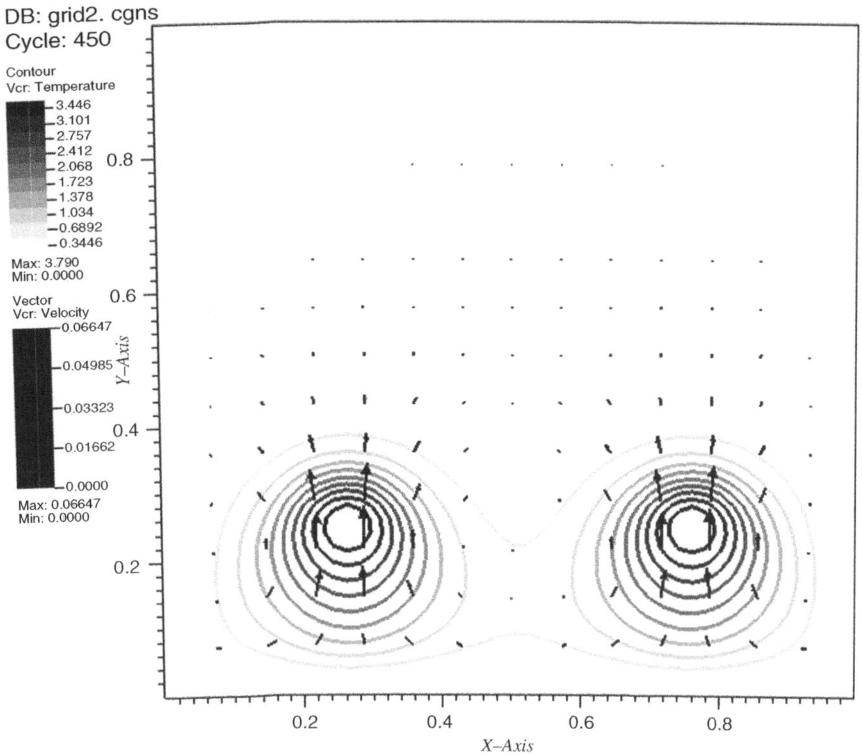

Figure 5.30 Unstable double-thermal system (Frank-Kamenetskii parameter δ above critical). $Ra = 6 \cdot 10^7$; $\theta_{\max} = 6.0$. Isolines of non-dimensional temperature and non-dimensional velocity field. $\tau = 3.86$. Isoline values and velocity magnitude are shown on the scale panel on the left-hand side of the figure.

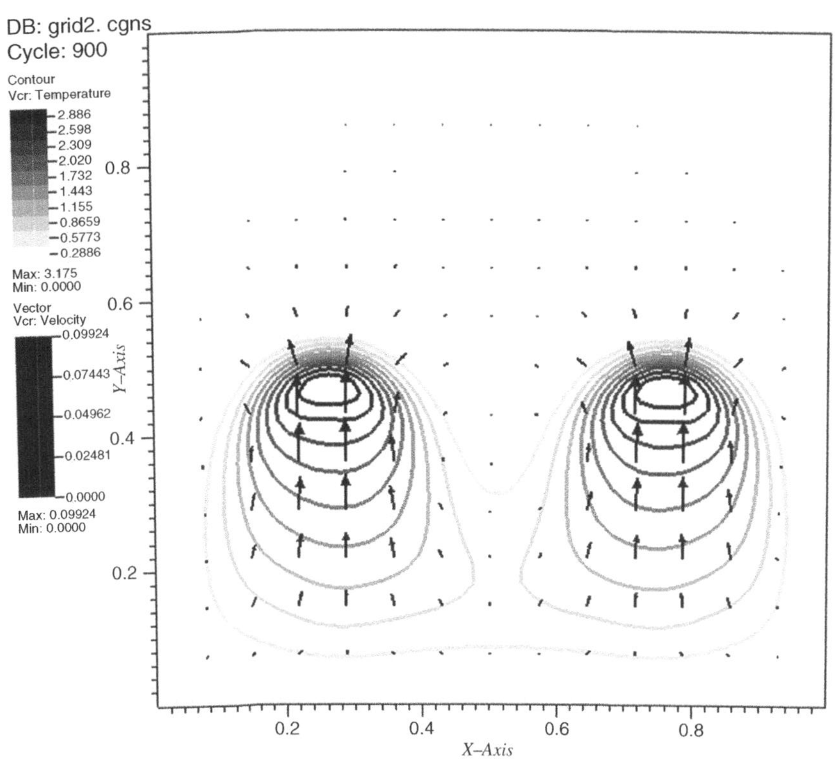

Figure 5.31 Unstable double-thermal system (Frank-Kamenetskii parameter δ above critical). $Ra = 6 \cdot 10^7$; $\theta_{\max} = 6.0$. Isolines of non-dimensional temperature and non-dimensional velocity field. $\tau = 7.71$. Isoline values and velocity magnitude are shown on the scale panel on the left-hand side of the figure.

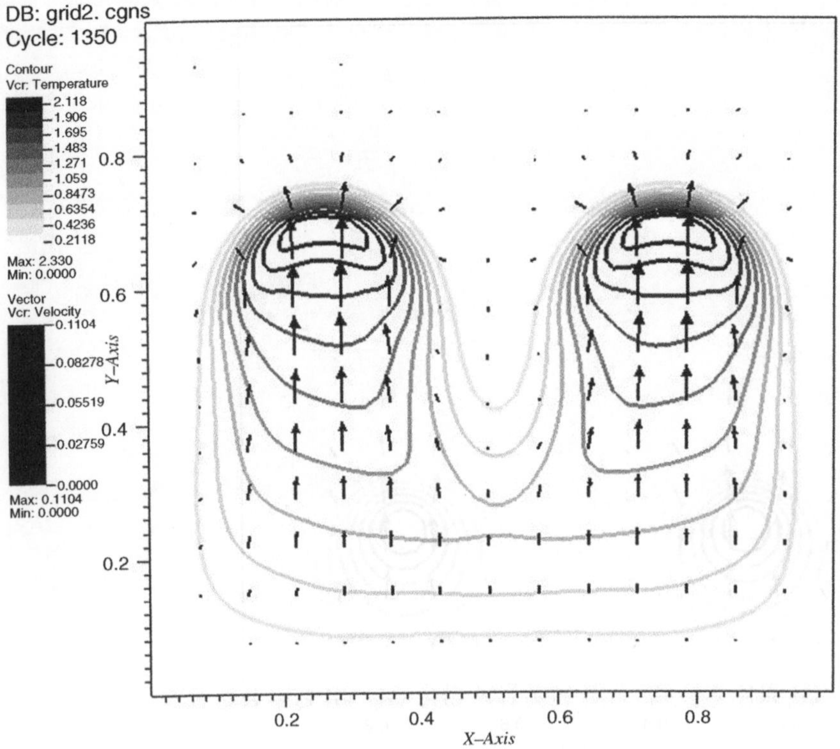

Figure 5.32 Unstable double-thermal system (Frank-Kamenetskii parameter δ above critical). $Ra = 6 \cdot 10^7$; $\theta_{max} = 6.0$. Isolines of non-dimensional temperature and non-dimensional velocity field. $\tau = 11.57$. Isoline values and velocity magnitude are shown on the scale panel on the left-hand side of the figure.

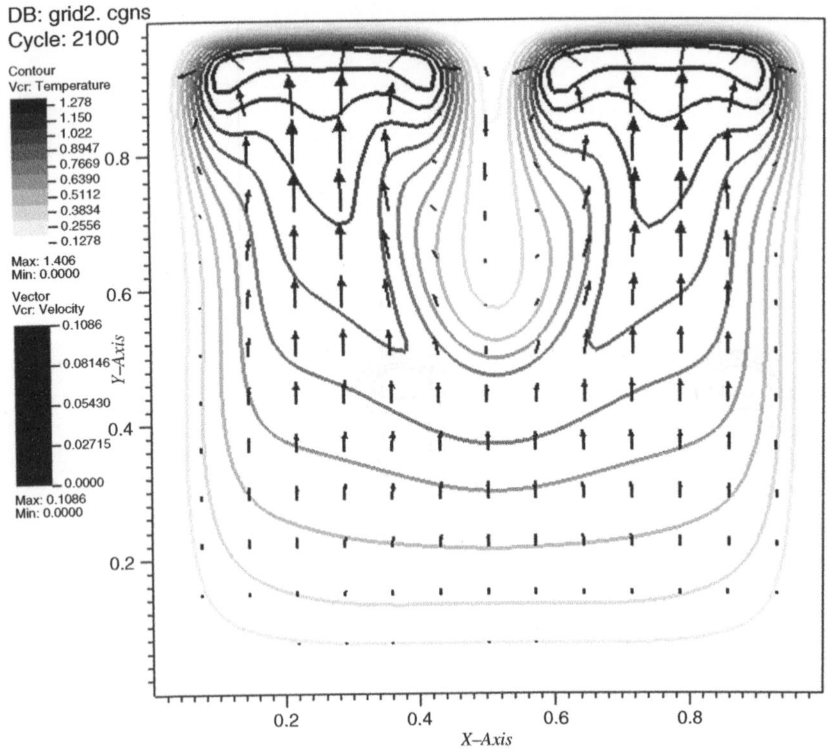

Figure 5.33 Unstable double-thermal system (Frank-Kamenetskii parameter δ above critical). $Ra = 6 \cdot 10^7$; $\theta_{max} = 6.0$. Isolines of non-dimensional temperature and non-dimensional velocity field. $\tau = 18.00$. Isoline values and velocity magnitude are shown on the scale panel on the left-hand side of the figure.

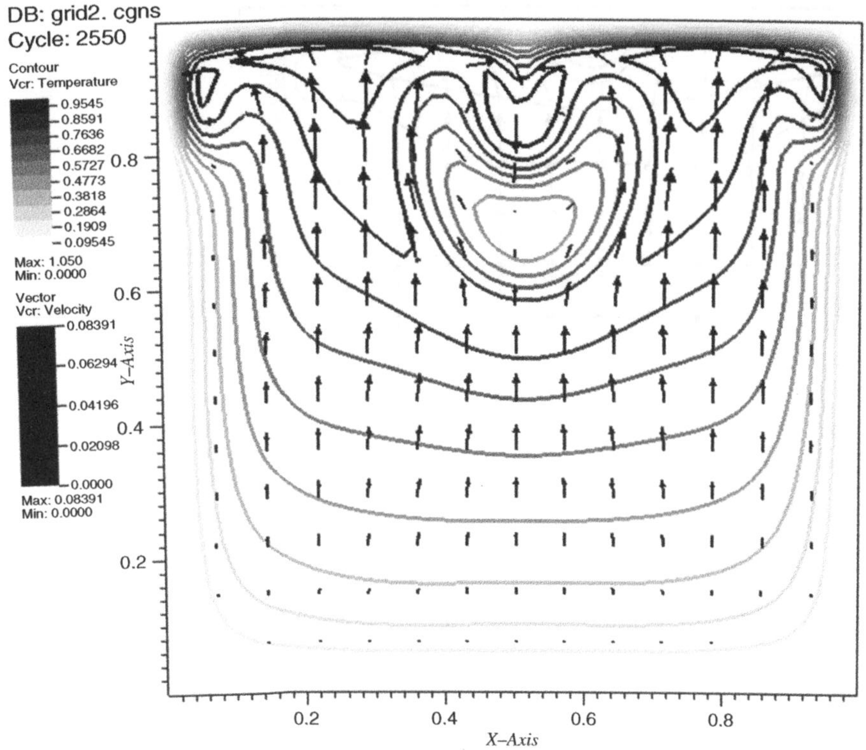

Figure 5.34 Unstable double-thermal system (Frank-Kamenetskii parameter δ above critical). $Ra = 6 \cdot 10^7$; $\theta_{max} = 6.0$. Isolines of non-dimensional temperature and non-dimensional velocity field. $\tau = 21.85$. Isoline values and velocity magnitude are shown on the scale panel on the left-hand side of the figure.

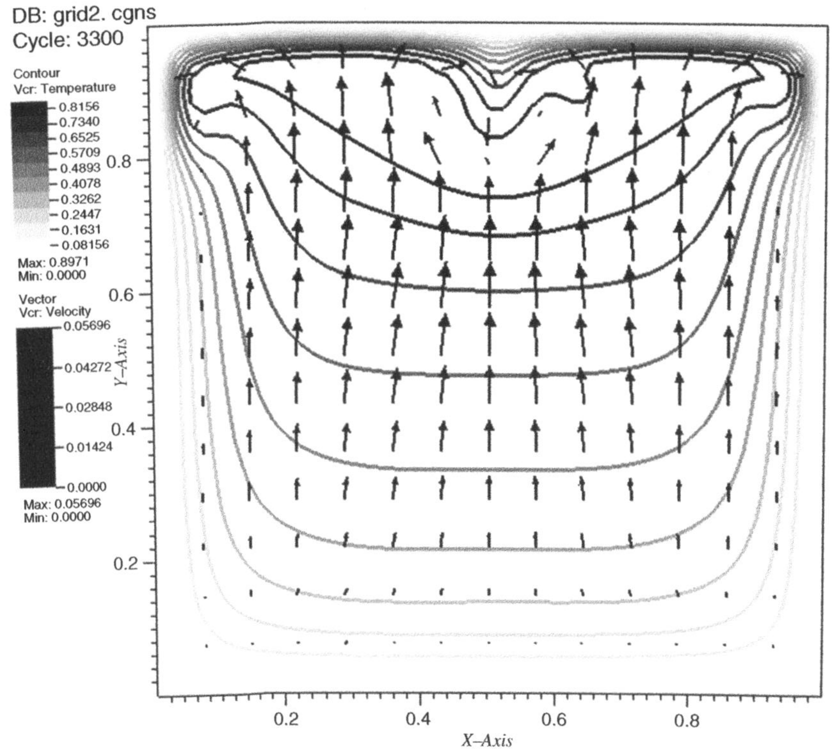

Figure 5.35 Unstable double-thermal system (Frank-Kamenetskii parameter δ above critical). $Ra = 6 \cdot 10^7$; $\theta_{max} = 6.0$. Isolines of non-dimensional temperature and non-dimensional velocity field. $\tau = 28.28$. Isoline values and velocity magnitude are shown on the scale panel on the left-hand side of the figure.

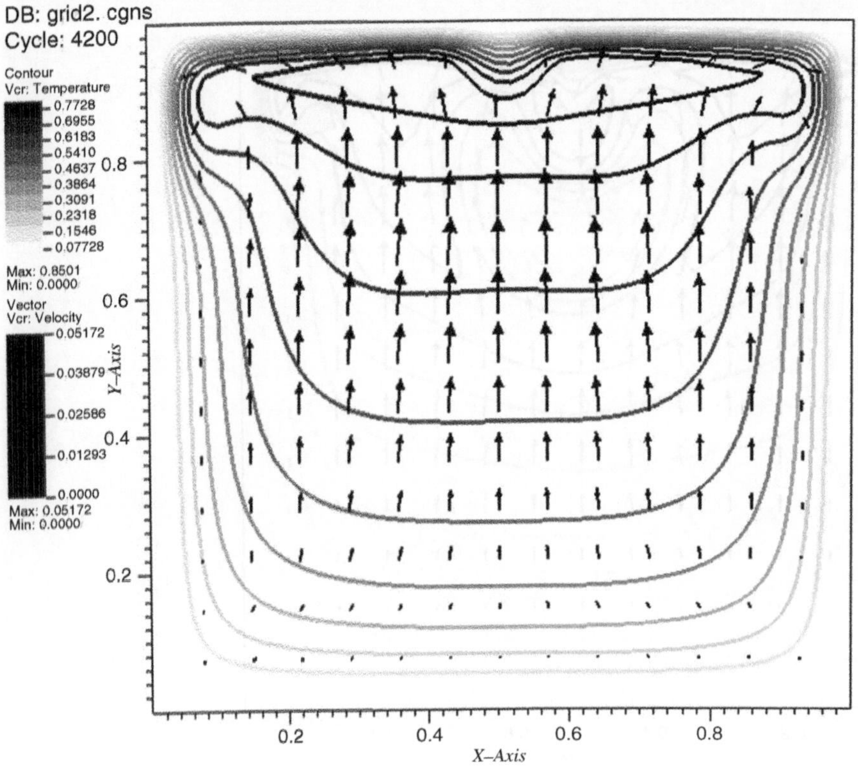

Figure 5.36 Unstable double-thermal system (Frank-Kamenetskii parameter δ above critical). $Ra = 6 \cdot 10^7$; $\theta_{max} = 6.0$. Isolines of non-dimensional temperature and non-dimensional velocity field. $\tau = 36.00$. Isoline values and velocity magnitude are shown on the scale panel on the left-hand side of the figure.

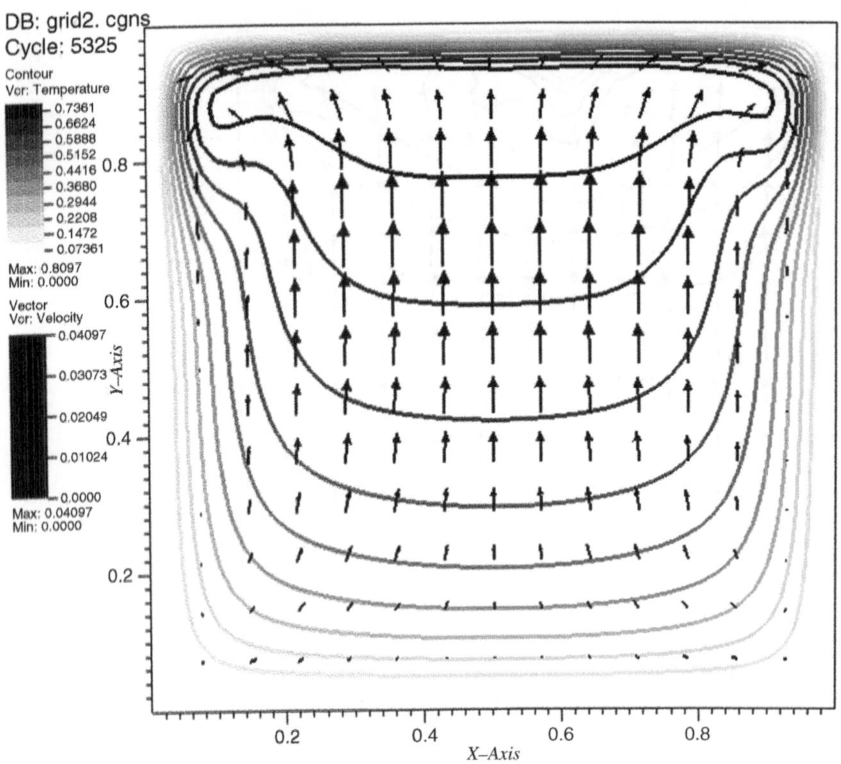

Figure 5.37 Unstable double-thermal system (Frank-Kamenetskii parameter δ above critical). $Ra = 6 \cdot 10^7$; $\theta_{max} = 6.0$. Isolines of non-dimensional temperature and non-dimensional velocity field. $\tau = 45.64$. Isoline values and velocity magnitude are shown on the scale panel on the left-hand side of the figure.

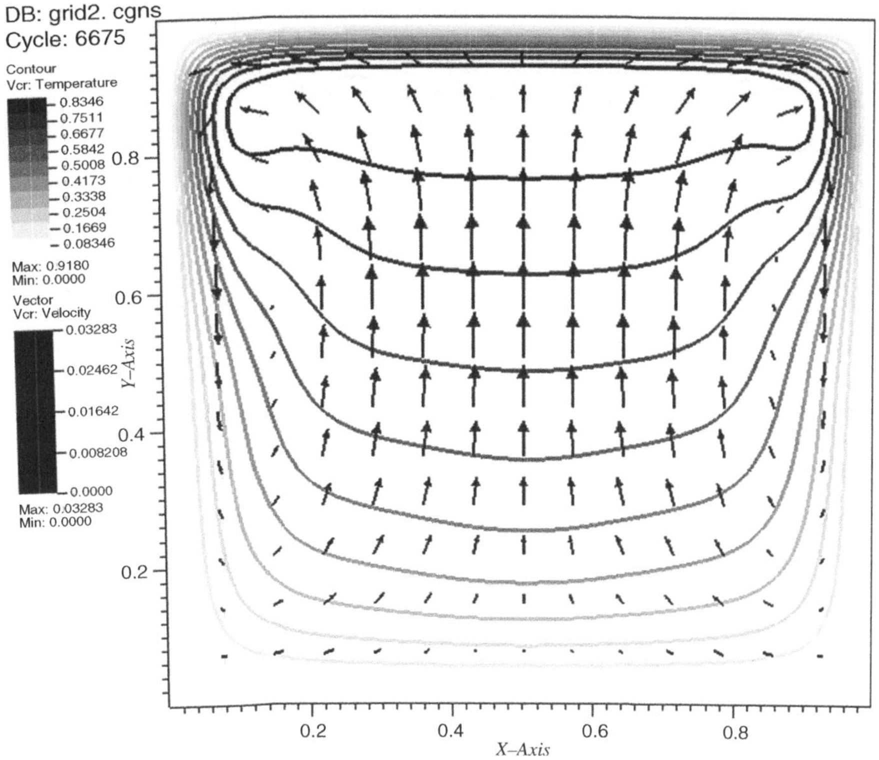

Figure 5.38 Unstable double-thermal system (Frank-Kamenetskii parameter δ above critical). $Ra = 6 \cdot 10^7$; $\theta_{max} = 6.0$. Isolines of non-dimensional temperature and non-dimensional velocity field. $\tau = 57.20$. Isoline values and velocity magnitude are shown on the scale panel on the left-hand side of the figure.

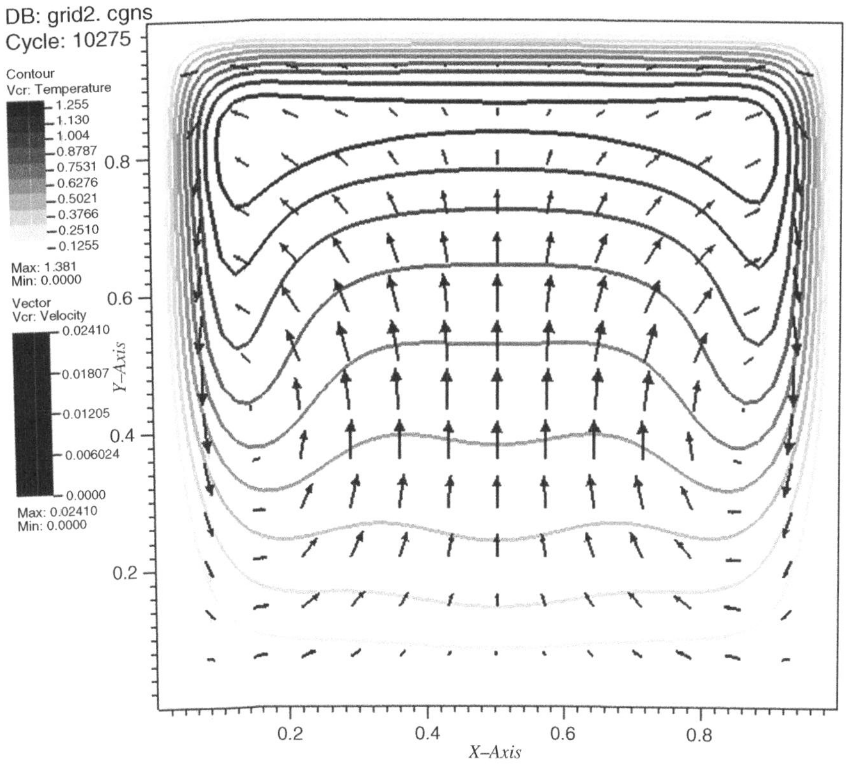

Figure 5.39 Unstable double-thermal system (Frank-Kamenetskii parameter δ above critical). $Ra = 6 \cdot 10^7$; $\theta_{max} = 6.0$. Isolines of non-dimensional temperature and non-dimensional velocity field. $\tau = 88.06$. Isoline values and velocity magnitude are shown on the scale panel on the left-hand side of the figure.

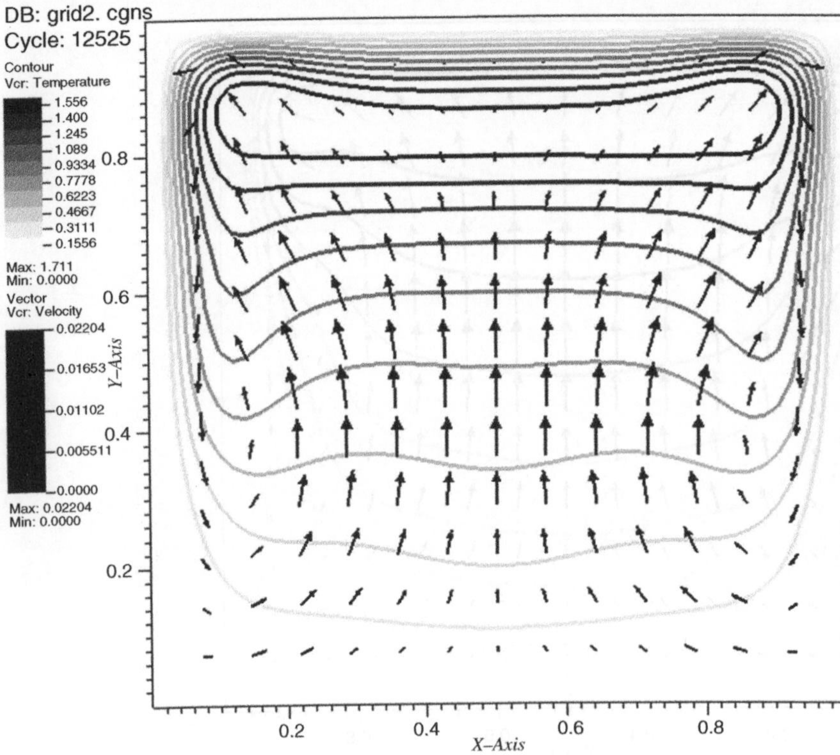

Figure 5.40 Unstable double-thermal system (Frank-Kamenetskii parameter δ above critical). $Ra = 6 \cdot 10^7$; $\theta_{max} = 6.0$. Isolines of non-dimensional temperature and non-dimensional velocity field. $\tau = 107.34$. Isoline values and velocity magnitude are shown on the scale panel on the left-hand side of the figure.

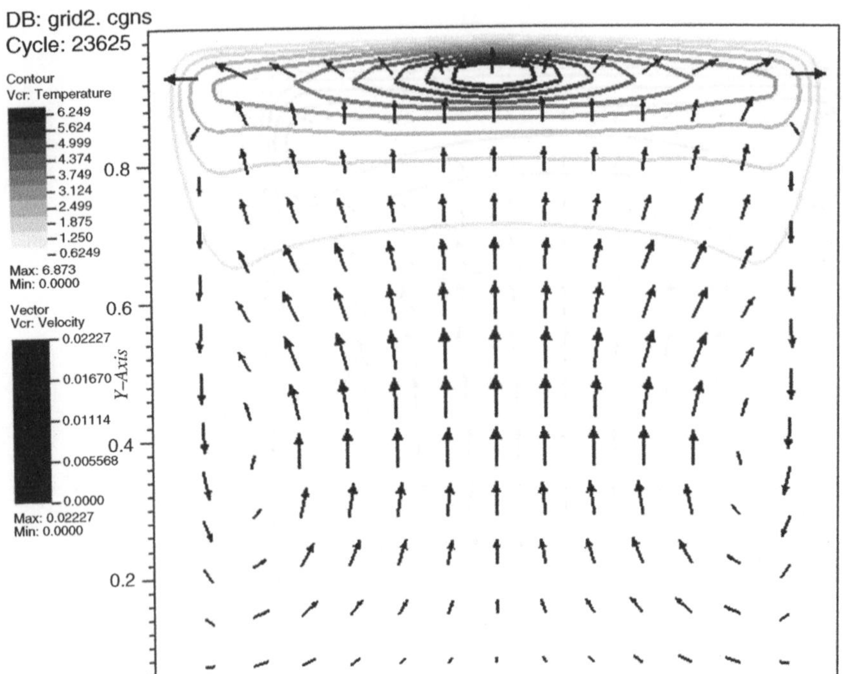

Figure 5.41 Unstable double-thermal system (Frank-Kamenetskii parameter δ above critical). $Ra = 6 \cdot 10^7$; $\theta_{max} = 6.0$. Isolines of non-dimensional temperature and non-dimensional velocity field. $\tau = 202.47$. Isoline values and velocity magnitude are shown on the scale panel on the left-hand side of the figure.

opposite ends of the covered region. This allows the role of natural convection intensity to be examined properly. While the majority of cases (Figures 5.11–5.41) are presented for high Rayleigh numbers ($Ra \sim 10^7$) where the influence of natural convection on critical conditions is substantial, one of the cases (Figures 5.5–5.10) is presented for a low Rayleigh number, for comparison.

Two cases represent single thermal development, and another three show the evolution of a pair of thermals.

In the following description of scenarios, the term stable refers to a non-explosive case where the mixture asymptotically (at $\tau \rightarrow \infty$) approaches a state of mechanical and thermal equilibrium, while unstable refers to an explosive case where the temperature in the domain grows infinitely at $\tau \rightarrow \infty$. Velocity magnitude means the absolute value of velocity.

Flow dynamics and the development of thermal explosion are distinctively different in the cases of low and high Rayleigh numbers.

At low Rayleigh numbers, natural convection plays a very limited role; conduction is the dominant mode of heat transfer. Indeed, low Rayleigh numbers correspond to high viscosity and therefore, since the Prandtl number is of the order of unity, to high thermal conductivity.

Mixture dynamics is determined by the two competing mechanisms, that is, heat generation by chemical reaction and heat dissipation through thermal diffusivity.

In sub-critical (below the explosion limit) mixtures, the cloud diffuses completely before the state of mechanical and thermal equilibrium is achieved.

Above the limit (super-critical mixtures), thermal explosion development depends on how large is the Frank-Kamenetskii parameter, compared to the critical one. If this difference is large, that is the mixture is far above the limit, thermal explosion develops in straightforward way with maximum temperature of the thermal growing monotonically and rapidly. Closer to (but still above) the limit thermal explosion development may be more peculiar. Dynamics of maximum temperature may be non-monotonic. It may drop first as thermal conductivity levels the temperature out across the domain. Dropping temperature gradients result in a decrease of the energy removal rate from the central region. Consequently, after some induction period, the energy generation rate due to chemical reaction starts to prevail, and the temperature increases again. It may increase monotonically further to infinity (thermal explosion).

This scenario of development is illustrated in Figures 5.5–5.10. Here, developing convective flow is very weak. The thermal maximum temperature drops dramatically at the beginning of the process, from ~3.6 to ~0.6 (Figures 5.5 and 5.6; $\tau = 0 - 40$), after which it starts to rise gradually (Figures 5.6 and 5.7; $\tau = 40 - 150$), and later more and more rapidly (Figures 5.7–5.10; $\tau = 150 - 478$), leading eventually to thermal instability.

In the opposite case of high Rayleigh numbers, the natural convection is intensive, shapes the flow dynamics and may become a dominant mode of heat transfer.

Nevertheless, explosion development far below and far above critical conditions remains qualitatively similar to the case of low Rayleigh numbers in the following sense. Below the critical limit thermal (or a pair of thermals) diffuses to a state of mechanical and thermal equilibrium. The difference with the case of low Rayleigh numbers is that convection plays dominant role in this process. Above the limit different scenarios are possible, but these are also qualitatively similar to the case of low Rayleigh numbers. Temperature may grow rapidly in the centre of the thermal, leading to thermal instability straightaway, or explosion may be delayed over a long induction period, and temperature varies non-monotonically until it eventually grows infinitely (this scenario is similar to the one presented in Figures 5.5–5.10 for low Rayleigh numbers).

Consider several examples.

Figures 5.11–5.16 present the evolution of the stable (below the critical condition) thermal. Emerging flow structure (Figures 5.12 and 5.13) is typical for an ascending buoyant cloud (or for a fireball) with colder air entrainment from the sides and bottom of the cloud. Once the hot region hits the upper wall of the enclosure (Figure 5.14), it breaks down into a pair of vortices. Further evolution of the flow (Figure 5.15) leads to the formation of a well-stratified mixture with warm layer just below the upper boundary. The maximum temperature, observed in the centre of the thermal, decreases monotonically in time during the early stages of the process (Figures 5.11–5.15; $\tau = 0 - 70$). At later stages (Figure 5.16), it increases slightly again, which means that the overall temperature decreases to the state of thermal equilibrium in an oscillatory manner. The system approaches the state of thermal and mechanical equilibrium asymptotically at $\tau \to \infty$. In this case, energy generation inside the thermal (due to chemical reaction) is not sufficient to overcome the cooling effect of the entrained colder mixture. Thermal explosion does not occur.

Further examples involve double-thermal systems. In principle, compared to a single thermal case, the major mechanisms and features of natural convection (and their influence on critical conditions of explosion) remain the same in this case. An additional complexity is added to the flow structure by interaction between the two thermals, which may occur either at early (initially closely located thermals) or later (initially substantially separated thermals) stages of flow evolution.

Another example of non-explosive mixture behaviour is presented by the double-thermal system (Figures 5.17–5.22). The thermals are substantially separated initially. Due to this initial condition, at the early stages of the process (Figure 5.17–5.19; $\tau = 0 - 9$), the thermals develop essentially separately. Evolution of each of the thermals is similar to a single thermal evolution (for example, to the one in Figures 5.11–5.13).

Once the thermals hit the upper boundary, a region of interaction is clearly seen in the upper central part of the domain (Figure 5.20). Later, this interaction leads to the development of the reverse, descending jet in the centre (Figure 5.21). Descending jets also develop near the left and the right walls of the domain. Later evolution of the flow results in the well-stratified structure (Figure 5.22) with the warm layer being formed beneath the upper boundary. This structure is similar to the one observed in Figure 5.16. Similar to that case, the temperature of the upper layer remains bounded at $\tau \to \infty$, ensuring that thermal instability (explosion) does not occur.

Explosive scenarios for high *Ra* are illustrated for double-thermal systems.

In the case presented in Figures 5.23–5.28, the system is substantially above the critical limit. The initial distance between the thermals is small, and the maximum temperature in their centres is rather high ($\theta_{max} = 8.0$). The thermals easily merge (Figures 5.24 and 5.25) and their further evolution, again, is essentially the same as the evolution of the single thermal (Figures 5.12 and 5.13). This is a straightforward case of thermal explosion development. Maximum temperature in the domain drops only slightly (Figures 5.23 and 5.24; $\tau = 0 - 0.9$) before it starts to increase again. Maximum temperature grows progressively and rapidly (Figures 5.24–5.28; $\tau = 0.9 - 4.5$) as the hottest zone in the centre of the thermal becomes more and more pronounced. After it reaches rather high values ($\theta_{max} \sim 12$; Figure 5.28), thermal explosion becomes inevitable. In principle, apart from more vigorous ascending natural convection flow, the present scenario would be similar to the evolution of a hot single thermal far above the critical limit for lower Rayleigh numbers.

Finally, a more complicated case of delayed thermal explosion is illustrated in Figures 5.29–5.41. Here the system is above but close to critical limit for explosion. During the induction period, flow dynamics is similar to the case presented earlier in Figures 5.17–5.22, where mixture conditions are below critical.

Indeed, there is a stage of initial separated thermals evolution (Figures 5.29–5.32; $\tau = 0 - 11.57$; similar to Figures 5.17–5.19), followed by the stage of thermal interaction below the upper boundary, including formation of the downward jet (Figures 5.33–5.34; $\tau = 18 - 21.85$; similar to Figures 5.20–5.21). At later stages, a well-stratified structure is formed (Figures 5.35–5.40; $\tau = 28.28 - 107.34$). Again, this configuration with a warm upper layer (Figure 5.40) is very similar to the previous case (Figure 5.22). However, in contrast to the former (Figures 5.17–5.19) case where the temperature remains bounded, in the present case, the upper layer transforms progressively into a hot spot (Figure 5.41), where the explosion eventually occurs.

Critical Frank-Kamenetskii parameter is shown as a function of Rayleigh number in Figure 5.42 for the initial condition in the form of a single thermal. In exactly the same manner as is the case for uniform initial distribution (Figure 5.4), the influence of natural convection on explosion critical conditions is maximal at high Rayleigh numbers. In this region, critical Frank-Kamenetskii parameter depends most strongly on the Rayleigh number. This figure also demonstrates the effect of initial maximum temperature of the thermal. At very high values of this temperature, which is observed in the centre of the thermal (for example, $\theta_{\max} = 10.0$; Figure 5.42), the influence of the Rayleigh number on the process is very small. The reason for this is the following. The maximum temperature does not enter the definition of the Frank-Kamenetskii parameter (5.20); on the other hand, the Rayleigh number (5.23) is proportional to this temperature. High values of temperature in the localized zone (central region of the thermal) trigger quick thermal explosion (as energy generation rate is high in that region and also convection does not have enough time to develop during very short induction period) even if the Frank-Kamenetskii parameter is only slightly above its value in the absence of natural convection. At the same time, the Rayleigh number, due to the dependence (5.23), becomes rather high at high values of $\theta_{\max}$. Therefore, sufficiently high values of the Rayleigh number Ra correspond to the values of the Frank-Kamenetskii parameter δ_{cr} only slightly larger than the value of δ_{cr} at $Ra = 0$. Therefore, the curve corresponding to $\theta_{\max} = 10.0$ in Figure 5.42 is very flat.

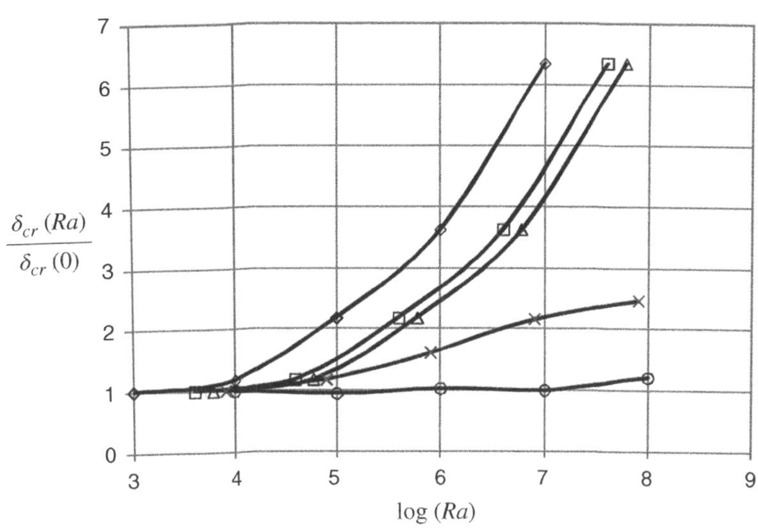

Figure 5.42 Dependence of critical Frank-Kamenetskii parameter (normalized by its value in the absence of natural convection) on Rayleigh number. Initial temperature distribution in the form of hot thermal. Effect of initial maximum temperature. ◊ – $\theta_{\max} = 1.0$; □ – $\theta_{\max} = 4.0$; Δ – $\theta_{\max} = 6.0$; X – $\theta_{\max} = 8.0$; ○ – $\theta_{\max} = 10.0$. *Source:* Reproduced with permission from Novozhilov and Byrne (2013).

Here, the effect of non-uniformity in the initial distribution can be seen clearly. For comparison, in the case of uniform initial distribution, there is also a region (region 4, Figure 5.4) where the induction period is shorter than the characteristic time of convection development. However, the boundary of that region depends strongly on the Rayleigh number. This is because, due to the absence of any region with initially elevated temperature, a higher value (at given Rayleigh number *Ra*) of the Frank-Kamenetskii parameter is required for explosion, compared to the non-uniform case presented in Figure 5.42.

In short, the Frank-Kamenetskii parameter is proportional to the reaction rate at the initial temperature. Peak with elevated temperature in the initial distribution provides an additional mechanism for rapid energy generation. Therefore, lower values of the critical parameter δ_{cr} are observed, at the same Rayleigh number *Ra*, in this case, compared with the case of uniform initial temperature distribution.

For progressively smaller values of the initial peak temperature θ_{max}, (Figure 5.42), the effect of Rayleigh number of critical conditions becomes more and more pronounced. The critical curve approaches the one shown in Figure 5.4 for uniform initial distribution.

Figure 5.43 shows the effect of the initial distribution width (for a single thermal system). According to the formula (5.22), larger values of the parameter *b* correspond to more narrow ('peak') distributions and smaller values to more wide ('flat') distributions. For the same value of the maximum temperature θ_{max}, narrow distributions contain less overall amount of energy than wider distributions. Due to this reason, a larger contribution of chemical reaction is required, in the former case, at critical conditions for an explosion. That means, that although all the curves in Figure 5.43 increase with Rayleigh number *Ra*, the curves corresponding to narrow distributions (for example, the curve for $b = b_0$) rise more rapidly.

Results for double-thermal systems may be presented in the same manner. Figure 5.44, analogous to Figure 5.42, shows the critical Frank-Kamenetskii parameter as a function of Rayleigh number for this case. Results are presented for different distances $d = \left|\vec{x}_1 - \vec{x}_2\right|$ between the centres of the two thermals. Qualitatively, Figure 5.44 demonstrates the same trends as Figure 5.42. Given,

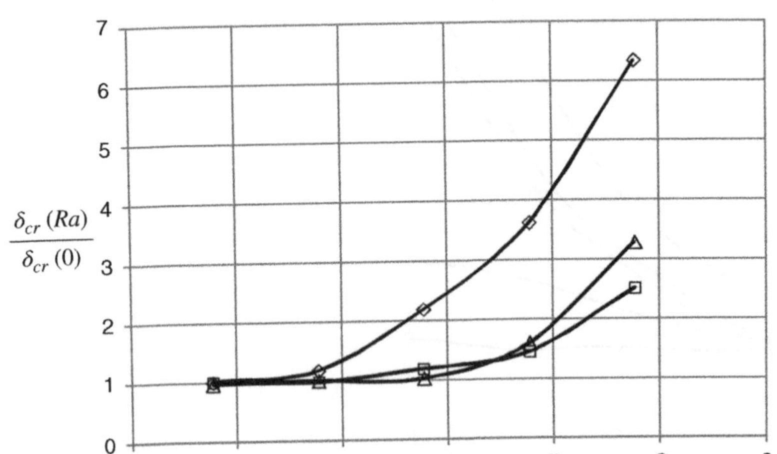

Figure 5.43 Dependence of critical Frank-Kamenetskii parameter (normalized by its value in the absence of natural convection) on the Rayleigh number. Initial temperature distribution in the form of hot thermal. Effect of initial temperature distribution width. $\theta_{max} = 6.0$. ◇ – $b = b_0 = 11.5$; Δ – $b = 5b_0$; □ – $b = 3b_0$. *Source:* Reproduced with permission from Novozhilov and Byrne (2013).

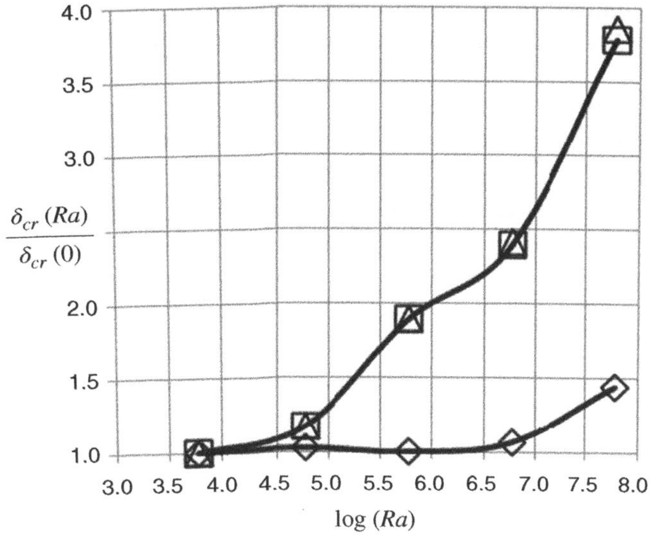

Figure 5.44 Dependence of critical Frank-Kamenetskii parameter (normalized by its value in the absence of natural convection) on Rayleigh number. Initial temperature distribution in the form of a pair of hot thermals. Effect of distance between centres of the thermals. θ_{max} = 6.0. ◇ – d = 0.1; □ – d = 0.3; Δ – d = 0.5. *Source:* Reproduced with permission from Novozhilov (2014).

however, that the overall energy content in the domain is larger (at the same value of peak temperature θ_{max}) in the case of pair of thermals, the curves in Figure 5.44 rise slower than corresponding curves in Figure 5.42. This can be seen by comparing the curve corresponding to $\theta_{max} = 6.0$ in Figure 5.42 with the both curves in Figure 5.44. Naturally, the larger amount of energy stored in the system initially requires smaller contribution to energy production from chemical reaction (which is described by the parameter δ_{cr}) to achieve critical conditions for thermal explosion.

Effect of initial separation between centres of the two thermals is also evident in this figure. Short separation distance (curve $d = 0.1$; Figure 5.44) slows down the rate of cooling of each of the clouds as they interact from the beginning of the process (reducing temperature gradients in the region between the clouds) and quickly merge. Reduced cooling rate results in low values of the critical Frank-Kamenetskii parameter and weak dependence of the latter on Rayleigh number. As the distance increases, thermals evolve essentially separately, with a very weak interaction, at the beginning of the process. Interaction occurs only at later stages of the induction period. The dependence of the critical parameter on distance $\delta_{cr}(d)$ very quickly pproaches the limit at $d \to \infty$. As Figure 5.44 demonstrates, already from $d = 0.3$, critical parameter becomes independent of the distance (curves $d = 0.3$ and $d = 0.5$ in Figure 5.44 coincide).

Novozhilov (2014) also considered the case of a more general initial distribution with the temperature varying harmonically in each spatial direction

$$\theta = \frac{\theta_{max}}{2}\left[1 + \prod_{i=1}^{3} \sin\left(\frac{2\pi}{\lambda}\left(x_i - \frac{1}{2}\right) - \frac{\pi}{2}\right)\right] \tag{5.24}$$

with λ being a wavelength of such variation.

Such distribution has a form of lattice (i.e. regular three-dimensional array) of thermals within the flow domain. The number and the size of each thermal are determined by the wavelength λ.

Critical conditions for explosion in this configuration are determined by the pair $(\lambda, \theta_{\max})$, as well as the Rayleigh number, that is, the critical Frank-Kamenetskii parameter is a function of the latter three parameters $\delta_{cr} = \delta_{cr}(\lambda, \theta_{\max}, Ra)$.

Results for this case are shown in a slightly different from the previous cases form. Instead of presenting critical value δ_{cr} as a function of the other three parameters, the values of the parameters δ and Ra are fixed; then the critical conditions are given by the pair of parameters $\left(\lambda^{cr}, \theta_{\max}^{cr}\right)$.

The value δ is fixed at some point below the critical conditions for uniform initial distribution. The degree of subcriticality is chosen by fixing the value of the parameter

$$S_{cr} = \frac{(\delta_{cr} - \delta)}{\delta_{cr}} \tag{5.25}$$

where $\delta < \delta_{cr}$ is a given fixed value, and δ_{cr} is the critical value of the Frank-Kamenetskii parameter for uniform initial temperature distribution in the absence of natural convection ($Ra = 0$). Thus, the subcriticality parameter S_{cr} measures the proximity of the system to critical conditions at uniform initial temperature.

Using the definition (5.25), fixing the degree of subcriticality is equivalent to fixing the value of the Frank-Kamenetskii parameter δ.

Critical curves for the initial distribution (5.24) are presented in Figure 5.45. The curves consist of critical points $\left(\lambda^{cr}, \theta_{\max}^{cr}\right)$ such that the region below each curve corresponds to stable (leading to thermal equilibrium) conditions, while the regions above correspond to thermal explosion conditions. Values of Rayleigh number for different points on curves 1, 2 and 3 in Figure 5.45 are fixed in such a way that, at the fixed value of $\theta_{\max}$, Rayleigh number increases by a factor of 10 from curve 1 to curve 2 and then by a factor of 10 again from curve 2 to curve 3.

Figure 5.45 demonstrates that for a fixed value of the maximum temperature $\theta_{\max}^{cr}$, the critical wavelength λ_{cr} decreases very quickly with the increase of the Rayleigh number. The reason for this is that natural convection quickly cools hot regions in the mixture at high Rayleigh numbers.

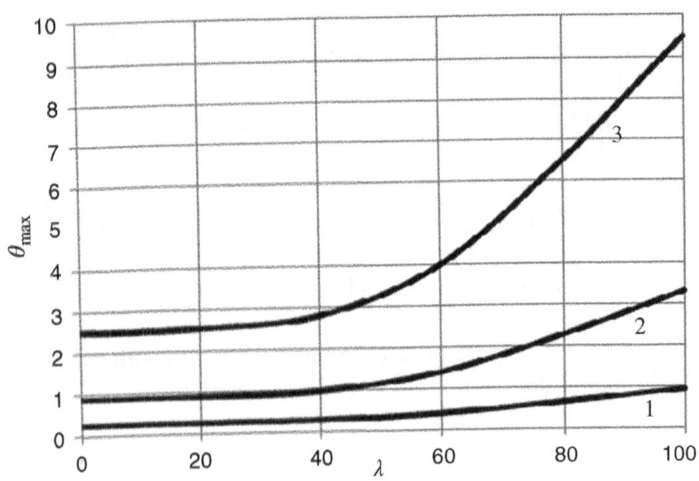

Figure 5.45 Critical thermal explosion conditions $\left(\lambda_{cr}, \theta_{\max}^{cr}\right)$. Initial temperature distribution in the form of a lattice of hot thermals. $Ra_3/Ra_2 = Ra_2/Ra_1 = 10$ (fixed $\theta_{\max}$). $S_{cr} = 0.0055$.

Consequently, tighter spacing between the areas with elevated temperature is required to achieve thermal explosion critical conditions.

5.3 Thermal Explosion Under Forced Convection Conditions

Along with natural, another type of convective flow that may exist in a reacting mixture is forced convection. External forces different from gravity must act on a system in order to develop this type of flow. In real systems, forced convection usually emerges for technological reasons. The motivation for such technological solutions is the intensification of heat and mass transfer processes in order to make a positive impact on the performance of practical combustion systems. Such intensification aims to achieve two major goals. The first is to enhance the mixing rate and uniformity of the reactants mixing. This may result in higher reaction rates within the mixture and have a positive effect on, for example, the efficiency of a chemical reactor. The second is to adjust heat exchange between the system and the surroundings in a desired way. From the general point of view, forced convection, as well as natural one, contributes to additional cooling of reactive media. Therefore, it may slow down or prevent entirely development of thermal explosion. This, obviously, also makes a positive effect on the performance of technological devices such as chemical reactors.

It is clear from the above remarks that, similarly to natural convection, forced convection also have an inevitable effect on critical conditions of potential thermal explosion in the mixture.

Since reactants in the present chapter are considered fully premixed anyway (and therefore it is assumed that the goal of uniform mixing has been achieved), the only effect of forced convection on thermal explosion conditions, which is being considered below, is an alteration of the heat exchange process between the system and its surroundings. In that respect, an analysis of the present section is analogous to that in Section 5.2.

One of the examples involving forced convection has, in fact, already been considered. This is the model homogeneous chemical reactor (or PSR) investigated in Section 5.1. Indeed, the perfect stirring and uniform flow properties assumed in the latter model are the result of the forced mixture injection into the reactor volume (Figure 5.1). That is, the convective flow (leading to uniform properties within the volume) is caused by external forces acting on a stream of mixture being injected.

Perfect mixing, as a result of forced convection, is of course an idealization, that is a limiting case. Generally, a distinctive flow structure develops as a result of forced convection.

Despite the undeniable importance, the analysis of thermal explosion development under forced convection conditions has received rather limited attention.

The problem may be considered in different formulations.

Kagan *et al.* (1997) considered the theoretical arrangement of an infinite periodic array of eddies with a prescribed identical flow configuration. This configuration is fixed by the choice of a particular stirring flow-field stream function.

The problem in a configuration of closed-volume chemical reactor has been considered by Novozhilov *et al.* (2002), Novozhilov *et al.* (2005) and Kunakov *et al.* (2013).

Process of stirring in the latter three studies may be thought as being caused by a number of mechanical stirrers each having a number of blades. Positions of the stirrers are prescribed, as well as the exact flow field configuration emerging due to stirring in each particular case.

Novozhilov *et al.* (2002) considered the simplest case of cylindrical reactor with the flow pocessing cylindrical symmetry. As such, all the variables depend on only one (radial) coordinate, that is the distance from the axis of symmetry of the cylinder, and geometrical arrangement, therefore, is identical to the one considered in Sub-section 2.3.2. All the principal assumptions of that sub-section, as well as definitions of non-dimensional temperature and Frank-Kamenetskii parameter, apply as well.

The effect of forced convection is assumed to be limited to a cylindrical layer bounded by the cylindrical surfaces at the radii $\tilde{r}_1$ and $\tilde{r}_2$, where $0 \leq \tilde{r}_1 \leq \tilde{r}_2 \leq \tilde{r}_0$, and $\tilde{r}_0$ is the radius of the cylindrical reactor. Forced convection manifests itself by increasing thermal diffusivity in the layer from its molecular value $\tilde{\kappa}_0$ to the larger effective value $\tilde{\kappa}_0 + \tilde{\kappa}_c$, where the addition, $\tilde{\kappa}_c$, is caused by forced convection. The quantity $\tilde{\kappa}_c$ will be called convective thermal diffusivity. Convection is supposed to be strong enough to force effective thermal diffusivity to be uniform (i.e. constant) throughout the stirred cylindrical layer $\tilde{r}_1 \leq \tilde{r} \leq \tilde{r}_2$. This is also in line with the principal assumptions of the model, discussed in Sub-section 2.3.2.

With the assumption of forced stirring, the flow area is being divided into the three regions, that is, $0 \leq \tilde{r} \leq \tilde{r}_1$, $\tilde{r}_1 \leq \tilde{r} \leq \tilde{r}_2$ and $\tilde{r}_2 \leq \tilde{r} \leq \tilde{r}_0$. Properties (thermal diffusivity) of the middle region are different from the properties of the other two.

It is clear, considering the scaled variables

$$\xi = \frac{\tilde{r}}{\tilde{r}_0}; r_1 = \frac{\tilde{r}_1}{\tilde{r}_0}; r_2 = \frac{\tilde{r}_2}{\tilde{r}_0} \tag{5.26}$$

that in each of the regions the equation governing temperature distribution is, in non-dimensional variables, of the form of Eq. (2.27) for the case of cylindrical symmetry (i.e. with the symmetry index equal to 1).

The only difference from Eq. (2.27) arises in the central region $r_1 \leq r \leq r_2$, where the equation must contain the ratio of the effective and molecular thermal diffusivities.

The problem takes, therefore, the form

$$\begin{aligned} \frac{d^2\theta_1}{d\xi^2} + \frac{1}{\xi}\frac{d\theta_1}{d\xi} &= -\delta\exp(\theta_1), 0 \leq \xi \leq r_1 \\ \frac{d^2\theta_2}{d\xi^2} + \frac{1}{\xi}\frac{d\theta_2}{d\xi} &= -s^2\delta\exp(\theta_2), r_1 \leq \xi \leq r_2 \\ \frac{d^2\theta_3}{d\xi^2} + \frac{1}{\xi}\frac{d\theta_3}{d\xi} &= -\delta\exp(\theta_3), r_2 \leq \xi \leq 1 \end{aligned} \tag{5.27}$$

Here the indices 1, 2, 3 in the notation of non-dimensional temperature refer to the three considered regions, and the parameter s is defined as

$$s = \frac{1}{\sqrt{m}}, m = \frac{\tilde{\kappa}_0 + \tilde{\kappa}_c}{\tilde{\kappa}_0} \tag{5.28}$$

The boundary conditions are

$$\frac{d\theta_1}{d\xi}(0) = 0; \theta_3(1) = 0 \tag{5.29}$$

Besides, an appropriate auxiliary conditions (continuity of temperature and energy flux distributions) must be fulfilled at the boundaries between the regions, that is

$$\begin{aligned} \theta_1(r_1) = \theta_2(r_1); s^2\frac{d\theta_1}{d\xi}(r_1) &= \frac{d\theta_2}{d\xi}(r_1) \\ \theta_2(r_2) = \theta_3(r_2); \frac{d\theta_2}{d\xi}(r_2) &= s^2\frac{d\theta_3}{d\xi}(r_2) \end{aligned} \tag{5.30}$$

In principle, the problem (5.27–5.30) may be investigated by the means of numerical analysis. Instead, Novozhilov *et al.* (2002) used the approximate method, described in Section 2.5, to determine the critical value of the Frank-Kamenetskii parameter δ_{cr}. In this method, the critical parameter is taken as being proportional to the minimum eigenvalue of the Laplacian.

Here, this method is generalized further to the case of conjugate heat transfer problem involving the regions with different properties (Eq. (5.27)).

The conjugate eigenvalue problem consists of the set of equations (all the eigenvalues in this case are positive as explained in Section 2.5)

$$\begin{aligned}
&\frac{d^2\theta_1}{d\xi^2} + \frac{1}{\xi}\frac{d\theta_1}{d\xi} + \mu^2\theta_1 = 0; 0 \le \xi \le r_1 \\
&\frac{1}{s^2}\left[\frac{d^2\theta_2}{d\xi^2} + \frac{1}{\xi}\frac{d\theta_2}{d\xi}\right] + \mu^2\theta_2 = 0; r_1 \le \xi \le r_2 \\
&\frac{d^2\theta_3}{d\xi^2} + \frac{1}{\xi}\frac{d\theta_3}{d\xi} + \mu^2\theta_3 = 0; r_2 \le \xi \le 1
\end{aligned} \tag{5.31}$$

with the same boundary and auxiliary conditions (5.29, 5.30).

Once the minimum eigenvalue μ is found, the critical Frank-Kamenetskii parameter may be calculated as

$$\delta_{cr} = \gamma\frac{\mu^2}{e}; \gamma \approx 0.93 \tag{5.32}$$

Here, the non-dimensional counterpart of the formula (2.80) is being used with $r = 1$.

The general solution of the set of Eq. (5.31), subjected to the first (at $\xi = 0$) of the boundary conditions (5.29) (taking also into account that eigenfunction is determined to within multiplication by the arbitrary constant), has the form

$$\begin{aligned}
&\theta_1 = J_0(\mu\xi) \\
&\theta_2 = AJ_0(\mu s\xi) + BY_0(\mu s\xi) \\
&\theta_3 = CJ_0(\mu\xi) + DY_0(\mu\xi)
\end{aligned} \tag{5.33}$$

Here, J_n and Y_n are the Bessel functions of the first kind, and the second kind, respectively; *A*,*B*,*C* and *D* are arbitrary integration constants.

The second (at $\xi = 1$) of the boundary conditions (5.29) and the auxiliary conditions (5.30) result in the following set of algebraic equations that must be fulfilled by the integration constants

$$\begin{aligned}
&J_0(\mu r_1) = AJ_0(\mu s r_1) + BY_0(\mu s r_1) \\
&sJ_1(\mu r_1) = AJ_1(\mu s r_1) + BY_1(\mu s r_1) \\
&AJ_0(\mu s r_2) + BY_0(\mu s r_2) = CJ_0(\mu r_2) + DY_0(\mu r_2) \\
&AJ_1(\mu s r_2) + BY_1(\mu s r_2) = s[CJ_1(\mu r_2) + DY_1(\mu r_2)] \\
&CJ_0(\mu) + DY_0(\mu) = 0
\end{aligned} \tag{5.34}$$

Here, the following properties of the cylinder functions have been used

$$\frac{dJ_0(x)}{dx} = -J_1(x); \frac{dY_0(x)}{dx} = -Y_1(x) \tag{5.35}$$

The set (5.34) consists of five equations for unknowns (integration constants) *A*, *B*, *C* and *D*. Therefore, its solution exists at certain values of the parameter μ (eigenvalues of the problem (5.28–5.31).

Arithmetic manipulation of a set of Eq. (5.34) leads to the following equation for the eigenvalues

$$\frac{G_1(\mu, r_1, r_2, s)}{G_2(\mu, r_1, r_2, s)} = \frac{G_3(\mu, r_1, r_2, s) - G_4(\mu, r_1, r_2, s)}{G_5(\mu, r_1, r_2, s) - G_6(\mu, r_1, r_2, s)} \tag{5.36}$$

where

$$\begin{aligned}
G_1(\mu, r_1, r_2, s) &= J_0(\mu r_1)Y_1(\mu s r_1) - sJ_1(\mu r_1)Y_0(\mu s r_1)\\
G_2(\mu, r_1, r_2, s) &= J_0(\mu s r_1)sJ_1(\mu r_1) - J_1(\mu s r_1)J_0(\mu r_1)\\
G_3(\mu, r_1, r_2, s) &= Y_1(\mu s r_2)[J_0(\mu r_2)Y_0(\mu) - Y_0(\mu r_2)J_0(\mu)]\\
G_4(\mu, r_1, r_2, s) &= Y_0(\mu s r_2)s[J_1(\mu r_2)Y_0(\mu) - Y_1(\mu r_2)J_0(\mu)]\\
G_5(\mu, r_1, r_2, s) &= J_0(\mu s r_2)s[J_1(\mu r_2)Y_0(\mu) - Y_1(\mu r_2)J_0(\mu)]
\end{aligned} \tag{5.37}$$

The set of Eqs (5.36) and (5.37) allows the minimum eigenvalue μ to be found. The critical Frank-Kamenetskii parameter δ_{cr} is obtained then from Eq. (5.32).

The dependence of the critical condition on the effect of stirring intensity is demonstrated in Figure 5.46. Critical Frank-Kamenetskii parameter increases with the increase of the effective thermal diffusivity (i.e. the parameter m) as the cooling rate of the mixture is enhanced. This effect, however, saturates once the convective thermal diffusivity becomes approximately two orders of magnitude larger than the molecular one ($m \sim 10^2$). For larger values of the stirring intensity (the parameter m), the critical Frank-Kamenetskii parameter (and the cooling rate) are only weakly dependent on this parameter. Such an effect of saturation is well known in chemical engineering.

Here, $p = \sqrt{r_2^2 - r_1^2}$ is square root of the ratio of volumes of the stirred region and the whole cylinder. Below, a maximum possible value of the internal radius r_1, which corresponds to a fixed value of p, that is $r_{1,\max} = \sqrt{1 - p^2}$, is also being used.

It should be noted that the regimes with $m \gg 1$ (that is, $\tilde{\kappa}_c \gg \tilde{\kappa}_0$) correspond to realistic conditions in practical stirred reactors. Indeed, by the order of magnitude, the convective thermal diffusivity is $\tilde{\kappa}_c \sim \tilde{d}^2 \tilde{r}_m$, where $\tilde{d}$ is the stirrer diameter, and $\tilde{r}_m$ is the stirrer rotational speed (in revolutions per unit time). In real reactors, this value is much higher than the molecular thermal diffusivity $\tilde{\kappa}_0 \sim 10^{-3}$ cm^2/s.

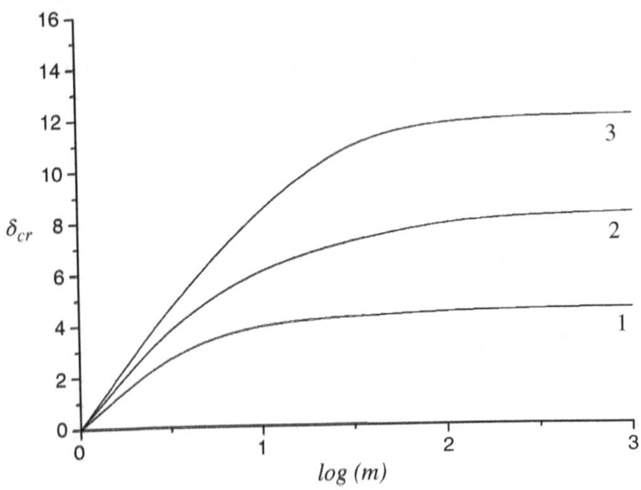

Figure 5.46 Critical Frank-Kamenetskii parameter as a function of stirring intensity. 1 – $r_1 = 0.5$, $p = 0.5$; 2 – $r_1 = 0.4$, $p = 0.7$; 3 – $r_1 = 0.3$, $p = 0.8$.

The limit $m \gg 1$ ($s \ll 1$) deserves, therefore, closer examination, and more specific results may be obtained in this limit.

In this limit ($s \to 0$), Novozhilov *et al.* (2002), using the following well-known (Abramowitz and Stegun 1964) asymptotic expansions at $x \to 0$

$$
\begin{aligned}
&J_0(x) \underset{x \to 0}{\sim} 1 - \frac{x^2}{4}; Y_0(x) \underset{x \to 0}{\sim} \frac{2}{\pi} \ln\left(\frac{x}{2}\right) \\
&J_1(x) \underset{x \to 0}{\sim} \frac{x}{2}; Y_1(x) \underset{x \to 0}{\sim} -\frac{2}{\pi x}
\end{aligned}
\tag{5.38}
$$

obtained the much reduced version of Eq. (5.36)

$$
\begin{aligned}
&\mu\left(r_2^2 - r_1^2\right) + 2r_1 \frac{J_1(\mu r_1)}{J_0(\mu r_1)} \\
&= 2r_2 \frac{Y_0(\mu)J_1(\mu r_2) - J_0(\mu)Y_1(\mu r_2)}{Y_0(\mu)J_0(\mu r_2) - J_0(\mu)Y_0(\mu r_2)}
\end{aligned}
\tag{5.39}
$$

Using Eq. (5.39) instead of the set of Eqs. (5.36) and (5.37) allows the critical Frank-Kamenetskii parameter to be calculated for the limiting case of infinite convective thermal diffusivity $s \to 0$.

Figure 5.47 shows the critical parameter as a function of the smaller stirred region radius r_1. As expected, increase of fractional volume occupied by stirred medium results in the increase of the critical Frank-Kamenetskii parameter (the curves with higher values of the parameter p lie above the curves with lower values, Figure 5.47).

For a fixed value of the parameter p, the function $\delta_{cr}(r_1)$ has a maximum that indicates position of the stirred region, providing the least favourable conditions for thermal explosion. This is a quite non-trivial result having obvious practical implications. Such position of the stirred region ensures an optimal (from the safety point of view) operating conditions.

In an exact way, the dependence of the location and the magnitude of maximum critical parameter on fractional volume of the stirred region is presented in Figure 5.48 (curves 1 and 3). For a

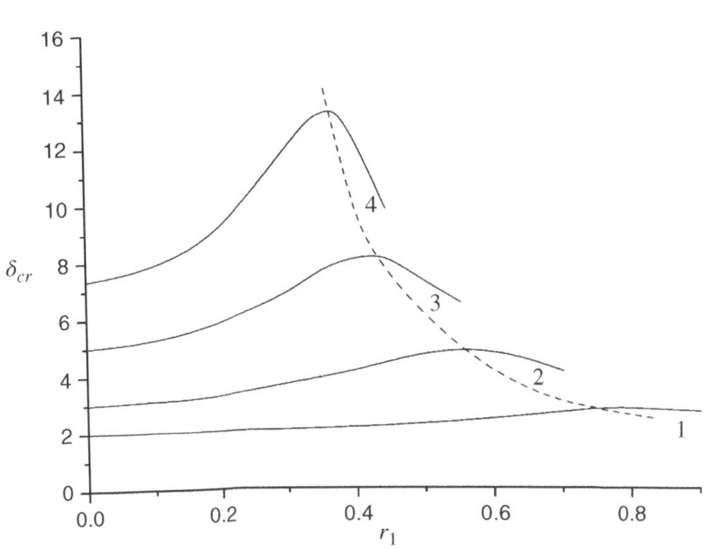

Figure 5.47 Critical Frank-Kamenetskii parameter as a function of internal radius of the stirred region. 1 – $p = 0.2$; 2 – $p = 0.5$; 3 – $p = 0.7$; 4 – $p = 0.8$. Dashed line – locus of maximums of the curves $\delta_{cr}(r_1)$.

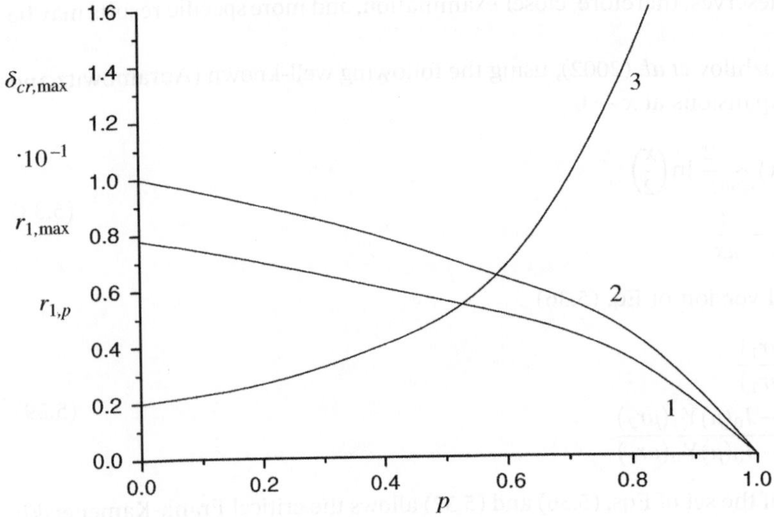

Figure 5.48 Maximum critical Frank-Kamenetskii parameter and geometrical parameter of the stirred region. 1 – optimal value of the internal radius $r_{1,\,p}$; 2 – maximum value of the internal radius $r_{1,\,max}$; 3 – maximum critical Frank-Kamenetskii parameter $\delta_{cr,\,max} \cdot 10^{-1}$.

comparison, the maximum possible, at the fixed fractional volume, internal stirred region radius is shown as curve 2. It is clear that the optimum position of the stirred region is always in the outer, rather close to the side surface of the cylinder, part of the volume.

A more complicated flow structure arises if a finite number of stirrers (agitators) are placed into a flow domain. Such configurations have been considered by Novozhilov *et al.* (2005) and Kunakov *et al.* (2013) for reactors of cylindrical shape. Flow structure in these studies is prescribed using an analytically specified stream function in polar coordinates. Such formulation mathematically correctly specifies a particular case of plane-parallel motion of incompressible inviscid fluid (Landau and Lifshitz 1987; Lavrentiev and Shabat 1977).

The stream function may be taken in the non-dimensional form

$$\psi(r,\varphi) = C\sin(\pi r^{m_r})\cos\left(\frac{n\varphi}{2}\right), 0 \le r \le 1;\ -\pi \le \varphi \le \pi \qquad n = 2k; k \in \mathbb{N} \tag{5.40}$$

Here (r, φ) are polar coordinates (the radial coordinate is non-dimensionalized using the radius $\tilde{r}_0$ of the cylindrical reactor as a scale), n is number of stirrers (agitators) and m_r is parameter prescribing (as will be clear below) the radial distance between the position of the stirrer and the axis of symmetry $r = 0$, $C \neq 0$ is arbitrary fixed constant.

Even number of agitators is considered as otherwise unstable tangential discontinuities emerge in the flow.

According to (5.40), the velocity components are given by

$$u_r = -\frac{1}{r}\frac{\partial\psi}{\partial\varphi};\ u_\varphi = \frac{\partial\psi}{\partial r} \tag{5.41}$$

Consider flow pattern enforced by the stream function (5.40). Due to angular periodicity of the stream function, the map of trajectories breaks into congruent (to within the direction of flow rotation, clockwise or counter-clockwise) sectors with the angles $\frac{2\pi}{n}$, containing one stirrer each.

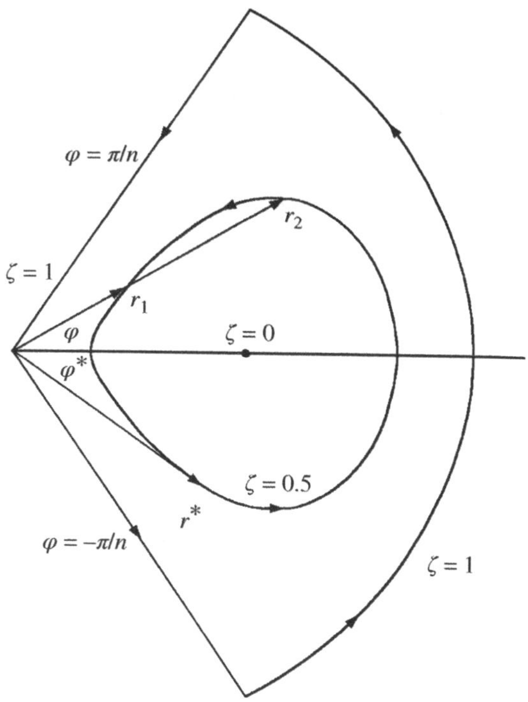

Figure 5.49 Sectorial subdivision of flow domain and geometry of streamlines. $C = -1$; $\varsigma(r, \varphi) = 1 + \psi(r, \varphi)$. *Source:* Reproduced with permission from Novozhilov *et al.* (2005) / Springer Nature.

The sector $-\frac{\pi}{n} \leq \varphi \leq \frac{\pi}{n}$ may be chosen as a representative (Figure 5.49). Angular component of velocity turns into zero along the boundaries of the sector. Stirrer is at the point $\left(r = r_{st} = 2^{-\frac{1}{m_r}}, \varphi = 0\right)$ where velocity is zero.

Examples of flow structure are also presented in Figure 5.50.

In the following analysis, $C = \frac{1}{\pi m_r}$.

The only remaining governing equation is that for energy. This equation is similar to the steady-state form of the third of Eq. (5.21). The only modifications made to that equation here are the following. Instead of convective velocity scale (5.20), a particular fixed velocity magnitude $\left|\vec{\tilde{u}}_0\right|$ (which may be chosen in arbitrary way) is used. Cylindrical reactor radius, as usual, is chosen as spatial scale. Frank-Kameneskii parameter is slightly redefined.

It is assumed further that the temperature dependence of the pre-exponent is negligible, i.e. $\alpha = 0$, and the Frank-Kamenetskii transformation (Chapter 2) is being used.

Specifically, with these modifications

$$Pe = \frac{\tilde{r}_0\left|\vec{\tilde{u}}_0\right|}{\tilde{\kappa}}; \delta = \frac{\tilde{r}_0^2\tilde{k}_0\tilde{Q}}{Ar\tilde{\lambda}\tilde{T}_0} \tag{5.42}$$

Peclet number Pe characterizes the intensity of stirring.

Boundary conditions for the temperature field include the standard fixed temperature condition at the surface of the cylinder, as well as the symmetry conditions at the side boundaries of the sector.

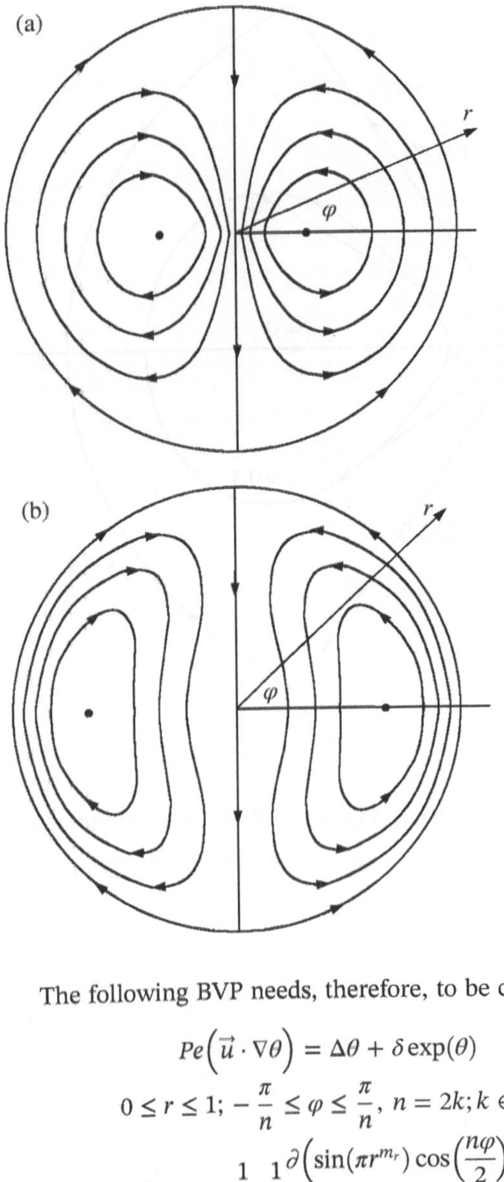

Figure 5.50 Streamlines in the case of a pair of stirrers. $n = 2$. (a) $r_{st} = 1/3$; (b) $r_{st} = 2/3$. *Source:* Reproduced with permission from Novozhilov *et al.* (2005) / Springer Nature.

The following BVP needs, therefore, to be considered

$$
\begin{aligned}
&Pe\left(\vec{u}\cdot\nabla\theta\right) = \Delta\theta + \delta\exp(\theta)\\
&0 \le r \le 1;\ -\frac{\pi}{n} \le \varphi \le \frac{\pi}{n},\ n = 2k; k \in \mathbb{N}\\
&u_r = -\frac{1}{\pi r^{m_r}}\frac{1}{r}\frac{\partial\left(\sin(\pi r^{m_r})\cos\left(\frac{n\varphi}{2}\right)\right)}{\partial\varphi}\\
&u_\varphi = \frac{1}{\pi r^{m_r}}\frac{\partial\left(\sin(\pi r^{m_r})\cos\left(\frac{n\varphi}{2}\right)\right)}{\partial r}\\
&\theta(1,\varphi) = 0; \frac{\partial\theta}{\partial\varphi}(r,\varphi)\bigg|_{\varphi=\pm\frac{\pi}{n}} = 0
\end{aligned}
\tag{5.43}
$$

Consider first the limiting case $Pe \to \infty$.

In this limit, the time of particle revolution over the streamline is zero; therefore streamlines are also isolines of temperature. Heat transfer occurs in the direction normal to streamlines only.

Formally, this fact is derived from the following consideration. In the limit $Pe \to \infty$, 1 all the terms of the HTE (5.43), including the one on the LHS, must remain finite. Therefore, the scalar product $\left(\vec{u}\cdot\nabla\theta\right)$ turns in this limit into zero. This product, however, is proportional to the Jacobian $J=\dfrac{D(\psi,\theta)}{D(r,\varphi)}$. Vanishing of this Jacobian means that stream function and non-dimensional temperature are functionally dependent. Temperature is constant on the isolines (streamlines) of the stream function and is a particular function $\theta = F(\psi)$.

Defining the parameter $\varepsilon=\dfrac{\pi m_r}{Pe}$ (so that $Pe\to\infty$ is equivalent to $\varepsilon\to 0$) and introducing the function

$$\Psi(r,\varphi) = -\pi m_r\psi(r,\varphi) \tag{5.44}$$

proportional to the stream function, the following operator of differentiation in the direction tangential to the streamline may be defined

$$L = -\frac{1}{r}\frac{\partial\Psi}{\partial\varphi}\frac{\partial}{\partial r} + \frac{1}{r}\frac{\partial\Psi}{\partial r}\frac{\partial}{\partial\varphi} \tag{5.45}$$

The PDE (5.43) may now be written as

$$\frac{1}{\varepsilon}L\theta = -\Delta\theta - \delta\exp(\theta) \tag{5.46}$$

The non-dimensional temperature θ may be sought in the form of an asymptotic expansion

$$\theta = \theta_0 + \varepsilon\theta_1 + \varepsilon^2\theta_2 + \cdots + \varepsilon^n\theta_n + \cdots \quad \varepsilon\to 0 \tag{5.47}$$

Balancing the terms containing the two leading (at $\varepsilon\to 0$) powers of ε gives the set of equations

$$\begin{aligned} L\theta_0 &= 0 \\ L\theta_1 &= -\Delta\theta_0 - \delta\exp(\theta_0) \end{aligned} \tag{5.48}$$

As noted earlier, the solution of the problem at $Pe\to\infty$ θ_0 assumes constant value on each of the streamlines and is, therefore, the function of the stream function only

$$\theta_0 = f(\Psi) \tag{5.49}$$

This fact may be seen now again from the first of Eq. (5.48).

The function $\Psi(r,\varphi)$ assumes the values $-1\le\Psi\le 0$. The set of streamlines is parameterized, therefore, by the parameter $-1\le s\le 0$.

For any $s\in[-1, 0]$, since

$$\begin{aligned} Lr &= -\frac{1}{r}\frac{\partial\Psi}{\partial\varphi} \\ L\varphi &= \frac{1}{r}\frac{\partial\Psi}{\partial r} \end{aligned} \tag{5.50}$$

the set of equations

$$\begin{aligned} \frac{\partial r}{\partial\alpha} &= -\frac{1}{r}\frac{\partial\Psi}{\partial\varphi} \\ \frac{\partial\varphi}{\partial\alpha} &= \frac{1}{r}\frac{\partial\Psi}{\partial r} \end{aligned} \tag{5.51}$$

with the initial conditions

$$\varphi(0) = 0; r(0) = \left(\frac{1}{\pi}\arcsin(-s)\right)^{\frac{1}{m}} \tag{5.52}$$

defines the dynamics of motion along the streamline $\Psi(r(\alpha), \varphi(\alpha)) = s$.

If $2A(s)$ is the period of trajectory of the dynamical system (5.52), then

$$2A(s) = \int_0^{2A(s)} d\alpha; \text{or } 2A(s) = 2\int_{r_{\min}}^{r_{\max}} \left(\frac{dr}{d\alpha}\right)^{-1} dr \tag{5.53}$$

where

$$r_{\min} = \left(\frac{1}{\pi}\arcsin(-s)\right)^{\frac{1}{m}} \tag{5.54}$$

and

$$r_{\max} = \left(1 - \frac{1}{\pi}\arcsin(-s)\right)^{\frac{1}{m}} \tag{5.55}$$

are the minimum and the maximum values of the radial coordinate along the streamline, respectively.

The explicit form of the function $\Psi(r, \varphi)$ provides

$$A(s) = \frac{2}{n}\int_{r_{\min}}^{r_{\max}} \frac{r}{\left(\sin^2 \pi r^{m_r} - s^2\right)^{\frac{1}{2}}} dr \tag{5.56}$$

Further, for any function $u(\alpha)$ on the streamline

$$Lu = \frac{du}{d\alpha}; \quad \int_0^{2A(s)} Lud\alpha = 0 \tag{5.57}$$

Application of these formulae to the function $\theta_1(r, \varphi) \equiv \theta_1(\alpha)$ (Eqs. (5.47) and (5.48)) yields the restriction

$$\int_0^{2A(s)} [\Delta f(\Psi) + \delta \exp(f)]d\alpha = 0 \tag{5.58}$$

Using explicit forms of the Laplace operator and the function $\Psi(r, \varphi)$, and making also change of the integration variable from α to r, it may be seen that the restriction (5.58) implies that the function $f(\Psi) \equiv f(s)$ (where $\Psi = s$) must satisfy the following differential equation

$$a(s)\frac{\partial^2 f}{\partial s^2} + b(s)\frac{\partial f}{\partial s} + \delta \exp(f) = 0 \quad s \in [-1, 0] \tag{5.59}$$

The coefficients $a(s)$ and $b(s)$ are given by the following formulae

$$a(s) = \frac{1}{A(s)} \int_{r_{\min}}^{r_{\max}} \frac{\left(\pi^2 m^2 r^{2m_r - 2} \mathrm{ctg}^2 \pi r^{m_r} - \dfrac{n^2}{4r^2}\right) s^2 + \dfrac{n^2}{4r^2} \sin^2 \pi r^{m_r}}{(\sin^2 \pi r^{m_r} - s^2)^{\frac{1}{2}}} \frac{2r}{n} dr$$

$$b(s) = \frac{s}{A(s)} \int_{r_{\min}}^{r_{\max}} \frac{\left(-\pi^2 m^2 r^{2m_r - 2} + \pi m^2 r^{2m_r - 2} \mathrm{ctg}^2 \pi r^{m_r} - \dfrac{n^2}{4r^2}\right)}{(\sin^2 \pi r^{m_r} - s^2)^{\frac{1}{2}}} \frac{2r}{n} dr \quad (5.60)$$

Boundary conditions for Eq. (5.59) are as follows. The condition $\theta(1, \varphi) = 0$ implies $f(0) = 0$. Further, at $s = -1$, the coefficient $a(s)$ turns into zero ($r_{\min} = r_{\max}$). Therefore, the solution f of Eq. (5.59) must satisfy the following two boundary conditions

$$f(0) = 0$$
$$b(-1)\frac{\partial f}{\partial s}(-1) + \delta \exp(f(-1)) = 0 \quad (5.61)$$

where

$$b(-1) = \frac{n^2 + \pi^2 m^2}{2^{\frac{(2m_r - 2)}{m_r}}} \quad (5.62)$$

The problem of finding critical conditions for thermal explosion in the limit $Pe \rightarrow \infty$ is reduced, therefore, to BVP (5.59–5.62), which determines the function $f(s)$.

The results for $Pe \rightarrow \infty$ (Novozhilov *et al.* 2005; and Kunakov *et al.* 2013) are presented in Figure 5.51. As expected, the critical Frank-Kamenetskii parameter increases with an increase of the number of stirrers. A larger number of stirrers enhances heat transfer rates between different portions of the mixture and consequently through its boundary.

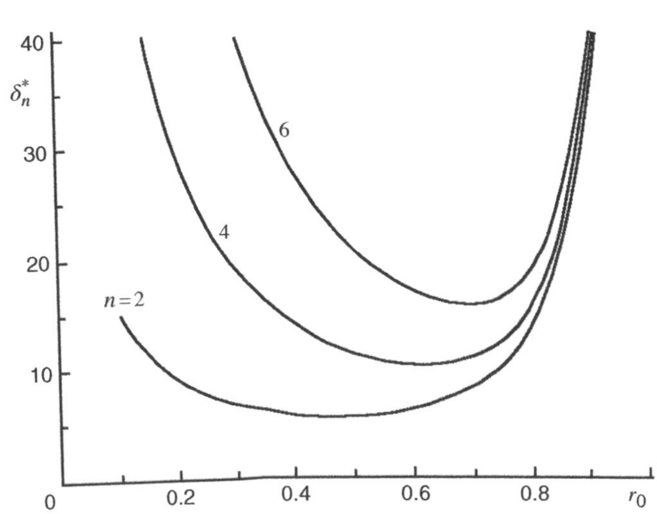

Figure 5.51 Critical Frank-Kamenetskii parameter $\delta_n^* \equiv \delta_{cr}^{(n)}$ as a function of the number of stirrers and the distance (r_0 stands for r_{st}, i.e. $r_0 \equiv r_{st}$ in this figure) from the centre of the reactor to the stirrer centreline. *Source:* Reproduced with permission from Novozhilov *et al.* (2005) / Springer Nature.

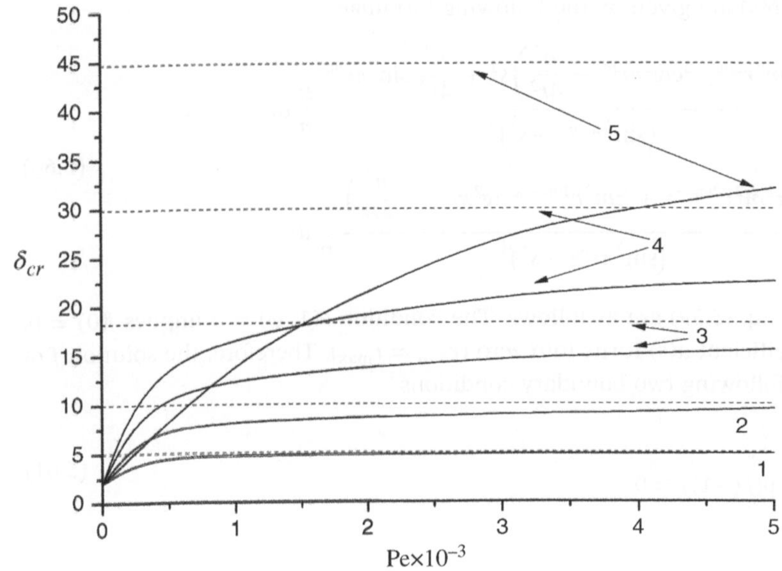

Figure 5.52 Critical Frank-Kamenetskii parameter as a function of Peclet number. $m_r = 1$ 1 – $n = 2$; 2 – $n = 4$; 3 – $n = 6$; 4 – $n = 8$; 5 – $n = 10$. Dashed lines – corresponding limits for curves 1, 2, 3, 4, 5 at $Pe \to \infty$.

It is also evident that forced convection significantly increases critical Frank-Kamenetskii parameters, compared to the standard (quiescent) case where $\delta_{cr}^{(0)} = 2$. From a practical point of view, this means that forced convection of the mixture makes reactor's operation safer.

Figure 5.51 also shows that the most advantageous (resulted in significantly increased critical Frank-Kamenetskii parameters) positions of stirrers are either closer to the cylindrical boundary of the reactor or closer to its centre. In the former case, the stirrer intensifies heat exchange in the close vicinity of the boundary and also increases the temperature gradient across the mixture. Both these mechanisms force excessive energy transfer to the surroundings. In the latter case, the stirrer intensively transfers energy from the hottest (close to the centre of the reactor) region of the flow towards its outer layers and eventually to the boundary. This cascading mechanism of energy transfer results again in the augmented rate of heat loss through the surface of the reactor.

For the stirrer positions around half-way from the centre of the reactor, the overall energy loss rate from the mixture is significantly lower, which means that thermal explosion may occur at much lower values of the Frank-Kamenetskii parameter.

In the case of finite Peclet number, the problem (5.43) has to be solved by numerical methods. In this way, the BVP (5.43) was investigated by Kunakov *et al.* (2013).

It should be noted that at finite $Pe > 0$, the isolines of the temperature $\theta(r, \varphi)$ do not coincide with the streamlines. Such a coincidence takes place only in the limit $Pe \to \infty$.

The principal results are presented in Figure 5.52. For any number of stirrers, as expected, the critical parameter increases with increasing Peclet number. The two limiting cases may also be seen in this figure. In the absence of convection ($Pe = 0$), all the curves assume the classical value $\delta_{cr} = 2$. In the limit $Pe \to \infty$, the curves approach limiting values shown by the dashed lines.

The remarkable feature of the model with forced convection, described by the problem (5.43), is the existence, at certain values of governing parameters, of four steady-state solutions, in contrast to two steady-state solutions, which are common for thermal explosion problems. Similarly to the common case, exactly one half (i.e. two out of four) of these solutions are unstable. This phenomenon was discovered and investigated by Kunakov *et al.* (2013).

6

Thermal Explosion and Fire Dynamics

6.1 Compartment Fire Flashover: Problem Description

Flashover is best described as a transition from slowly growing to fully developed compartment fire (Drysdale 2011; Quintiere 2016).

Obviously, this rather informal definition requires clarification and quantification.

Building fires normally start in slowly growing mode. Their further development is influenced by many factors: flammability properties of materials in the building, building design (size of the building, its layout and compartmentation), heating/ventilation/air conditioning (HVAC) systems and fire protection measures (both passive and active), which are put in place. In addition, there are more 'random' circumstances involving chance and human factor: failure of fire detection and/or automatic active protection systems, particularly unfortunate traffic conditions delaying access by fire brigades, misjudgement of the severity and growth rate of fire, etc.

As a result of the combination of the above factors, the evolution of fire may be dramatically different. It may continue in slowly growing mode (eventually rendering itself to easy extinguishment) and be suppressed (fully or partially) by active fire protection systems. In more unfortunate circumstances, however, it experiences a very rapid transition to the state where all combustible materials within a particular compartment of the building get involved in fire (this situation is referred to as a fully developed fire).

Such a very rapid transition in fire behaviour is called flashover. Needless to say, flashover represents the most dangerous scenario of fire development as it may, for example, suddenly block the evacuation process.

Transition from slowly growing to fully developed fire normally also implies transition from fuel-controlled to ventilation-controlled combustion regime in the compartment (Thomas 1995; Drysdale 2011; Quintiere 2016). The former regime, observed at early stages of fire development, implies that solid combustible material vapours (volatiles) are in deficiency in relation to oxygen and burn immediately as they are being released in the fuel gasification process. The latter regime develops as fire progresses to the fully developed stage, where the situation reverses itself, and deficiency of oxygen starts to be observed due to its overwhelmingly quick consumption.

The fully developed fire normally breaks the window glasses due to rising internal pressure in the compartment. Flames start to emerge from broken windows, as at this point fire starts to fill the oxygen starvation.

The onset of such flame behaviour may be taken as a definition of the moment of flashover. Visual impression of flashover is shown in Figure 6.1.

Thermal Explosion: Theory and Application, First Edition. Vasily B. Novozhilov.

Figure 6.1 Visual impression of flashover.

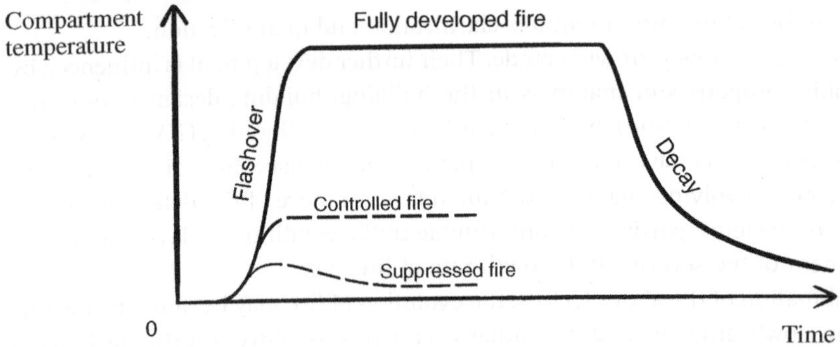

Figure 6.2 Average compartment temperature evolution in different fire scenarios. Solid curve – unintervened growing fire with flashover. Dashed curves – fire with intervention (i.e. active fire suppression): bounded compartment temperature (controlled fire) or temperature approaching asymptotically ambient value (suppressed fire).

In a more detailed, qualitative, way evolution of compartment fire through different stages of its development is demonstrated in Figure 6.2. Here average compartment temperature is drawn against time, starting at fire ignition. The solid line represents behaviour where fire progresses to the fully developed stage due to the combination of relevant factors (see above) and the absence of intervention (i.e. active suppression of fire). Flashover is shown as a steep rise of this curve at a particular moment in time. Dashed curves (relevant for later chapters) show possible fire behaviour under active fire suppression conditions: either with finite, but restricted heat release rate (HRR), or with asymptotically diminishing HRR.

The possibility of an enclosure fire flashover exists due to the influence of solid boundaries (walls, ceiling, etc.) on compartment fire development. Even a ceiling, without side walls, erected above the fire results in dramatic increase in its burning rate (Friedman 1975). The latter experiment featured burning of a slab of polymethylmethacrylate (PMMA). The reasons for this increase are the formation of a ceiling jet containing hot smoke and subsequent thermal radiation by that smoke back to fire source, as well as direct reflection of fire radiation by the ceiling towards fire base.

Correspondingly, flashover may occasionally occur even in similar largely open spaces, where solid boundaries essentially amount to the roof of the structure. The Bradford Football Stadium fire in May 1985 (Drysdale 2011) is an example of this kind of flashover.

In a more tightly closed compartment, in the presence of walls and with relatively small area of openings, the situation becomes much more dramatic. In such an environment, the rate of dissipation of heat generated by fire drops tremendously. Therefore, disbalance between the rates of energy generation and dissipation develops. It is easy to see that such disbalance is qualitatively completely identical to the description of classical thermal explosion.

There are, however, some important quantitative differences in the way flashover develops in compartments, compared to the classical models.

First, the combustion regime is different; as in compartment fires energy is predominantly being released in diffusion flames, while nearly all thermal explosion models developed to date have considered premixed chemical reactions. (The exception is the diffusion thermal explosion considered in Section 4.6.) Further, the previous models considered either quiescent environment or laminar flow dynamical models (Chapter 5). Flow and reaction rate distributions in such models follow rather regular patterns. In compartment fires, in contrast, there are a number of distinctive highly turbulent regions (such as fire plume and ceiling jet), separated from other flow regions (e.g. bottom cold layer containing clean air). Therefore, both flow structures and reaction rate distributions are highly irregular.

Based on a large volume of real fire observations and experimental data (Thomas 1995; Drysdale 2011; Quintiere 2016 and further references in these monographs), the contemporary view is that compartment fire flashover develops by the following mechanism.

Upon ignition, fire (in case it ends up in flashover) evolves in a slowly growing mode. During this phase, fire plume develops and, once it hits the ceiling, gives rise to ceiling jet. Since lateral spread of ceiling jet is restricted by compartment walls, ceiling jet flow forms smoke layer. The thickness of this layer gradually grows while its temperature increases.

The formation of hot sooty smoke layer underneath the ceiling of the compartment results in intensive thermal radiation fluxes within the compartment. In particular, strong thermal radiation is felt on the floor and side walls of the compartment, in the vicinity of which majority of combustible materials within the compartment are normally located.

Initially, fire size is restricted, therefore, the majority of combustible materials in the compartment are not burning, i.e. represent 'virgin' fuel. As the level of radiation emitted from the smoke layer grows, such virgin materials eventually ignite. Fire size and its intensity therefore grow, and this, in turn, provides more intensive energy and soot input into the smoke layer. Positive feedback loop is created: growing fire results in thicker and hotter smoke layer, and the later, in turn, results in ignition of more and more new portions of virgin fuel, contributing to further fire growth.

In such a positive feedback loop, in the absence of comparably fast heat dissipation mechanisms, energy progressively and quickly accumulates in the compartment. Eventually, all the available virgin fuel ignites, manifesting flashover, and fire progresses to fully developed stage.

As in classical thermal explosion, unrestricted energy accumulation via positive feedback loop mechanism is the essence of fire flashover.

The flashover mechanism description provided earlier suggests that the occurrence of flashover must be strongly correlated with two parameters: the temperature of the upper (smoke) layer and the intensity of radiation fluxes (in particular, on the floor) within the compartment. Indeed, these correlations may be quantified (Thomas 1995; Drysdale 2011; Quintiere 2016; Waterman 1966, 1968; Babrauskas 1979; Hägglund *et al.* 1974; Fang 1975).

In summary, the occurrence of flashover may be judged by one or more of the following attributes:

- flames start to emerge from compartment openings;
- temperature of the smoke layer grows to the values of about 600 °C;
- thermal radiation flux on the floor of compartment reaches approximately 20 kW/m^2.

These values are, of course, indicative. For example, in some cases, flashover may not occur until smoke layer temperature rises to above 673 °C (Budnick and Klein 1978).

It should be noted that, due to high complexity of flashover mechanism, certain simplifications are unavoidable in developing theoretical models. Further developments in the present chapter follow zone modelling approach to fire dynamics, which is reviewed briefly in Section 6.2.

6.2 One-Variable Thermal Explosion Models of Fire Flashover

In Sections 6.2, 6.3 and 6.4, a parallelepiped compartment with the length $\tilde{L}$, width $\tilde{W}$ and height $\tilde{H}$ is being considered.

The first model of flashover, based on the principles of thermal explosion theory, was proposed by Graham *et al.* (1995, 1999). Quantitative description of flashover mechanism, described qualitatively in Section 6.1, requires analysis of fire dynamics within compartment. Figure 6.3 presents a sketch of typical fire environment within compartment during flashover development. Figure 6.3 represents pre-flashover stage of fire, where there is a smoke leak from compartment, but flames do not spread through the windows (as in Figure 6.1).

Description of fire dynamics in all the models considered in the present chapter is performed in the framework of the so-called 'zone' fire modelling approach. This approach is widely known (Cox 1995; Karlsson and Quintiere 2022) in application to fire modelling, as well as to some other problems in heat transfer and fluid mechanics. It is impractical to discuss here all the underlying

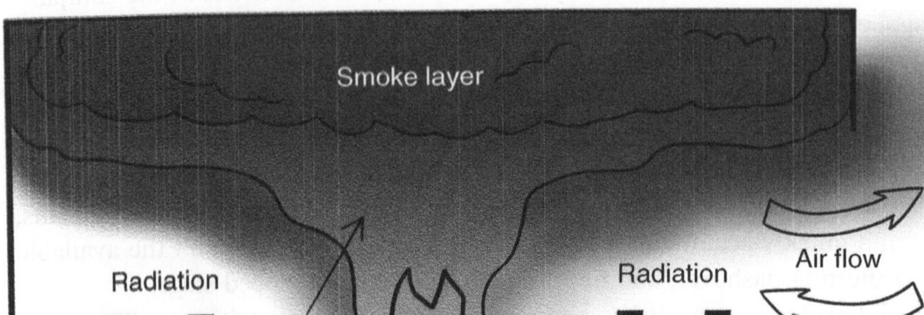

Figure 6.3 Cross-section of compartment with essential features of flashover zone model.

assumptions of the zone method, as long as its nuances. Reader interested in such a comprehensive discussion may refer to the above-referenced sources.

It is sufficient to note that the essence of the zone method is the following. Compartment volume is divided into a number of zones (in most typical and simple case, hot smoke layer and cold bottom, fresh air layer) in such a way that each of the zones may (with reasonable accuracy) be characterized by a finite number of key parameters. For example, smoke layer is typically characterized by its average temperature and its thickness (both the parameters being, generally speaking, time dependent). Conservation laws for each of the zones, as well as for the whole compartment, are being written as a set of equations (typically first-order nonlinear differential equations with respect to time), which contain zone parameters as unknowns. Solution of this set of equations provides a full description of compartment fire dynamics in terms of evolution of parameters describing all the zones.

In application to the model considered in present section (Graham *et al.* 1995), the zone approach is applied as follows.

It is assumed that at the start of the process, fire covers area $\tilde{A}_f$, and this area does not change. Fire intensity may, however, change as will be discussed below.

Thickness and mass of the smoke layer are assumed to be constant during flashover development. This is only possible if smoke continuously leaves compartment, as shown in Figure 6.3. An additional simplifying assumption is that initial smoke layer temperature is equal to environment temperature.

Smoke layer temperature $\tilde{T}$ serves as the only state variable in the dynamical model. That is, the developed model is a one-variable model.

The energy balance equation for the upper (smoke) layer is written in the form

$$\tilde{m}\tilde{c}_p \frac{d\tilde{T}}{d\tilde{t}} = \dot{\tilde{G}} - \dot{\tilde{L}} \tag{6.1}$$

where $\dot{\tilde{G}}$ is the rate of energy gain by the smoke layer, and $\dot{\tilde{L}}$ is the rate of heat losses from the layer.

The initial condition for Eq. (6.1) is

$$\tilde{T}(0) = \tilde{T}_0 \tag{6.2}$$

In line with the general concept of non-steady thermal explosion theory, flashover (thermal instability) is associated with infinite growth of temperature, as a solution of Eq. (6.1).

The EGD function on the RHS of Eq. (6.1) needs to be specified.

The heat gain rate term is taken in the following form:

$$\dot{\tilde{G}} = \chi \frac{\tilde{Q}}{\tilde{H}_{vap}} \tilde{A}_f \left[\dot{\tilde{q}}'' + \varepsilon_g \tilde{\sigma}\left(\tilde{T}^4 - \tilde{T}_0^4\right)\right] \tag{6.3}$$

This is the simplest reasonable model accounting for the variable fire HRR in the course of flashover development. The ratio $\tilde{Q}/\tilde{H}_{vap}$ of the heat of reaction to the amount of energy required to vaporize a given amount of fuel is the so-called combustibility ratio. The first term on the RHS of Eq. (6.3) corresponds to the steadily burning (at the initial rate) fire. The burning area of this initial fire may be assumed to be a square of the area equal to the actual fire area $\tilde{A}_f$. The second term describes augmentation of the initial burning rate by a radiative flux from the smoke layer of variable temperature.

Energy losses from the smoke layer are due to convection, conduction and radiation mechanisms.

The energy loss rate term assumes (upon evident (Incropera and DeWitt 2002) simplifications regarding emissivities and view factors) the following form:

$$\begin{aligned}\dot{\tilde{L}} = \tilde{m}_{out}\tilde{c}_p(\tilde{T}-\tilde{T}_0) + \left[\tilde{A}_U - \tilde{A}_V^{(2)}\right]\tilde{h}_c(\tilde{T}-\tilde{T}_w) + \tilde{A}_V^{(2)}\tilde{h}_V(\tilde{T}-\tilde{T}_0)\\ + \varepsilon_g\tilde{\sigma}\left[\tilde{A}_U - \tilde{A}_V^{(2)}\right]\left(\tilde{T}^4-\tilde{T}_w^4\right) + \varepsilon_g\tilde{\sigma}\tilde{A}_V^{(2)}\left(\tilde{T}^4-\tilde{T}_0^4\right)\\ + \varepsilon_g\tilde{\sigma}\left[\tilde{A}_L - \tilde{A}_f\right]\left(\tilde{T}^4-\tilde{T}_0^4\right) + \varepsilon_g\tilde{\sigma}\tilde{A}_f\left(\tilde{T}^4-\tilde{T}_f^4\right)\end{aligned} \tag{6.4}$$

The first term on the RHS of Eq. (6.4) describes energy loss due to the flow of smoke leaving the compartment (Figure 6.3). It will be specified precisely below.

The second and third terms describe energy losses by conduction. The second term accounts for conductive heat loss from the smoke layer to compartment walls in contact with the upper (hot) layer. The temperature of this upper part of the walls is denoted as $\tilde{T}_w$. This temperature is dependent on smoke temperature, and an assumed corresponding relation will be presented below. The third term specifies conductive energy loss through the upper part of the opening.

The rest of the terms on the RHS of Eq. (6.4) are radiative energy losses from the upper layer to the walls in contact with smoke, to the surroundings through the opening and through the walls surrounding the bottom layer and, finally, to the fire area.

The energy balance Eq. (6.1) can now be written as

$$\begin{aligned}\tilde{m}\tilde{c}_p\frac{d\tilde{T}}{dt} &= \chi\frac{\tilde{Q}}{\tilde{H}_{vap}}\tilde{A}_f\left[\tilde{q}'' + \varepsilon_g\tilde{\sigma}\left(\tilde{T}^4-\tilde{T}_0^4\right)\right]\\ &- \tilde{m}_{out}\tilde{c}_p(\tilde{T}-\tilde{T}_0) - \left[\tilde{A}_U-\tilde{A}_V^{(2)}\right]\tilde{h}_c(\tilde{T}-\tilde{T}_w) - \tilde{A}_V^{(2)}\tilde{h}_V(\tilde{T}-\tilde{T}_0)\\ &- \varepsilon_g\tilde{\sigma}\left[\tilde{A}_U-\tilde{A}_V^{(2)}\right]\left(\tilde{T}^4-\tilde{T}_w^4\right) - \varepsilon_g\tilde{\sigma}\left[\tilde{A}_L+\tilde{A}_V^{(2)}-\tilde{A}_f\right]\left(\tilde{T}^4-\tilde{T}_0^4\right)\\ &- \varepsilon_g\tilde{\sigma}\tilde{A}_f\left(\tilde{T}^4-\tilde{T}_f^4\right)\end{aligned} \tag{6.5}$$

Flow rate of hot gases leaving the compartment is approximated as (Graham *et al.* 1995)

$$\tilde{m}_{out} = \frac{2}{3}c_d\tilde{\rho}_0\tilde{A}_V\sqrt{2\tilde{g}\tilde{H}_V^{(2)}\frac{\tilde{T}_0}{\tilde{T}}\left(1-\frac{\tilde{T}_0}{\tilde{T}}\right)} \tag{6.6}$$

modifying slightly the conventionally used arguments originally due to Rockett (1976).

Further, using the approximation

$$\frac{\tilde{T}_0^{1/2}(\tilde{T}-\tilde{T}_0)^{3/2}}{\tilde{T}} \approx \frac{(\tilde{T}-\tilde{T}_0)}{2} \tag{6.7}$$

the energy loss term, due to convective flow of smoke, is written as

$$\tilde{m}_{out}\tilde{c}_p(\tilde{T}-\tilde{T}_0) \approx \frac{1}{3}c_d\tilde{\rho}_0\tilde{A}_V\sqrt{2\tilde{g}\tilde{H}_V^{(2)}}(\tilde{T}-\tilde{T}_0) \tag{6.8}$$

It is easy to show that the approximation (6.7) has a relative error below 20% for the temperatures of up to 1,500 K, which well exceeds the range of upper layer temperatures during flashover (Section 6.1).

The other additional assumptions are made.

The term accounting for radiative energy loss to the fire area is approximated as

$$\varepsilon_g\tilde{\sigma}\tilde{A}_f\left(\tilde{T}^4-\tilde{T}_f^4\right) \approx \varepsilon_g\tilde{\sigma}\tilde{A}_f\left(\tilde{T}^4-\tilde{T}_0^4\right), \tag{6.9}$$

which results in a relatively small error assuming that fire area is small compared to floor area of the compartment.

It is further assumed that the temperature of the compartment walls in contact with the smoke layer may be written in the form

$$\tilde{T}_w = \tilde{T}_0 + \beta(\tilde{T} - \tilde{T}_0); \quad 0 \leq \beta \leq 1 \tag{6.10}$$

where parameter β measures thermal inertia of the walls.

At small values of this parameter $\beta \approx 0$ (walls of high thermal inertia), the wall temperature is close to the external (ambient) temperature $\tilde{T}_w \approx \tilde{T}_0$. At high values $\beta \approx 1$ (walls of low thermal inertia), the temperature of the walls is close to the temperature of the upper layer.

Using the assumptions (6.8–6.10), Eq. (6.5), after tedious algebraic manipulations, may be transformed to the following form:

$$\begin{aligned}
\tilde{m}\tilde{c}_p \frac{d\tilde{T}}{d\tilde{t}} &= \chi \frac{\tilde{Q}}{\tilde{H}_{vap}} \tilde{A}_f \tilde{\dot{q}}'' + \left[\frac{1}{3}\tilde{c}_d \tilde{\rho}_0 \tilde{A}_V \sqrt{2\tilde{g}\tilde{H}_V^{(2)}} \right. \\
&\quad + \tilde{A}_V^{(2)} \tilde{h}_V + \left[\tilde{A}_U - \tilde{A}_V^{(2)}\right] \tilde{h}_c (1-\beta) \Big] \tilde{T}_0 - \left[\chi \frac{\tilde{Q}}{\tilde{H}_{vap}} \tilde{A}_f \varepsilon_g \tilde{\sigma} - \varepsilon_g \tilde{\sigma} \left[\tilde{A}_L + \tilde{A}_V^{(2)}\right] \right. \\
&\quad \left. - \varepsilon_g \tilde{\sigma} \left(\tilde{A}_U - \tilde{A}_V^{(2)}\right)(1-\beta)^4 \right] \tilde{T}_0^4 \\
&\quad + \left[\chi \frac{\tilde{Q}}{\tilde{H}_{vap}} \tilde{A}_f \varepsilon_g \tilde{\sigma} - \varepsilon_g \tilde{\sigma} \left[\tilde{A}_L + \tilde{A}_V^{(2)}\right] - \varepsilon_g \tilde{\sigma} \left(\tilde{A}_U - \tilde{A}_V^{(2)}\right)\left(1-\beta^4\right) \right] \tilde{T}^4 \\
&\quad - \left[\frac{1}{3}\tilde{C}_d \tilde{\rho}_0 \tilde{A}_V \sqrt{2\tilde{g}\tilde{H}_V^{(2)}} + \tilde{A}_V^{(2)} \tilde{h}_V + \left[\tilde{A}_U - \tilde{A}_V^{(2)}\right] \tilde{h}_c (1-\beta) \right. \\
&\quad \left. - \varepsilon_g \tilde{\sigma} \left(\tilde{A}_U - \tilde{A}_V^{(2)}\right) 4\beta(1-\beta)^3 \tilde{T}_0^3 \right] \tilde{T} \\
&\quad + \varepsilon_g \tilde{\sigma} \left(\tilde{A}_U - \tilde{A}_V^{(2)}\right) 4(1-\beta)\beta^3 \tilde{T}_0 \tilde{T}^3 \\
&\quad + \varepsilon_g \tilde{\sigma} \left(\tilde{A}_U - \tilde{A}_V^{(2)}\right) 6(1-\beta)^2 \beta^2 \tilde{T}_0^2 \tilde{T}^2
\end{aligned} \tag{6.11}$$

Equation (6.11) may be transformed to non-dimensional form in the following way.

Non-dimensional temperature is defined in the present chapter simply as

$$\theta = \frac{\tilde{T}}{\tilde{T}_0} \tag{6.12}$$

since the characteristic temperature scale of combustion reaction does not play an essential role here, in contrast to the classical thermal explosion problem. Rather, it is important to follow how fire HRR increases over the characteristic scale of temperature change of the upper layer.

Further, the non-dimensional time is

$$\tau = \frac{\tilde{t}}{\tilde{t}_0} \tag{6.13}$$

where the characteristic time scale is introduced as

$$\tilde{t}_0 = \frac{b_{-1}\tilde{m}\tilde{c}_p \tilde{T}_0}{\tilde{\dot{Q}}_0} \tag{6.14}$$

where

$$\tilde{\dot{Q}}_0 = \chi \frac{\tilde{Q}}{\tilde{H}_{vap}} \tilde{A}_f \tilde{\dot{q}}'' \tag{6.15}$$

is the HRR of self-sustained fire, and b_{-1} is a parameter that will be defined later.

Using the non-dimensional variables (6.12, 6.13), Eq. (6.11) is written in the following form:

$$\frac{d\theta}{b_{-1}d\tau} = b_0^{(-1)} + b_4^{(-1)}\theta^4 - b_1^{(-1)}\theta + b_3^{(-1)}\theta^3 + b_2^{(-1)}\theta^2 \tag{6.16}$$

where

$$\begin{aligned}
b_0^{(-1)} &= 1 + \tilde{\dot{Q}}_0^{-1}\Bigg[\left[\frac{1}{3}\tilde{c}_d\tilde{\rho}_0\tilde{A}_V\sqrt{2\tilde{g}\tilde{H}_V^{(2)}} + \tilde{A}_V^{(2)}\tilde{h}_V + \left[\tilde{A}_U - \tilde{A}_V^{(2)}\right]\tilde{h}_c(1-\beta)\right]\tilde{T}_0 \\
&\quad - \left[\chi\frac{\tilde{Q}}{\tilde{H}_{vap}}\tilde{A}_f\varepsilon_g\tilde{\sigma} - \varepsilon_g\tilde{\sigma}\left[\tilde{A}_L + \tilde{A}_V^{(2)}\right] - \varepsilon_g\tilde{\sigma}\left(\tilde{A}_U - \tilde{A}_V^{(2)}\right)(1-\beta)^4\right]\tilde{T}_0^4\Bigg] \\
b_4^{(-1)} &= \tilde{\dot{Q}}_0^{-1}\tilde{T}_0^4\left[\chi\frac{\tilde{Q}}{\tilde{H}_{vap}}\tilde{A}_f\varepsilon_g\tilde{\sigma} - \varepsilon_g\tilde{\sigma}\left[\tilde{A}_L + \tilde{A}_V^{(2)}\right] - \varepsilon_g\tilde{\sigma}\left(\tilde{A}_U - \tilde{A}_V^{(2)}\right)(1-\beta^4)\right] \\
b_1^{(-1)} &= \tilde{\dot{Q}}_0^{-1}\tilde{T}_0\Bigg[\frac{1}{3}\tilde{c}_d\tilde{\rho}_0\tilde{A}_V\sqrt{2\tilde{g}\tilde{H}_V^{(2)}} + \tilde{A}_V^{(2)}\tilde{h}_V + \left[\tilde{A}_U - \tilde{A}_V^{(2)}\right]\tilde{h}_c(1-\beta) \\
&\quad - \varepsilon_g\tilde{\sigma}\left(\tilde{A}_U - \tilde{A}_V^{(2)}\right)4\beta(1-\beta)^3\tilde{T}_0^3\Bigg] \\
b_3^{(-1)} &= \tilde{\dot{Q}}_0^{-1}\tilde{T}_0^3\varepsilon_g\tilde{\sigma}\left(\tilde{A}_U - \tilde{A}_V^{(2)}\right)4(1-\beta)\beta^3\tilde{T}_0 \\
b_2^{(-1)} &= \tilde{\dot{Q}}_0^{-1}\tilde{T}_0^2\varepsilon_g\tilde{\sigma}\left(\tilde{A}_U - \tilde{A}_V^{(2)}\right)6(1-\beta)^2\beta^2\tilde{T}_0^2
\end{aligned} \tag{6.17}$$

Equation (6.16) can be written in the following form:

$$\begin{aligned}
\frac{d\theta}{b_{-1}d\tau} &= b_0^{(-1)} + b_4^{(-1)} - b_1^{(-1)} + b_4^{(-1)}(\theta^4 - 1) - b_1^{(-1)}(\theta - 1) \\
&\quad + b_3^{(-1)}\theta^3 + b_2^{(-1)}\theta^2
\end{aligned} \tag{6.18}$$

Setting

$$b_{-1} = \left(b_0^{(-1)} + b_4^{(-1)} - b_1^{(-1)}\right)^{-1} \tag{6.19}$$

and defining further

$$\begin{aligned}
a_1 &= \frac{b_4^{(-1)}}{\left(b_0^{(-1)} + b_4^{(-1)} - b_1^{(-1)}\right)} \\
a_2 &= \frac{b_1^{(-1)}}{\left(b_0^{(-1)} + b_4^{(-1)} - b_1^{(-1)}\right)} \\
a_3 &= \frac{b_3^{(-1)}}{\left(b_0^{(-1)} + b_4^{(-1)} - b_1^{(-1)}\right)} \\
a_4 &= \frac{b_2^{(-1)}}{\left(b_0^{(-1)} + b_4^{(-1)} - b_1^{(-1)}\right)}
\end{aligned} \tag{6.20}$$

The following final equation is obtained:

$$\frac{d\theta}{d\tau} = 1 + a_1(\theta^4 - 1) - a_2(\theta - 1) + a_3\theta^3 + a_4\theta^2 \tag{6.21}$$

Note that physically relevant sets of input parameters result in a_1, $a_2 > 0$, a_3, $a_4 \geq 0$.

Equation (6.21) allows the critical conditions for thermal explosion (flashover) to be established. The description of a critical boundary in a four-dimensional parametric space (a_1, a_2, a_3, a_4) is a rather difficult task.

Important specific cases where analysis simplifies dramatically are the cases of walls of infinitely high thermal inertia ($\beta = 0$) and of walls of infinitely low thermal inertia ($\beta = 1$). In both these cases, according to (6.17) and (6.20), the coefficients a_3 and a_4 vanish, that is, $a_3, a_4 = 0$.

Eq. (6.21) reduces to

$$\frac{d\theta}{d\tau} = 1 + a_1\left(\theta^4 - 1\right) - a_2(\theta - 1) \tag{6.22}$$

Critical conditions for thermal explosion correspond to the disappearance of real root(s) of the EGD function

$$F(\theta) = 0; F(\theta)F(\theta) = a_1\left(\theta^4 - 1\right) - a_2(\theta - 1) + 1 \tag{6.23}$$

Rewriting the latter equation in the form

$$\widehat{F}(\theta) = 0; \quad \widehat{F}(\theta) = \frac{a_1}{a_2}\left(\theta^4 - 1\right) + 1 + \frac{1}{a_2} - \theta \tag{6.24}$$

allows the considerations of the Legendre transformation, outlined in Chapter 2, to be applied again. Critical conditions are given by the requirement

$$g(1) = 0 \tag{6.25}$$

which, in turn, translates into the following two (completely identical to (2.16)) conditions

$$\begin{aligned} &\frac{d\widehat{F}}{d\theta}(\theta_{cr}) = 0 \\ &\widehat{F}(\theta_{cr}) = 0 \end{aligned} \tag{6.26}$$

for the unknown critical thermal explosion (flashover) temperature θ_{cr}.

Specifically, Eq. (6.26) read

$$\begin{aligned} &4\frac{a_1}{a_2}\theta_{cr}^3 = 1 \\ &\frac{a_1}{a_2}\left(\theta_{cr}^4 - 1\right) + 1 + \frac{1}{a_2} - \theta_{cr} = 0 \end{aligned} \tag{6.27}$$

Elimination of the critical temperature from the above equations leads to the following implicit dependence $a_1 = a_1(a_2)$ for the critical boundary

$$\frac{a_2}{4a_1} = \left(\frac{4}{3}\right)^3\left(\frac{1 - a_1 + a_2}{a_2}\right)^3 \tag{6.28}$$

In parametric form, the critical boundary may be represented as

$$\begin{aligned} a_1 &= \frac{1}{\left(3\theta_{cr}^4 - 4\theta_{cr}^3 + 1\right)} \\ a_2 &= \frac{4\theta_{cr}^3}{\left(3\theta_{cr}^4 - 4\theta_{cr}^3 + 1\right)} \end{aligned} \tag{6.29}$$

Here critical temperature θ_{cr} is a parameter that varies from 1 to ∞.

The dependence (6.28) (or (6.29)) is presented in Figure 6.4.

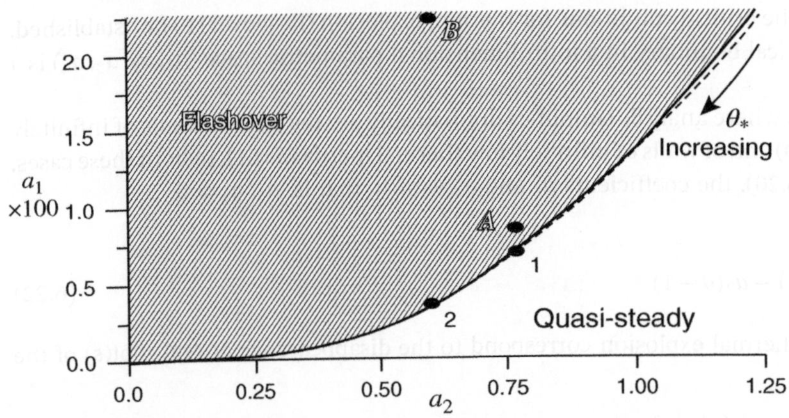

Figure 6.4 Critical boundary for flashover. $\beta = 0$ or $\beta = 1$. Points 1 and 2 belong to the critical boundary. Points A and B describe flashover conditions for the increased values of the parameter a_1 and for the same values of the parameter a_2 as for points 1 and 2, respectively. Dashed curve – approximation (6.30). *Source:* Reproduced with permission from Graham *et al.* (1999) / Elsevier.

For controlling parameters (a_1, a_2) below the critical boundary, the fire burns in a quasi-steady manner; above the boundary, flashover occurs.

Note that the limit $a_1 \ll 1$ is well described by the simple explicit expression

$$a_1 \approx \frac{27}{256} \frac{a_2^4}{(1 + a_2)^3} \tag{6.30}$$

(dashed curve in Figure 6.4).

Possible steady-state solutions (for $\beta = 0$ or $\beta = 1$) are presented in Figure 6.5. There may exist zero, one or two equilibrium points of Eq. (6.22).

Figure 6.5 is completely analogous to Figure 2.3 for the Semenov model. Curve 1 corresponds to such parameters (a_1, a_2) that the EGD function has no real roots, so that the upper layer temperature grows infinitely. Curve 2 corresponds to the critical conditions (exactly one root where the EGD function is tangent to the horizontal, θ-axis). This single steady-state solution is unstable. Finally, Curve 3 corresponds to the case of two real roots. The equilibrium point corresponding to the least of these two roots is stable; the other one is unstable.

The effect of variable thermal inertia of walls may be illustrated by the specific case (Graham *et al.* 1999), where the coefficients a_1, a_2 may be expanded around the point $\beta = 1$ to the zeroth and first order of $1 - \beta$, respectively, and the last two terms on the RHS of Eq. (6.21) are negligible

$$a_1 \approx a_1|_{\beta=1}; a_2 \approx a_2|_{\beta=1} + \alpha_1(1-\beta); a_3\theta^3 \ll 1; a_4\theta^2 \ll 1 \tag{6.31}$$

Variation of the position of critical boundary with varying thermal inertia of walls β is presented in Figure 6.6. Points 1 and 2 from Figure 6.4 are also plotted here.

Upper layer temperature time history curves have been presented by Graham *et al.* (1995, 1999) and turned out (not very surprisingly) to be topologically identical to the analogous curves for Semenov and Frank–Kamenetskii models (Figures 2.9 and 2.11).

Graham *et al.* (1995, 1999) also considered induction period of flashover and provided approximate formula for induction period for the case of parameters close to critical conditions.

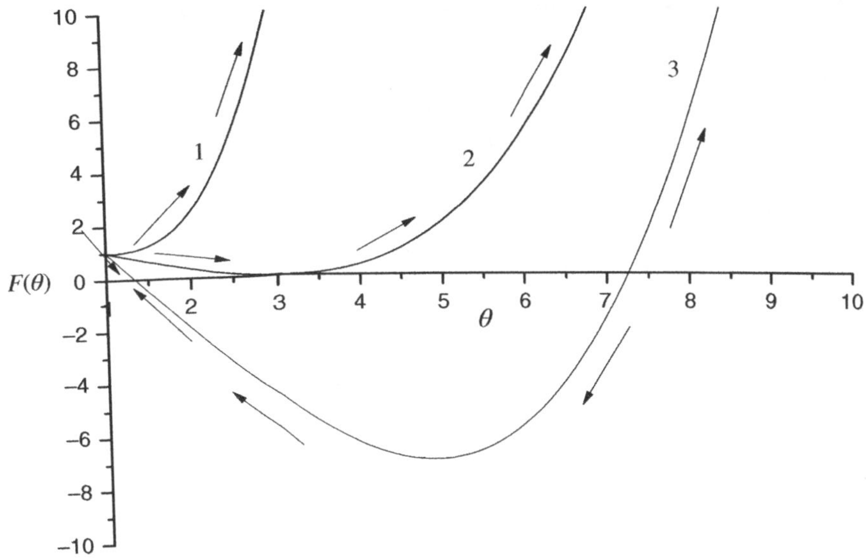

Figure 6.5 EGD function of the Graham *et al.* (1995, 1999) model in the vicinity of equilibrium points and the system evolution. $\beta = 0$ or $\beta = 1$. 1 – $a_1 = 0.15$, $a_2 = 0.75$; 2 – $a_1 = 0.64 \cdot 10^{-2}$, $a_2 = 0.75$; 3 – $a_1 = 0.64 \cdot 10^{-2}$, $a_2 = 3.00$. Arrows – directions of system evolution in different regions of phase space.

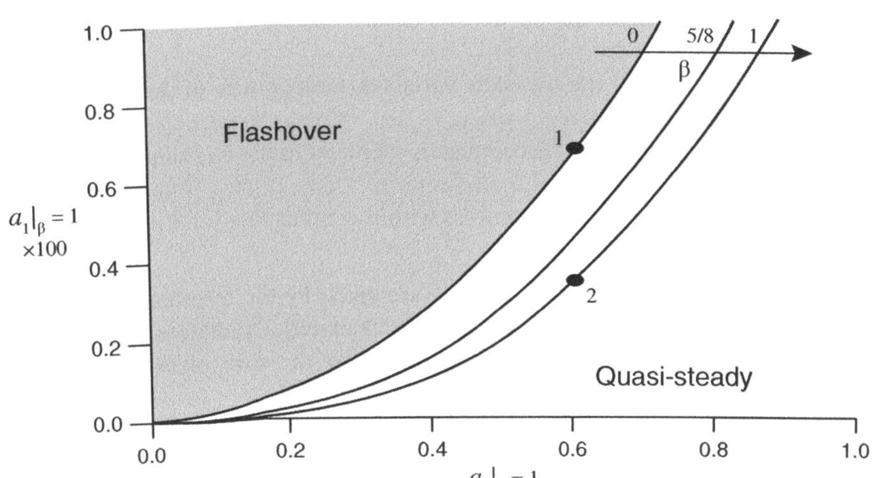

Figure 6.6 Effect of variation of thermal inertia of walls on critical boundary. $\alpha_1 = 0.165$. *Source:* Reproduced with permission from Graham *et al.* (1999) / Elsevier.

One-variable models (with the same state variable, i.e. smoke layer temperature) have also been considered by Bishop *et al.* (1993) and Beard *et al.* (1995). The former study includes formally the second equation for fire radius, but it is trivially integrable, resulting essentially in *a'priori* chosen law of fire radius growth and one-variable model.

6.3 Two-Variable Thermal Explosion Models of Fire Flashover

The model of the previous section described conditions of uncontrolled energy accumulation in a compartment. This model has two clear assumptions. First, the fire area remains the same although the fire burning rate grows. Second, quantitatively, the growth mechanism is given by Eq. (6.3), which is really the simplest model of solid fuel burning under variable external heat flux. In particular, this model does not take into account the exact mechanism of heat transfer within the solid material. The model (6.3) is equivalent to the assumption that all the energy supplied to the fuel is accumulated within a thin layer (which is being vaporized), adjacent to the surface. As such, this model is zero-dimensional with respect to combustible solid material. Both the discussed assumptions are not realistic in a real flashover scenario.

The next level of complexity is represented by one-dimensional model, which considers HTE within the solid material while also allowing fire to spread over initially inert portions of fuel (which are marked in Figure 6.3 as combustible materials). The model of this kind was constructed by Novozhilov (2010).

The Novozhilov (2010) model considers solid fuel as thermally thick; therefore one-dimensional HTE may be considered on the semi-infinite interval $(-\infty, 0]$, where the origin of the coordinate system is at the fuel's surface.

The model describes ignition of virgin fuel (which manifests rapid fire spread, i.e. flashover) explicitly. The simplest criterion, conventional in classical fire dynamics, of fixed ignition surface temperature is being used. Consequently, the second state variable, namely material surface temperature, which is being tracked in time, emerges.

Flashover is associated with ignition of all combustible materials existing on the floor of the compartment (Figure 6.3).

The model consists of two equations for the two state variables, temperature of the upper layer and fuel surface temperature.

The first equation is ODE, which is taken to be completely identical to the Graham *et al.* (1995, 1999) model (6.21).

The second equation describes the heat transfer process within combustible solid material, subjected to relevant boundary conditions.

Non-dimensional (scaled with $\dot{\tilde{Q}}_0/\tilde{\kappa}\tilde{t}_0$, where $\tilde{t}_0$ and $\dot{\tilde{Q}}_0$ are given by the formulae (6.14) and (6.15), respectively, and $\tilde{\kappa}$ is thermal diffusivity of solid combustible material) radiative flux received by virgin fuel from smoke layer (Figure 6.3) is approximated by the case of two radiation-exchanging long parallel planes (Incropera and DeWitt 2002)

$$\dot{q}''_{rad} = -V(\varepsilon_1, \varepsilon_2)\frac{\tilde{\sigma}\tilde{\kappa}\tilde{t}_0\tilde{T}_0^4}{\dot{\tilde{Q}}_0}(\theta^4 - \theta_s^4) \tag{6.32}$$

where $V(\varepsilon_1, \varepsilon_2) = \left(\left(\frac{1}{\varepsilon_1}\right) + \left(\frac{1}{\varepsilon_2}\right) - 1\right)^{-1}$. The minus sign reflects the direction of the x–axis from $-\infty$ to 0.

The exact view factor for parallel rectangles can easily be incorporated but is not essential for the present analysis.

Using the following scaling and non-dimensional numbers

$$\begin{aligned} &\Theta = \frac{\tilde{\Theta}}{\tilde{T}_0}, x = \frac{\tilde{x}}{\sqrt{\tilde{\kappa}\tilde{t}_0}}; \tau = \frac{\tilde{t}}{\tilde{t}_0}, Bi = \sqrt{\tilde{\kappa}\tilde{t}_0}\frac{\tilde{h}_s}{\tilde{\lambda}} \\ &Bi_R = V(\varepsilon_1, \varepsilon_2)\frac{\sqrt{\tilde{\kappa}\tilde{t}_0}\tilde{\sigma}\tilde{T}_0^3}{\tilde{\lambda}} \end{aligned} \tag{6.33}$$

the BVP for temperature distribution through the solid fuel (in the direction orthogonal to its surface) is written in the form

$$\begin{aligned}
&\frac{\partial \Theta}{\partial \tau} = \frac{\partial^2 \Theta}{\partial x^2}, \qquad -\infty < x \le 0 \\
&\frac{\partial \Theta}{\partial x} = Bi_R\left(\theta^4 - \theta_s^4\right) - Bi(\theta_s - 1), \quad x = 0 \\
&\frac{\partial \Theta}{\partial x} = 0, \quad x = -\infty \\
&\Theta(x) = 1, \quad \tau = 0
\end{aligned} \tag{6.34}$$

Note that the parameter Bi_R plays the role of radiative Biot number.

It is convenient to consider an integral equivalent of the PDE (6.34) (Landau and Lifshitz 1987)

$$\Theta(x, \tau) - 1 = \frac{d^{-\frac{1}{2}}}{d\tau^{-\frac{1}{2}}}\left[\frac{d}{dx}(\Theta(x, \tau))\right] \tag{6.35}$$

which holds for any $-\infty < x \le 0$ and where the fractional derivative is given by the Riemann–Liouville definition (Oldham and Spanier 2002)

$$\frac{d^{-\frac{1}{2}}}{ds^{-\frac{1}{2}}}[\phi(s)] = \frac{1}{\sqrt{\pi}}\int_0^s \frac{\phi(\xi)}{\sqrt{s-\xi}}d\xi \tag{6.36}$$

The advantage of using integral formulation (6.35) is that it can be applied just at the fuel's surface $x = 0$ to produce an equation explicitly for combustible material surface temperature, which serves as the second state variable in the model.

Consequently, combination of Eqs. (6.34) and (6.35) (at $x = 0$) gives the two-variable flashover model by Novozhilov (2010) in the form

$$\begin{aligned}
&\frac{d\theta}{d\tau} = 1 + a_1\left(\theta^4 - 1\right) - a_2(\theta - 1) + a_3\theta^3 + a_4\theta^2 \\
&\theta_s(\tau) - 1 = \frac{d^{-(1/2)}}{d\tau^{-(1/2)}}\left[Bi_R\left(\theta^4 - \theta_s^4\right) - Bi(\theta_s - 1)\right]
\end{aligned} \tag{6.37}$$

or in expanded form

$$\begin{aligned}
&\frac{d\theta}{d\tau} = 1 + a_1\left(\theta^4 - 1\right) - a_2(\theta - 1) + a_3\theta^3 + a_4\theta^2 \\
&\theta_s - 1 = \sqrt{\frac{\tau}{\pi}}\int_0^1 \frac{\left[Bi_R\left(\theta^4 - \theta_s^4\right) - Bi(\theta_s - 1)\right]}{\sqrt{1-s}}ds
\end{aligned} \tag{6.38}$$

The equations are one-way coupled, that is, the solution of the second equation is dependent on the solution of the first, but not vice versa. The temperature of the upper layer is completely determined by the first of Eq. (6.38). Surface temperature of the material is determined by both of the above equations.

In terms of this model, flashover is seen as ignition of all the available combustible loading in the compartment. This manifests fully developed fire (Figure 6.2). Flashover occurs, therefore, if the surface temperature of virgin fuel, predicted by Eqs. (6.37) (or (6.38)), reaches its *a'priori* prescribed critical value θ_s^{cr}. Quasi-steady fire occurs if the surface temperature stays below the critical value at any time.

Since fuel thermal properties are assumed uniform throughout the compartment, ignition of all the virgin fuel occurs in the model simultaneously. In principle, it is not difficult, however, to extend the model to the case of variable thermal properties.

It should be noted that the model formulation may be compressed into the single equation

$$\theta_s - 1 = \sqrt{\frac{\tau}{\pi}} \int_0^1 \frac{\left[Bi_R\left(\theta^4 - \theta_s^4\right) - Bi(\theta_s - 1)\right]}{\sqrt{1-s}} ds \tag{6.39}$$

where the function $\theta(\tau)$ is known and given by

$$\int_1^\theta \frac{1}{1 + a_1(s^4 - 1) - a_2(s - 1) + a_3 s^3 + a_4 s^2} ds = \tau \tag{6.40}$$

Equations (6.39) and (6.40) provide an alternative, fully equivalent to Eqs. (6.37) (or (6.38)), problem formulation.

Critical conditions for the Novozhilov (2010) model may be established in the following way. Consider again, for simplicity, the case $a_3 = a_4 = 0$ (walls of either infinitely high or infinitely low thermal inertia).

As discussed in Section 6.2, the critical boundary for the Graham *et al.* (1995, 1999) model (the first of Eq. (6.37)) is represented by a solid curve in Figure 6.4. This boundary is also shown in Figure 6.7 as curve 1. Addition of the second state variable (governed by the second of Eq. (6.37)) changes critical conditions for flashover.

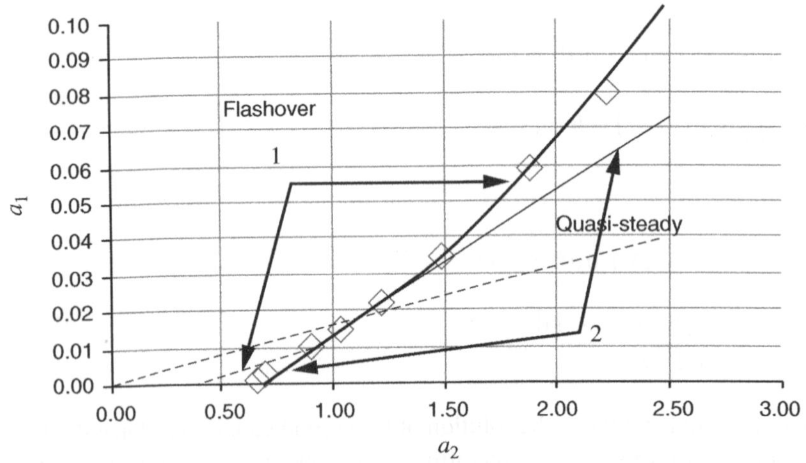

Figure 6.7 Comparison of flashover regions between the Graham *et al.* (1995, 1999) and the Novozhilov (2010) models. $\theta_s^{cr} = 1.9$; $Bi/Bi_R = 27.4$. 1 – the critical boundary according to the Graham *et al.* (1995, 1999) model; 2 – the boundary for the region described by the second of the inequalities (6.48). Dashed line – the boundary for the region described by the third of the inequalities (6.48). Bold curve – the critical boundary according to the Novozhilov (2010) model. Open diamond – the critical boundary for the Novozhilov (2010) model obtained by numerical simulation. *Source:* Novozhilov (2010) / Springer Nature.

There are two qualitatively different possible regimes of behaviour of solutions of the set of Eq. (6.37): either both variables θ and θ_s grow infinitely or both of them approach asymptotically finite values. This fact follows easily from the structure of the set of Eq. (6.37).

Flashover, according to the imposed ignition criterion, occurs at such values (a_1, a_2) that either

1) θ grows infinitely with time, or
2) θ approaches such value $\theta_{\lim} = \lim_{\tau \to \infty} \theta(\tau)$ that the corresponding value $\theta_s^{\lim}$ exceeds ignition threshold, that is, $\theta_s^{\lim} = \lim_{\tau \to \infty} \theta_s(\tau) > \theta_s^{cr}$.

Since surface temperature approaches finite value, temperature profile in the solid material flattens out asymptotically

$$Bi_R\left(\theta_{\lim}^4 - \left[\theta_s^{\lim}\right]^4\right) - Bi\left(\theta_s^{\lim} - 1\right) \underset{\tau \to \infty}{\to} 0 \tag{6.41}$$

or

$$\theta_{\lim}^4 - \left[\left[\theta_s^{\lim}\right]^4 + \frac{Bi}{Bi_R}\left(\theta_s^{\lim} - 1\right)\right] \underset{\tau \to \infty}{\to} 0 \tag{6.42}$$

Therefore

$$\theta_s^{\lim} > \theta_s^{cr} \text{ if } \theta_{\lim} > \sqrt[4]{\left[\theta_s^{cr}\right]^4 + \frac{Bi}{Bi_R}\left(\theta_s^{cr} - 1\right)} \tag{6.43}$$

Note that $\theta_s^{cr} > 1$, and therefore $\theta_{\lim} > 1$.

The two cases (1) and (2), specified earlier, can be described in terms of analysis of the first of the set of Eqs. (6.37). This equation can be rewritten (recalling the case $a_3 = a_4 = 0$ is currently being considered) as

$$\frac{d\theta}{d\tau} = F(\theta); F(\theta) = a_1\left(\theta^4 - 1\right) - a_2(\theta - 1) + 1\,(\theta \geq 1) \tag{6.44}$$

At the left boundary of the interval of interest, $F(\theta) = F(1) = 1 > 0$.

Case (1) corresponds to the absence of critical points of $F(\theta)$ (Eq. (6.44)). This case is described in terms of critical conditions (6.27), that is, case (1) occurs if

$$a_1 > \widehat{f}(a_2) \tag{6.45}$$

where the function $a_1 = \widehat{f}(a_2)$ (the critical boundary) is defined by (6.27).

If case (2) occurs, then $\theta_{\lim}$ will be equal to the least root of the polynomial $F(\theta)$ on $1 \leq \theta < \infty$. Therefore, there must exist such a root, and this root must be greater than $\sqrt[4]{\left[\theta_s^{cr}\right]^4 + \frac{Bi}{Bi_R}\left(\theta_s^{cr} - 1\right)}$. This happens if and only if the inequalities

$$\begin{aligned} &a_1 < \widehat{f}(a_2); F\left(\sqrt[4]{\left[\theta_s^{cr}\right]^4 + \frac{Bi}{Bi_R}\left(\theta_s^{cr} - 1\right)}\right) > 0 \\ &\frac{dF}{d\theta}\left(\sqrt[4]{\left[\theta_s^{cr}\right]^4 + \frac{Bi}{Bi_R}\left(\theta_s^{cr} - 1\right)}\right) < 0 \end{aligned} \tag{6.46}$$

are satisfied simultaneously.

When (6.45) and (6.46) are combined, the region of parameters (a_1, a_2) leading to flashover is described by the following conditions:

$$a_1 > \widehat{f}(a_2) \tag{6.47}$$

or

$$\begin{cases} a_1 < \widehat{f}(a_2) \\ a_1\left(\left[\theta_s^{cr}\right]^4 + \dfrac{Bi}{Bi_R}(\theta_s^{cr} - 1) - 1\right) - \\ - a_2\left(\sqrt[4]{\left[\theta_s^{cr}\right]^4 + \dfrac{Bi}{Bi_R}(\theta_s^{cr} - 1)} - 1\right) + 1 > 0 \\ 4a_1\left(\left[\theta_s^{cr}\right]^4 + \dfrac{Bi}{Bi_R}(\theta_s^{cr} - 1)\right)^{3/4} - a_2 < 0 \end{cases} \tag{6.48}$$

For a fixed ignition temperature, the conditions (6.47, 6.48) depend on the three parameters, namely, a_1, a_2 and Bi/Bi_R, and therefore the critical boundary can generally be written, for example, as

$$a_1 = f\left(a_2, \frac{Bi}{Bi_R}\right) \tag{6.49}$$

where

$$\frac{Bi}{Bi_R} = V^{-1}(\varepsilon_0, \varepsilon_1)\frac{h_s}{\sigma T_0^3} \tag{6.50}$$

The obtained result may be interpreted on the plane of the governing parameters (a_1, a_2). Realistic values of the Biot numbers Bi and Bi_R may be estimated from typical emissivities and correlation for a natural convective flow above a heated horizontal plate (Incropera and DeWitt 2002). The values are taken as $Bi = 4.0 \cdot 10^{-2}$ and $Bi_R = 1.46 \cdot 10^{-3}$, so that $\dfrac{Bi}{Bi_R} \approx 27.4$.

The flashover region and the critical boundary are plotted in Figure 6.7.

The region described by the inequality (6.47) lies to the left from the critical boundary of the Graham *et al.* (1995, 1999) model (6.28, 6.47) (curve 1, Figure 6.7). The region described by the first of the inequalities (6.48) lies to the right of this curve. The second of the conditions (6.48) is satisfied above line 2, Figure 6.7. Finally, the third of the conditions (6.48) is satisfied below the dashed line, Figure 6.7.

As a result, the critical flashover boundary, predicted by the conditions (6.47, 6.48), is shown by the bold curve in Figure 6.7. The region of controlling parameters leading to flashover lies above this curve.

Note few basic properties of Figure 6.7.

It is easy to demonstrate that the boundaries of the second, the third conditions in (6.48) and the critical boundary of Graham *et al.* (1995, 1999) model intersect at a single point. Indeed, at the point of intercession between the boundaries of the second and third inequalities in (6.48), exactly the same conditions are satisfied as the critical conditions defined by Graham *et al.* (1995, 1999) (curve 1, Figure 6.7).

Further, by calculating the tangent of the critical curve 1 from conditions (6.27), it becomes evident that the boundary of the region described by the second of conditions (6.48) has the same slope at the intersection point as the Graham *et al.* (1995, 1999) critical boundary.

The above results mean that the critical boundary for the Novozhilov (2010) model consists of the two sections: the section of the boundary described by the second of conditions (6.48) for smaller values of parameter a_1 and the part of curve 1 (the critical boundary for the Graham *et al.* (1995, 1999) model) for larger values of a_1. The two sections join smoothly at the intersection point.

Figure 6.7 thus shows the difference between flashover conditions predicted by the Graham *et al.* (1995, 1999) and the Novozhilov (2010) models. Results of numerical integration of the set of Eq. (6.38) are also shown and confirm the correctness of the critical boundary plotted in Figure 6.7 for the Novozhilov (2010) model.

As may be expected, Novozhilov (2010) model predicts a wider range of parameters leading to flashover, compared with the Graham *et al.* (1995, 1999) model. Therefore, the Graham *et al.* (1995, 1999) model underestimates the risk of flashover. The difference between the two model predictions is seen at the lower left part of a plot in Figure 6.7 (i.e. for small values of a_1 and a_2). This region is magnified in Figure 6.8.

The significance of the difference between the two models is due to the fact that the coefficients a_1, a_2 are inversely proportional to the HRR of the self-sustained fire (Eq. (6.17)). The magnitude of the underestimation by the Graham *et al.* (1995, 1999) model can be analyzed based on Figure 6.8.

With all the other parameters fixed, the parameters a_1, a_2 can be written as functions of $\tilde{\dot{Q}}_0$:

$$a_1 = \frac{k_1}{\tilde{\dot{Q}}_0}; a_2 = \frac{k_2}{\tilde{\dot{Q}}_0} \tag{6.51}$$

and therefore

$$a_1 = \frac{k_1}{k_2} a_2 \tag{6.52}$$

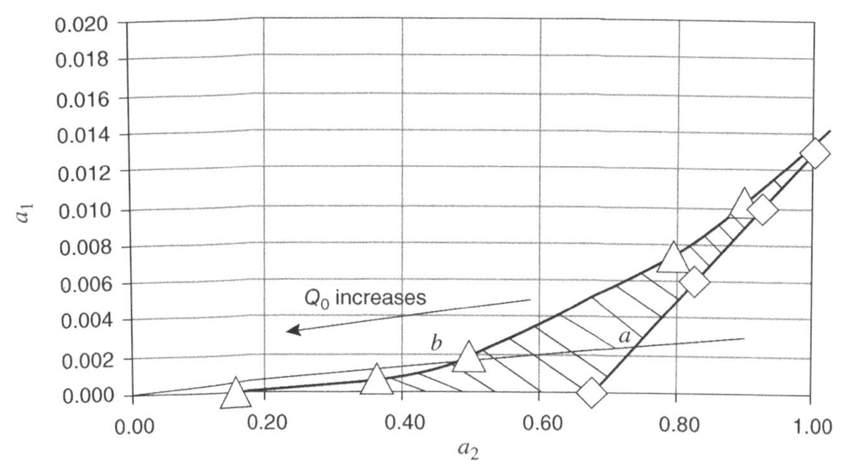

Figure 6.8 Comparison of flashover regions between the Graham *et al.* (1995, 1999) and Novozhilov (2010) models (magnified). $\theta_s^{cr} = 1.9$; $Bi/Bi_R = 27.4$. Open triangle – the critical boundary according to the Graham *et al.* (1995, 1999) model. The flashover region is to the left of the open triangle curve. Open diamond – the critical boundary according to the Novozhilov (2010) model. The flashover region is to the left of the open diamond curve. The dashed region represents the difference in predicted flashover regions between the two models. Line *ab* – the linear relation between a_1 and a_2 parameterized by $\tilde{\dot{Q}}_0$ (with all the other parameters fixed). *Source:* Reproduced with permission from Novozhilov (2010) / Springer Nature.

The line (6.52) is shown in Figure 6.8; $\tilde{\dot{Q}}_0$ increases in the direction of decreasing (a_1, a_2). As the HRR increases, the critical boundaries of the Novozhilov (2010) and the Graham *et al.* (1995, 1999) models are intersected by the line (6.52) at points a and b, respectively. The actual slope of the line (6.52) depends on the model parameters. In particular, if $k_2 \to \infty$, then $a_2(a) \approx 0.7$, $a_2(b) \approx 0.2$. Therefore, the difference in critical HRRs is $\tilde{\dot{Q}}_0(b) - \tilde{\dot{Q}}_0(a) \approx k_2 \left[\frac{1}{0.2} - \frac{1}{0.7}\right] \to \infty$. Therefore, the overestimation of the critical HRR by the Graham *et al.* (1995, 1999) model may be arbitrarily large. This is obviously unacceptable from the Fire Safety Design point of view.

A particularly important difference is that the Novozhilov (2010) model may predict flashover for $a_1 = 0$ (or arbitrarily small value of a_1) and finite value of a_2. In the Graham *et al.* (1995, 1999) model, on the other hand, $a_1 \to 0$ requires also $a_2 \to 0$ for flashover. The term involving parameter a_1 (Eqs. (6.22) and (6.24)) describes the difference between heat gains in the smoke layer, caused by burning rate enhancement, and radiative heat losses form the smoke layer. At the limit $a_1 \to 0$, the upper layer temperature stabilizes, and only heat generation from the initial (non-radiation enhanced) burning rate determines the stabilization temperature. The latter is given by $\theta = 1 + \frac{1}{a_2}$. Ignition of virgin fuel, on the other hand, is determined by this asymptotic temperature and the two Biot numbers, Bi and Bi_R. For certain, sufficiently small values of a_2, ignition of virgin fuel occurs. In contrast, in the Graham *et al.* (1995, 1999) model, smoke temperature stabilization is always interpreted as a quasi-steady fire. In situations with insufficient heat losses from the compartment, however, the limiting smoke temperature, $\theta = 1 + \frac{1}{a_2}$, may be high enough to cause radiative feedback sufficient for ignition. Therefore, the Novozhilov (2010) model describes real conditions for flashover far more accurately than the Graham *et al.* (1995, 1999) model.

Figure 6.9 shows temperature histories for a number of different cases. Each of the cases is characterized by a pair of curves, one corresponding to the upper layer temperature (upper curves), and another to the solid fuel surface temperature (lower curves). Pair 3 corresponds to conditions close to critical; the solid-phase surface temperature approaches the ignition temperature asymptotically.

The system behaviour may be described more conveniently on phase-space diagram involving both state variables. Figure 6.10 presents several phase trajectories for the different values of the

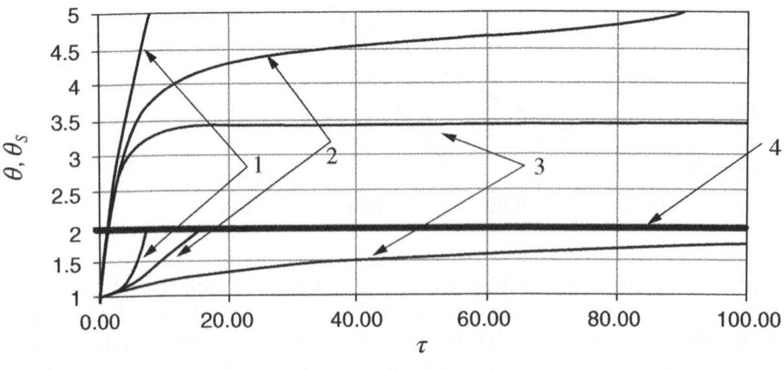

Figure 6.9 Temperature time histories for the smoke layer and the virgin fuel surface. $\theta_s^{cr} = 1.9$; $Bi = 4.0 \cdot 10^{-2}$; $Bi_R = 1.46 \cdot 10^{-3}$. 1 – $a_1 = 0.001$, $a_2 = 0.3$; 2 – $a_1 = 0.001$, $a_2 = 0.4$; 3 – $a_1 = 0.001$, $a_2 = 0.47$; 4 – bold horizontal line: ignition criteria $\theta_s = \theta_s^{cr}$. *Source:* Reproduced with permission from Novozhilov (2010) / Springer Nature.

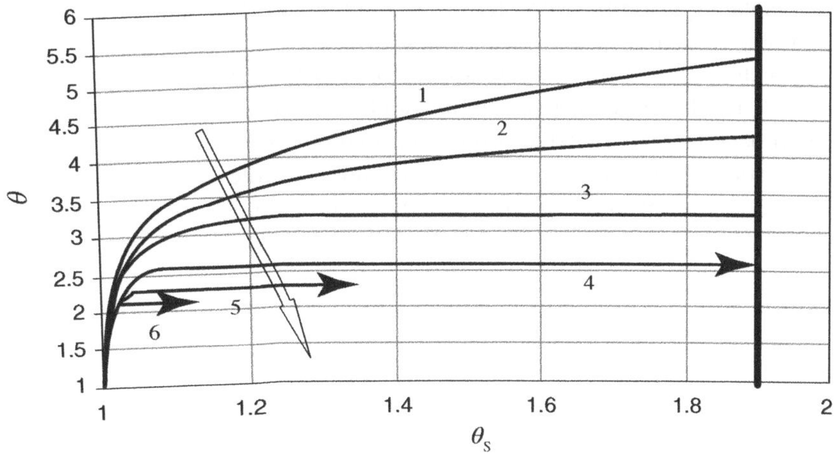

Figure 6.10 Phase space of the state variables: upper layer temperature θ and virgin fuel surface temperature θ_s and phase trajectories for several cases. $\theta_s^{cr} = 1.9$; $Bi = 4.0 \cdot 10^{-2}$; $Bi_R = 1.46 \cdot 10^{-3}$. 1 – $a_1 = 0.001$, $a_2 = 0.2$; 2 – $a_1 = 0.001$, $a_2 = 0.4$; 3 – $a_1 = 0.001$, $a_2 = 0.5$; 4 – $a_1 = 0.001$, $a_2 = 0.66$; 5 – $a_1 = 0.001$, $a_2 = 0.8$; 6 –$a_1 = 0.001$, $a_2 = 0.9$; bold vertical line: ignition criteria $\theta_s = \theta_s^{cr}$. Hollow arrow indicates direction of increase for the parameter a_2. *Source:* Reproduced with permission from Novozhilov (2010) / Springer Nature.

parameters (a_1, a_2). The fuel ignition temperature is represented by the solid vertical line. Arrows in Figure 6.10 indicate limiting points of trajectories as $\tau \rightarrow \infty$.

Flashover occurs upon intersection of a phase trajectory with the critical temperature line (bold vertical line, Figure 6.10). Critical conditions correspond to the trajectory that approaches the critical temperature line asymptotically in the normal direction (curve 4, Figure 6.10).

Note that the same critical conditions apply for the case of thermally thin fuel. Indeed, all the arguments presented above with respect to the derivation of those conditions will still be valid for such a case. Of course, although the critical conditions remain the same, a thermally thin fuel load will result in a much shorter induction period of flashover.

The model developed by Novozhilov (2010) is much more realistic, compared to other currently existing models, such as the Graham *et al.* (1995, 1999) model, and others. Ignition of fuel by radiative fluxes is an undisputable and routinely observed mechanism of flashover; therefore, this phenomenon must be taken into account.

Comparison of the Novozhilov (2010) model with experiment requires, however, much more accurate experimental data on flashover than is currently available.

Several possible generalizations of the model are discussed by Novozhilov (2010). These include the case of non-uniform fuel properties and the case of under-ventilated fires.

6.4 Pseudo – Three-Variable Models and Other Results in Thermal Explosion Modelling of Fire Flashover

The two state variables (the upper layer temperature and the solid fuel surface temperature), proposed by Novozhilov (2010), form a natural set of state variables. They are fully sufficient for fire flashover modelling from the thermal explosion standpoint. Critical conditions involve one or both of these two variables: the smoke layer temperature may grow infinitely and/or fuel surface temperature may reach the ignition threshold.

It is desirable, however, to extend the formulation by Novozhilov (2010) to such a model, where the fuel surface temperature θ_s is a function of both the time and the location on the floor of the compartment. Technical approach to such an extension is discussed below.

It should be noted that an attempt to increase the number of state variables to three has been made by Beard *et al.* (1994–1995) and by Beard (2001,2003,2010). These studies consider the set of formally three ODEs of the form

$$\begin{aligned}\frac{d\tilde{T}}{d\tilde{t}} &= \tilde{F}_1(\tilde{T},\tilde{Z},\tilde{R})\\ \frac{d\tilde{Z}}{d\tilde{t}} &= \tilde{F}_2(\tilde{T},\tilde{Z},\tilde{R})\\ \frac{d\tilde{R}}{d\tilde{t}} &= \tilde{F}_3(\tilde{T},\tilde{Z},\tilde{R})\end{aligned} \tag{6.53}$$

where $\tilde{T}$ is, as usual, the temperature of the upper layer, $\tilde{Z}$ is the upper layer thickness and $\tilde{R}$ is the fire radius.

However, the function $\tilde{F}_3$ is chosen as an *a priori* known empirical function of the radius only $\tilde{F}_3(\tilde{T},\tilde{Z},\tilde{R}) \equiv \tilde{F}_3(\tilde{R})$, so the third of Eq. (6.53) is trivially integrable. In essence, this means that a particular law of fire radius growth with time is assumed. This law is completely decoupled from the properties of the smoke layer, while this very coupling is responsible for flashover. Moreover, the law assumed by Beard *et al.* (1994–1995), Beard (2001,2003,2010) and Bishop *et al.* (1993) is such that fire radius grows up to the maximum size of combustible material on the floor of the compartment. This means that all the above-listed models predict that flashover always occurs eventually due to full involvement of entire combustible load. Needless to say, such a prediction is unphysical, incorrect and unsatisfactory.

As far as the thickness of the smoke layer $\tilde{Z}$ is concerned, it should be noted that this is really an auxiliary variable used to find the upper layer temperature. This variable is not suitable to describe critical conditions for flashover. For example, achievement of the maximum value of smoke layer thickness does not automatically constitute flashover.

The above are the reasons why the two-variable (upper layer temperature and fuel surface temperature) model developed by Novozhilov (2010) is the natural, consistent and complete model required to predict flashover. Criticality may occur through the behaviour of one, or both of these two variables.

Nevertheless, a simple extension of this model, making fuel surface temperature a function of both the time and the spatial location (on the floor of the compartment) $\theta_s = \theta_s(y, \tau)$, is desirable. This requires that ceiling jet temperature is also a function of the spatial location $\theta = \theta(y, \tau)$.

Such an extension can be made along the following lines (Novozhilov 2019).

Instead of assuming uniform temperature of the upper layer, this temperature is allowed to vary in the longitudinal, that is, along the ceiling (Figure 6.3) y-direction of the ceiling jet. Still, there is no variation in the transversal direction (i.e. across the thickness of the upper layer).

Longitudinal temperature variation in the steady-state ceiling jet may be predicted by the theoretical model of Tanaka and Novozhilov (2019). Their investigation shows an exponential decay of temperature downstream of the ceiling jet, which also conforms with the experiments (Tanaka and Novozhilov 2019) and the results of other studies. Specifically (in the far field, beyond the point $\tilde{y} = \frac{\tilde{W}}{2}$, where $\tilde{W}$ is the compartment width) (Tanaka and Novozhilov 2019)

$$\tilde{T}(\tilde{y}) = \tilde{T}_0 + \left(\tilde{T}\left(\frac{\tilde{W}}{2}\right) - \tilde{T}_0\right)\exp\left[-\frac{\tilde{h}\tilde{P}}{\tilde{\dot{m}}\tilde{c}_p}\left(\tilde{y} - \frac{\tilde{W}}{2}\right)\right], \quad \tilde{y} \geq \frac{\tilde{W}}{2} \tag{6.54}$$

where $\tilde{h}$ is the heat transfer coefficient for losses to compartment walls and $\tilde{P}$ is the part of compartment cross-sectional perimeter corresponding to locations of contact between walls and hot smoke.

Mass flow rate in fire plume $\tilde{\dot{m}}$ is given by

$$\tilde{\dot{m}} = \frac{(1-\gamma_r)\tilde{\dot{Q}}_0}{\tilde{c}_p\left(\tilde{T}(0) - \tilde{T}_0\right)} \tag{6.55}$$

The results (6.54) and (6.55) are applicable to steady-state ceiling jet.

Ceiling jet temperature distribution in the near field, closer to fire source $0 \leq \tilde{y} \leq \frac{\tilde{W}}{2}$, may be obtained from the well-known Alpert correlation (Evans 1995). This includes the temperature $\tilde{T}\left(\tilde{W}/2\right)$ required in the relationship (6.54).

For the purpose of the Novozhilov (2010) flashover model extension, one needs to consider the progressive involvement of virgin fuel in the fire. Fire size would grow with time, and the relationships (6.54) and (6.55) need to be modified.

Given the general nature of experimentally verified dependence (6.54), it is logical to assume the following temperature distribution in longitudinal direction of the ceiling jet:

$$\tilde{T}(\tilde{y},\tilde{t}) = \begin{cases} \tilde{T}(0,\tilde{t}), & 0 \leq \tilde{y} \leq \tilde{y}^{cr}(\tilde{t}) \\ \tilde{T}_0 + \left(\tilde{T}(\tilde{y}^{cr}(\tilde{t})) - \tilde{T}_0\right)\exp\left[-\frac{\tilde{h}\tilde{P}}{\tilde{\dot{m}}\tilde{c}_p}(\tilde{y} - \tilde{y}^{cr}(\tilde{t}))\right], & \tilde{y} \geq \tilde{y}^{cr}(\tilde{t}) \end{cases} \tag{6.56}$$

Here $\tilde{y}^{cr}$ is the coordinate of the edge of the fire region

$$\tilde{y}^{cr}(\tilde{t}) = \max_{\tilde{y}}\left[\tilde{y} : \tilde{T}_s(\tilde{y},\tilde{t}) = \tilde{T}_s^{cr}\right] \tag{6.57}$$

The temperature $\tilde{T}(0,\tilde{t})$ in the relationship (6.56) is obtained from the zone model, that is, (in non-dimensional form) from the first of Eq. (6.38). It is assumed that temperature in the longitudinal direction of the ceiling jet adapts quickly to the changing values of the fire region $\tilde{y}^{cr}(\tilde{t})$ and the jet temperature over that region $\tilde{T}(0,\tilde{t})$.

Surface temperature of the solid material $\tilde{T}_s(\tilde{y},\tilde{t})$ is obtained (in non-dimensional form) from the equation similar to the second of Eq. (6.38). This modified equation is discussed below.

Further, the mass flow rate in the plume becomes the function of time

$$\tilde{\dot{m}} = \frac{(1-\gamma_r)\left[\tilde{\dot{Q}}_0 + \tilde{W}\tilde{\dot{m}}_f''\tilde{Q}\tilde{y}^{cr}(\tilde{t})\right]}{\tilde{c}_p\left(\tilde{T}(0,\tilde{t}) - \tilde{T}_0\right)} \tag{6.58}$$

where $\tilde{\dot{m}}_f''$ is the mass burning rate, per unit area, of the ignited fuel and $\tilde{Q}$ is the heat of combustion of materials in the compartment.

Besides, due to the fact that temperature in the ceiling jet becomes non-uniform, the expression for the heat flux $\dot{q}_{rad}''$ needs to be modified.

The relevant expression, which replaces the expression (6.32) for the heat flux at any given compartment floor position $0 \leq \tilde{y} \leq \tilde{L}$, is

$$\tilde{\dot{q}}''_{rad}(\tilde{y},\tilde{t}) = -\varepsilon\tilde{\sigma}\frac{1}{\tilde{\tilde{L}}}\int_0^{\tilde{L}} F(\tilde{\xi},\tilde{y})\left(\tilde{T}^4 - \tilde{T}_s^4\right)d\tilde{\xi} \tag{6.59}$$

where $F(\tilde{\xi},\tilde{y})$ is the view factor between the strips of the width $\tilde{W}$, with infinitely small lengths, positioned at the horizontal locations $0 \le \tilde{\xi} \le \tilde{L}$ and $0 \le \tilde{y} \le \tilde{L}$, on the bottom and the upper boundaries of the compartment, respectively.

The latter view factor is constructed from the 'kernel' $\tilde{F}_k(\tilde{\xi},\tilde{y})$

$$\tilde{F}_k(\tilde{w}_1,\tilde{w}_2) = \frac{\cos(\theta_1)\cos(\theta_2)}{\pi\tilde{r}^2} \tag{6.60}$$

representing the 'density' of the view factor between the two infinitesimal elements, positioned at the locations (along the above described strips) $0 \le \tilde{w}_1 \le \tilde{W}$, $0 \le \tilde{w}_2 \le \tilde{W}$, at the distance $\tilde{r}$ from each other (Ozişik 1985; Modest 1993).

The construction has the form

$$F(\tilde{\xi},\tilde{y}) = \int_0^{\tilde{W}}\int_0^{\tilde{W}} \tilde{F}_k d\tilde{w}_1 d\tilde{w}_2 \tag{6.61}$$

In non-dimensional form (using the same scaling factors as in Sections 6.2 and 6.3)

$$\dot{q}''_{rad} = -\frac{\varepsilon_1\tilde{\sigma}\tilde{\kappa}\tilde{t}_0\tilde{T}_0^4}{\tilde{\dot{Q}}_0}\int_0^1 F(\varsigma,y)\left(\theta^4 - \theta_s^4\right)d\varsigma \tag{6.62}$$

Extended model can now be written, by modifying the formulation (6.38), in the form

$$\begin{aligned}
&\frac{d\theta_2}{d\tau} = 1 + a_1(\tau)\left([\theta_2]^4 - 1\right) - a_2(\tau)(\theta_2 - 1) + a_3(\tau)[\theta_2]^3 + a_4(\tau)[\theta_2]^2\\
&0 \le \tau \le \infty\\
&\theta_s(y,\tau) - 1\\
&= \sqrt{\frac{\tau}{\pi}}\int_0^1 \frac{\left[Bi_R^{(1)}\int_0^1 F(\varsigma,y)\left([\theta(\varsigma,s)]^4 - [\theta_s(\varsigma,s)]^4\right)d\varsigma - Bi(\theta_s(y,s) - 1)\right]}{\sqrt{1-s}}ds;\\
&y^{cr} \le y \le L\\
&\theta_s \equiv \theta_s^{cr}; 0 \le y \le y^{cr}(\tau)\\
&y^{cr}(\tau) = \max_{\tilde{y}}\left[y : \theta_s(y,\tau) = \theta_s^{cr}\right]\\
&\theta = \theta_2(\tau); 0 \le y \le y^{cr}(\tau)\\
&\theta(y,\tau) = 1 + (\theta_2(\tau) - 1)\exp\left[-\wp_0\frac{(\theta_2(\tau) - 1)}{[Q + \wp_1 y_{cr}(\tau)]}(y - y_{cr}(\tau))\right];\\
&y^{cr} \le y \le L\\
&A_f(0) = A_f^0 > 0; y^{cr}(0) = \sqrt{A_f^0}\\
&\theta_s(0) = \begin{cases}\theta_s^{cr}; 0 \le y \le y^{cr}(\tau)\\ 1; y^{cr} \le y \le L\end{cases}
\end{aligned} \tag{6.63}$$

where

$$Bi_R^{(1)} = \frac{\varepsilon_1 \sqrt{\tilde{\kappa}\tilde{t}_0}\tilde{\sigma}\tilde{T}_0^3}{\tilde{\lambda}}; L = \frac{\tilde{L}}{\sqrt{\tilde{\kappa}\tilde{t}_0}};$$
$$\wp_0 = \frac{\tilde{h}\tilde{P}\tilde{T}_0\sqrt{\tilde{\kappa}\tilde{t}_0}}{(1-\gamma_r)\tilde{\dot{Q}}_0}; \wp_1 = \frac{\tilde{W}\tilde{\dot{m}}_f''\tilde{Q}_c\sqrt{\tilde{\kappa}\tilde{t}_0}}{\tilde{\dot{Q}}_0} \tag{6.64}$$

The first of the above parameters, $Bi_R^{(1)}$, is the modified radiative Biot number (compare to (6.33)). The set of Eq. (6.63) has the following structure.

Both the ceiling jet temperature and the combustible material surface temperature are the functions of the longitudinal coordinate (i.e. along the compartment length) and time. The variable $y(\tau)$ is the edge of the fire area, so that the latter grows with time as $\tilde{A}_f = \left(\tilde{y}^{cr}(\tilde{t})\right)^2$. Consequently, the coefficients a_i in the first of Eq. (6.63) become functions of time $a_i = a_i(\tau)$, and corresponding changes must be introduced into the copies (from Section 6.2) of all the formulae for the coefficients a_i.

In the 'core' region of fire plume $0 \le y \le y^{cr}$, the ceiling jet and the material surface temperatures are assumed to be spatially constant. The first of them is equal to the solution of the first of Eq. (6.63), while the second is also constant in time and always equal to the critical ignition temperature θ_s^{cr}.

The first of the above assumption (regarding ceiling jet temperature) admits various formulations. For example, one can assume half-sum of ceiling jet temperatures at both ends of the compartment (i.e. at $y = 0$ and $y = y^{cr}$) to be equal to the solution of the first of Eq. (6.63). Ultimately, refinement of this assumption is a matter of comparison between prediction of the model and available experimental data.

It is important to note that the first and the second of Eq. (6.63) are two-ways coupled. This is a fundamental difference from Eqs. (6.38) of the original Novozhilov (2010) model, which are only one-way coupled.

As fire progresses, fire area and leading edge of fire region $y^{cr}(\tau)$ grow, and, at some conditions, the equality $\lim_{\tau \to \infty} y^{cr}(\tau) = L$ may be satisfied. Such conditions would represent critical conditions for flashover. In this way, using the arguments similar to those in Section 6.3, critical conditions may be written in explicit form although the analysis becomes much more involved.

Obviously, a predicted induction period of flashover will be affected by the model modification.

It is helpful to point out a necessary condition for initiation of ignition wave and, therefore, for flashover.

In order for combustion wave to be initiated, the radiative flux emitted by smoke layer must reach, at least at the initial fire location $y = 0$, a level sufficient for ignition of the virgin material.

The latter condition is met if there exists such $\tau^* < \infty$; that $\frac{d\theta_2}{d\tau} = 1 + a_1(\tau)\left([\theta_2]^4 - 1\right) - a_2(\tau)(\theta_2 - 1) + a_3(\tau)[\theta_2]^3 + a_4(\tau)[\theta_2]^2$;

$$0 \le \tau \le \tau^* \tag{6.65}$$

$$\theta_s^{cr}(y^{cr}(0), \tau^*) - 1 = \sqrt{\frac{\tau^*}{\pi}} \times$$
$$\times \int_0^1 \frac{\left[Bi_R^{(1)} \int_0^1 F(\varsigma, y^{cr}(0))\left([\theta(\varsigma, s)]^4 - [\theta_s(y^{cr}(0), s)]^4\right) d\varsigma - Bi(\theta_s(y^{cr}(0), s) - 1)\right]}{\sqrt{1-s}} ds$$

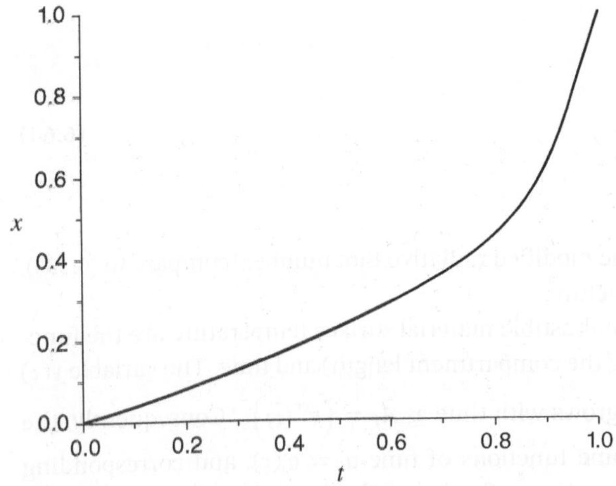

Figure 6.11 Estimated non-dimensional position of ignition front as a function of non-dimensional time. Fire Research Station (FRS, UK) experiment (Novozhilov 2019). *Source:* Reproduced with permission from Novozhilov (2019).

The extended model, in the form (6.63) and (6.64), may be validated against available experimental evidence.

Experimental data on flashover are rare due to necessity to create relatively well-controlled large-scale fires. Examination of available evidence reveals that ignition process during flashover occurs rapidly but not instantaneously. Such conclusion is derived, for example, from experimental tests conducted by the former Fire Research Station (FRS, UK) in the course of replicating the Stardust nightclub fire in Dublin in 1981.

FRS created an elongated compartment (similar to short tunnel with rectangular cross-section) with one end closed. The number of rows of seats was installed on the floor. The initial fire was initiated at the closed end of the compartment. As the initial fire develops and forms smoke layer, an ignition wave spreading through the seat rows was eventually observed. As the wave travelled, the magnitude of fire increased since ignited seats contribute to HRR.

Precise parameters of the experiment are not known. Basic estimation of the speed of propagation of ignition wave may be obtained, nevertheless from the experiment video (http://bufvc.ac.uk/dvdfind/index.php/title/781, Novozhilov 2019). The position of ignition wave front versus time is shown in Figure 6.11. The plot is presented in non-dimensional variables, where distance is normalized by the length of the compartment and time by the overall propagation time inferred from the video. Overall acceleration of ignition wave is evident. In particular, a very strong short-time acceleration occurs as the wave has travelled about a half of compartment length (at $x \approx 0.5$). Propagation up to this point takes majority of the overall wave spread time, $\Delta t \approx 0.85$, compared to just $\Delta t \approx 0.15$ required to spread through the other half-length of the compartment $\approx 0.5 < x \leq 1$.

This plot may be used for direct validation of the extended Novozhilov (2010) model.

7

Thermal Explosion in Granular Reacting Media, Biosolid Fuels and Electric Batteries

7.1 Experimental Data

The present chapter considers thermal explosion in versatile media, which are fundamentally different from homogeneous gas systems discussed in Chapters 2 and 3. The difference concerns both their physical structure and chemical nature.

The necessity of these considerations is mostly dictated by practical reasons, rather than theoretical advances.

Most fuels considered here are in the form of granulated (or porous) media. They form essentially two-phase reacting systems, where solid particles (of various, not necessarily small, sizes and various shapes) are surrounded by air.

Biosolids are a large class of versatile fuels and may include, for example, agricultural and forest residues; sawdust; forest, municipal and domestic waste; energy crops and other sources (Ibitoye *et al.* 2023).

Technically, all the above fuels may be classified as both granular and biomass solids.

For convenience, however, coal (as a fossil fuel) is separated into a dedicated section called 'Thermal Explosion in Granular Reacting Media'.

All the other fuels are discussed as biomass solid fuels (biosolids) (Section 7.3), which both conform better with conventional terminology and emphasize more precisely their nature as reacting systems (rather than being simply described as granular).

A specific separate topic that is also covered by the present chapter is thermal explosion in specific modern engineering devices, namely electrical batteries. These are discussed in Section 7.4.

In contrast to gaseous mixtures that have been known to be self-ignitable for over a century, self-ignition of biosolid fuel has been discovered and analyzed from a scientific point of view relatively recently.

Obtaining meaningful experimental data and its interpretation is a first fundamental step, which is required before any mathematical models of reacting biosolids may be developed.

Chemical kinetics of various gaseous mixtures has been studied scrupulously for decades. On the other hand, data concerning reacting behaviour of biomass solid fuels is much more limited.

Koseki (2011) and Murasawa and Koseki (2015) conducted evaluation of self-heating and self-ignition hazards presented by a number biosolid fuels, such as refuse-derived fuels (RDFs), refuse paper and plastic fuels (RPFs), wood chips and pellets, palm shell fuel and several others, using small samples and various advanced highly sensitive experimental techniques.

The principal finding of these studies was the fact that many biosolid fuels, especially those fabricated from waste, exhibit exothermic transformations due to microbiological activity. The latter

Thermal Explosion: Theory and Application, First Edition. Vasily B. Novozhilov.

comprise aerobic processes of oxygen consumption by microorganisms in the course of their metabolism.

Such exothermic activity is typically observed in the temperature range of 30 – 80 °C, with the peak of heat release at around 60 °C. These figures broadly agree with the earlier study by Rynk (2000). Elevated levels of moisture further promote biologically driven energy release.

Although heat generation rates are pretty small, of the order of milliwatts per gram, they provide essential link from the initial 'cold' state to higher temperatures where conventional oxidation reactions peak up. Such low-temperature 'link' mechanism is missing in purely gaseous mixtures. This is why most of the latter mixtures may be kept safely at low temperatures, and their autoignition temperatures are much higher. It is this additional heat generation mechanism that singles out the class of biomass solid fuel, makes them susceptible to self-ignition and, therefore, presents fire hazard at conventional storage temperatures.

Contribution of microbiological activity into the overall energy release within biosolid fuels has also been studied by a number of other researchers (Fengel and Wegener 1984; Springer and Hajny 1970; De Boer *et al.* 2005). Ferrero *et al.* (2009) performed respirometric tests with wet pine wood chips in order to assess intensity of microbiological activity. Respirometric tests allow microbial metabolic activity to be quantified on the basis of the amount of carbon dioxide production. Ferrero *et al.* (2009) observed rapid increase of CO_2 production, up to its maximum value, within the first 30 – 40 hours of their tests, implying very high metabolic activity and microorganism growth rate in wood samples. A second peak in carbon dioxide production was found in isotheral experiment at 30 °C after about 1,450 hours. This peak appeared as less intensive and was believed to be due either to biodegradation of different wood components or establishment of successional microbial communities or both (De Boer *et al.* 2005). Ferrero *et al.* (2009) also conducted a limited number of large scale pile heating tests, which essentially confirmed their laboratory observations.

Energy generation over the temperature range of 80 – 100 °C is believed to be caused by chemical transformation of fatty acid esters, typically contained in biosolid fuels (Li *et al.* 2006).

For higher temperatures, thermal decomposition reactions take over (Gray *et al.* 1984; Rychlý *et al.* 2004). A number of experimental techniques may be used to quantify kinetics of such reactions. These include, for example, thermogravimetry differential thermal analysis (TG-DTA), spontaneous ignition test (SIT) and calorimetry. Koseki (2011) and Murasawa and Koseki (2015) used a range of highly sensitive calorimeters: C80, MS80 or TAM. Typical data on thermal decomposition reactions and their temperature range is presented in Figures 7.1–7.5.

In particular, Figure 7.5 clearly shows transition to thermal explosion upon increasing ambient temperature. Autoignition temperature was estimated in this case by Koseki (2011) as about 160 °C (for the ambient temperature slightly above this value, explosion induction phase lasts for many hours, as it is evident from Figure 7.5). Curves of thermal explosion development presented in Figure 7.5 are qualitatively identical to predictions of unsteady Semenov model (with non-trivial initial condition, $Zn > 0$).

Thermal explosion of biomass products (leaves and shavings) was also observed in small-scale experiments by Jones and Raj (1988).

Based on the original range of small-scale experiments, as well as results of preceding studies, Koseki (2011) identified the following key stages in self-ignition of biomass solid fuels:

1) Weak heat generation in the range of temperatures 30 – 80 °C.
2) Combustion of fatty acid esters, contained in materials.
3) Combustion of cellulose or other organic materials.

In line with these findings, Murasawa and Koseki (2015) devised the qualitative timeline of self-ignition of biosolid fuels as shown in Figure 7.6.

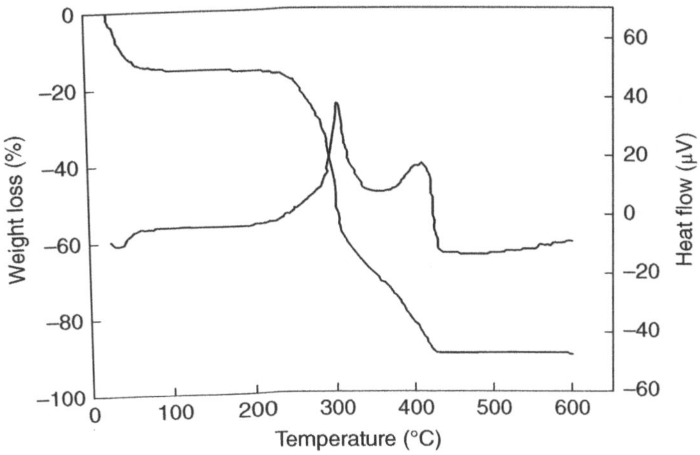

Figure 7.1 TG-DTA curves for wood chips. Sample mass 20 mg; heating rate 2 K/min. *Source:* Koseki (2011) / MDPI / CC BY 4.0.

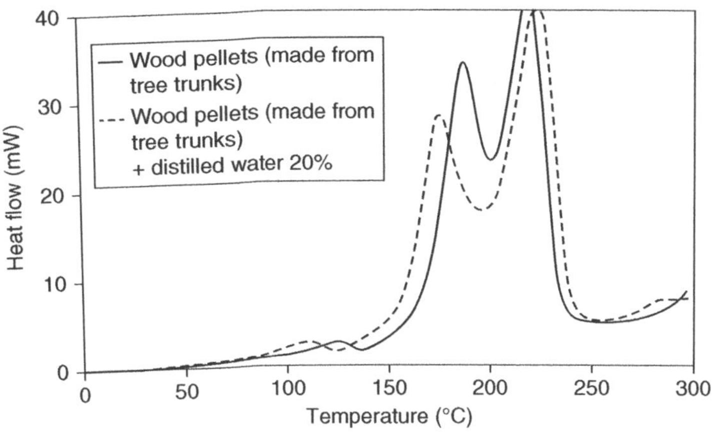

Figure 7.2 C80 calorimetry curves for wood pellets made of tree trunks. *Source:* Murasawa and Koseki (2015) / MDPI / CC BY 4.0.

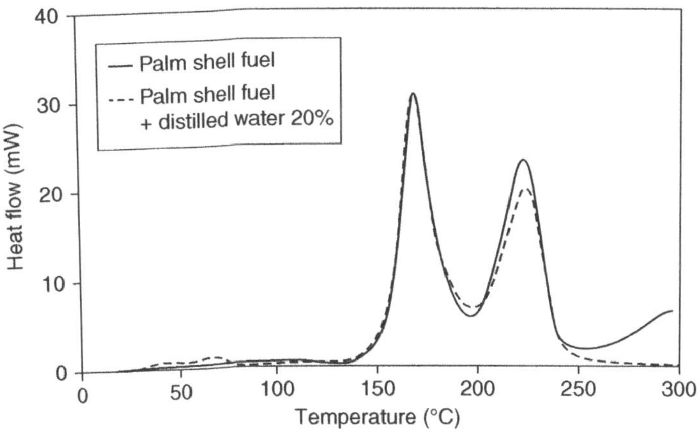

Figure 7.3 C80 calorimetry curves for palm shell fuel. *Source:* Murasawa and Koseki (2015) / MDPI / CC BY 4.0.

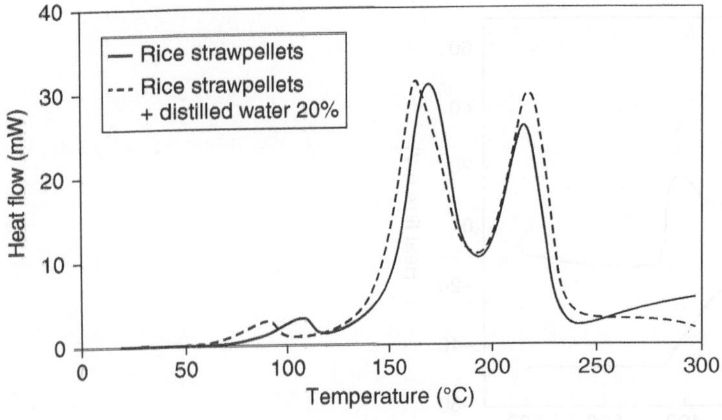

Figure 7.4 C80 calorimetry curves for rice straw pellets. *Source:* Murasawa and Koseki (2015) / MDPI / CC BY 4.0.

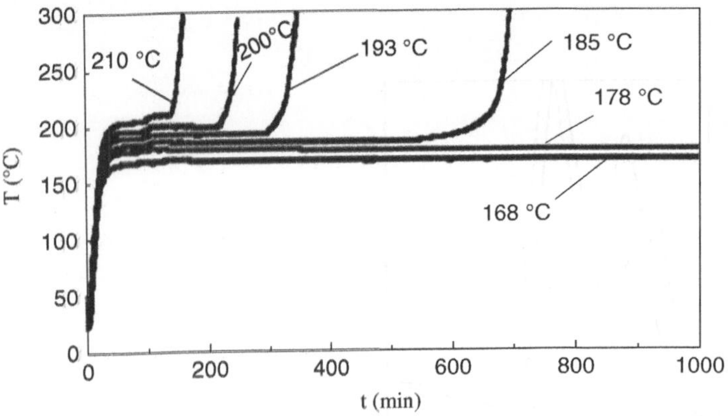

Figure 7.5 SIT results for RPF. *Source:* Koseki (2011) / MDPI / CC BY 4.0.

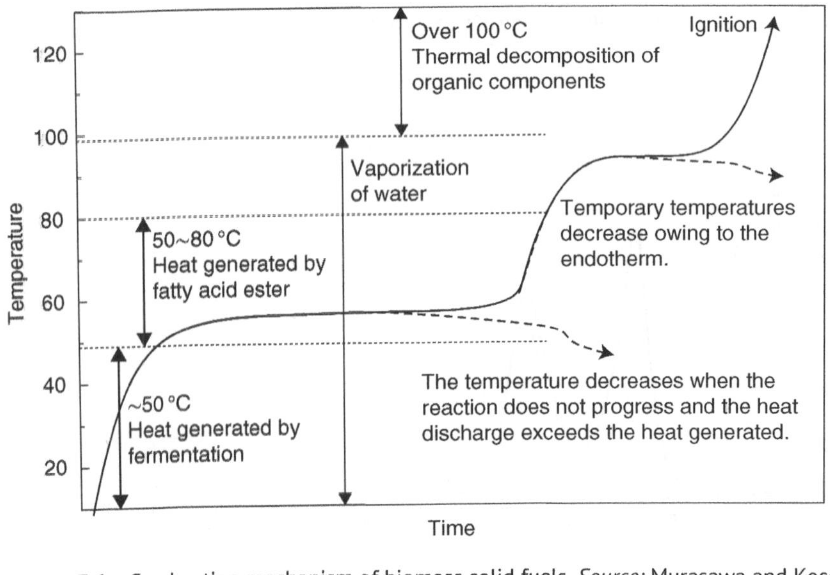

Figure 7.6 Combustion mechanism of biomass solid fuels. *Source:* Murasawa and Koseki (2015) / MDPI / CC BY 4.0.

7.2 Thermal Explosion in Granular Reacting Media

Many fuels exhibiting self-heating and self-ignition (thermal explosion) mechanisms may be stored in granular form. While biosolid mass fuels of such structure are discussed in Section 7.3, the present section focuses essentially on coal. Structural arrangements of this fuel that attracts most attention and concerns, from thermal explosion point of view, are coal stockpiles.

Carras and Young (1994) provided an extensive review on the problem, from both experimental and theoretical (modelling) perspectives. They point out that coal self-heating has a long history of being recognized as a hazard in the industry. Apart from storage in stockpiles, self-ignition hazards also affect the safety of vessels during coal transportation, as well as raise major fire safety concerns in underground mining, due to the self-heating of coal goafs.

Mechanisms, both physical and chemical, driving coal self-heating are complex but, in principle, are related to coal interaction with oxygen.

As coal stockpiles are fully open to the effects of environment, they are subject to complicated, multiscale heat and mass transfer processes, which include mechanisms of conduction, convection and (solar) radiation. Most importantly, these processes are responsible for the transport of oxygen and water vapour into and out of the stockpile. Propagating to small scales, transport of oxygen leads to the phenomenon known as physical adsorption of oxygen (Carras and Young 1994). Physical adsorption of oxygen also affects kinetics of coal oxidation as it directly influences oxygen concentration in the micropores of coal particles, as well as in the inter-particle space.

The other fundamental role of physical heat and mass transfer processes is that they control the heat gain/heat dissipation balance in the stockpile, which ultimately determines whether thermal explosion (self-ignition) occurs.

Propagation of oxygen to fine scales inside the stockpile allows oxidation processes to occur.

As far as the chemistry of oxidation is concerned, the two principal processes occurring are as follows:

1) *Chemical adsorption*. This mechanism leads to the formation of coal-oxygen complexes and oxygenated carbon species. The latter include, for example, carboxyl and carbonyl;
2) *Oxidation*. Oxidation reactions release gaseous products, mostly carbon monoxide, carbon dioxide and water (in the form of vapour).

Processes of chemical interaction between coal and oxygen have been studied extensively in the second half of the last century (Schmidt 1945; Yohe 1958; Nordon *et al.* 1979; Herring 1983; Schmal 1989). Experiments by Nordon *et al.* (1979) demonstrated that the oxidation rate of coal char is governed by both the Arrhenius and the so-called Elovich kinetics. The latter describes the decrease of the rate of sorption with increasing uptake (Khan *et al.* 1990; Carras and Young 1994).

Nordon *et al.* (1979) also provided a full description of the relevant laboratory methods used to study coal kinetics at laboratory scales. On a larger scale, in industrial tests, experimental data on coal self-ignition have been obtained by such methods as crossing point measurements, isothermal and adiabatic calorimetry, oxygen sorption and basket methods (Carras and Young 1994).

Extensive experimental research and accumuled understanding of the major processes controlling coal self-heating allowed a number of numerical models to be developed over time. These models range from simplified one-dimensional treatments to fairly comprehensive three-dimensional CFD models of fluid flow, mass and heat transfer, coupled with an appropriate phenomenological quantitative description of kinetic mechanisms.

Most of the models are fairly complicated and are not discussed here in detail. Examples of such models may be found in Nordon (1979), Brooks and Glasser (1986), Brooks *et al.* (1988), Schmal

(1989), Edwards (1990), Carras and Young (1994), Bradshaw *et al.* (1991), Chen (1992) and Krishnaswamy *et al.* (1996).

Comprehensive models allow the effects of different control parameters on thermal explosion to be investigated. It should be noted, however, that despite a considerable complexity of the described system, key dependencies extracted from those models are remarkably similar qualitatively to predictions of non-steady Semenov and Frank-Kamenetskii theories for gaseous mixtures. Examples can be found, for example, in the study by Nordon (1979). The author presents time histories of temperature within coal stockpiles for different stockpile dimensions and also for different velocities of external airflow. Both figures demonstrate a typical transition to thermal explosion (for pile size lengths exceeding critical and for velocities of external airflow below the critical value). Also, both these 'phase portraits' are qualitatively identical to Figures 2.9 and 2.11 from Chapter 2.

Li *et al.* (2018) used an approach different from the first physical principles utilized in CFD models. Instead, they had an array of sensors feeding real-time information on coal parameters to time series-type forecasting model. The model predicts the parameter that is called by the authors heat loss capacity (HLC) and represents an aggregated characteristic (index) of various factors inducing spontaneous combustion. The models of this type have a clear advantage of the use of real-time data. On the other hand, the form of HLC index proposed by Li *et al.* (2018) seems too simplistic to rely on in predictions of such complicated phenomenon as thermal explosion. The study of Li *et al.* (2018) focuses mostly on purely computational aspects of modelling, and direct comparison of the model with real data on coal self-ignition is still required.

Perhaps the most promising modelling approach, to be developed in the future, would be real-time data-informed CFD modelling.

Research on the spontaneous ignition of coal is actively developing. Valuable additional information may be found in the reviews by Onifade and Genc (2020) and Nelson and Chen (2007). The latter of these reviews covers specifically experimental methods in spontaneous coal combustion research.

7.3 Thermal Explosion of Biosolid Fuels

The present section considers one of the possible scenarios of thermal explosion involving large quantities of biosolid fuels. An analysis essentially follows the study by Novozhilov (2017).

Large volumes of biosolid fuels may typically be found in either dedicated storage facilities or, during transportation, in cargo compartments of ships.

Figure 7.7 shows a sketch of biosolid fuel packed in a large quantity.

Generally, whether it is a cargo compartment, or a storage, some clearance height $\tilde{H} > 0$ may exists between the compartment roof and the top surface of the fuel, with the area $\tilde{S}$.

As explained in Section 7.1, significant masses of biosolid fuels are prone to spontaneous ignition, via the mechanisms (1–4) described in the latter section. Such spontaneous ignition may give rise to the initial fire sketched in Figure 7.7. The later development of such small (initially) fire may lead to a catastrophic thermal explosion. This would have to occur via a flashover scenario considered in Chapter 6.

Correspondingly, the following analysis proceeds essentially along the lines of the latter chapter. There is, however, a significant difference. While the thickness of the smoke layer in a compartment was a fixed constant parameter in the models of Chapter 6, it is predicted dynamically here as a part of the overall problem solution. In this respect, the treatment provided below is more comprehensive than in Chapter 6.

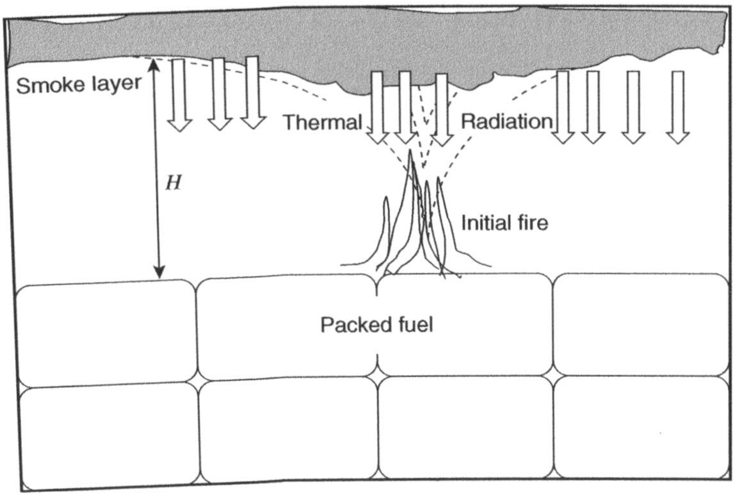

Figure 7.7 Thermal explosion (flashover) scenario involving a large volume of biosolid fuel, for example, in a cargo compartment of a ship. *Source:* Novozhilov (2017) / MDPI / CC BY 4.0.

To achieve the depth of analysis described above, the smoke filling process in the compartment must be described first. This is done using the two well-known and widely used in Fire Science (Karlsson and Quintiere 2022) phenomenological correlations for mass flow rate and centreline excess temperature in the fire plume

$$\tilde{\dot{m}}_p(\tilde{z}) = \beta\left(\frac{\tilde{\rho}_0^2\tilde{g}}{\tilde{c}_p\tilde{T}_0}\right)^{1/3} \cdot \tilde{\dot{Q}}_0^{1/3} \cdot \tilde{z}^{5/3} \tag{7.1}$$

$$\tilde{\theta}_p(\tilde{z}) = \beta^{-1}\left(\frac{\tilde{T}_0}{\tilde{g}\tilde{c}_p^2\tilde{\rho}_0^2}\right)^{1/3} \cdot \tilde{\dot{Q}}_0^{2/3} \cdot \tilde{z}^{-5/3} \tag{7.2}$$

Here, $\tilde{\dot{Q}}_0$ is the initial fire HRR; $\tilde{\rho}_0$, $\tilde{c}_p$, $\tilde{T}_0$ properties of ambient air (density, specific heat and temperature, respectively); $\tilde{g}$ is the gravity acceleration and $\beta \approx 0.02$ is the phenomenologically adjusted parameter.

The correlation (7.1) provides an estimation of the mass flow rate through the cross-section, at the height $\tilde{z}$ above the source of fire, in the fire plume. The correlation (7.2) estimates the excess (i.e. elevation above ambient) temperature at the centreline of the plume, again at the height $\tilde{z}$ above the fire source. Thus, $\tilde{\theta}_p(\tilde{z}) = \tilde{T}_p(\tilde{z}) - \tilde{T}_0$, where $\tilde{T}_p(\tilde{z})$ is the centreline plume temperature.

The following analysis makes use of appropriate scales in order to write governing equations in non-dimensional form. Specifically, non-dimensional variables are introduced as follows:

$$\begin{gathered} y = \frac{\tilde{z}}{\tilde{H}}; \quad \tau = \sqrt{\frac{\tilde{g}}{\tilde{H}}}\frac{\tilde{H}^2}{\tilde{S}}\tilde{t}; \quad p = \frac{\tilde{\mu}}{\tilde{R}\tilde{\rho}_0\tilde{T}_0}\tilde{p}; \quad \rho = \frac{\tilde{\rho}}{\tilde{\rho}_0}; \quad T = \frac{\tilde{T}}{\tilde{T}_0} \\ \dot{Q}_0 = \frac{\tilde{\dot{Q}}_0}{\tilde{c}_p\tilde{\rho}_0\tilde{T}_0\sqrt{\tilde{g}}\tilde{H}^{5/2}}; \quad \dot{m}_p = \frac{\tilde{\dot{m}}_p}{\tilde{\rho}_0\sqrt{\tilde{g}}\tilde{H}^{5/2}} \end{gathered} \tag{7.3}$$

Here $\tilde{R}$ is the universal gas constant, and $\tilde{\mu}$ is the molecular weight.

With the help of the variables (7.3), the correlations (7.1) and (7.2) are written in a very simple non-dimensional form

$$\dot{m}_p(y) = \beta \dot{Q}_0^{1/3} y^{5/3} \tag{7.4}$$

$$\theta_p(y) = \beta^{-1} \dot{Q}_0^{2/3} y^{-5/3} \tag{7.5}$$

with a straightforward consequence (identity)

$$\dot{Q}_0 = \dot{m}_p(y)\theta_p(y) \tag{7.6}$$

for any vertical location y, which has a transparent physical meaning: fire HRR is equal to the enthalpy flow in the plume (radiation losses are neglected).

Using the correlations (7.4) and (7.5), as well as just-introduced scaling (7.3), a set of equations involving averaged properties of the bottom (clean air) and upper (smoke) layers can be derived to predict the smoke-filling process in the compartment.

Below properties pertaining to the bottom layer are marked with subscript '1' and those for the upper layer with subscript '2'.

First of all, it is straightforward to write down equations of state for both layers (assuming the ideal gas law)

$$p = \rho_1(1 + \theta_1) \tag{7.7}$$

$$p = \rho_2(1 + \theta_2) \tag{7.8}$$

Next, equations describing conservation of mass may be obtained. Two independent equations are needed. In the present case, the first of these is written for the compartment as a whole and the second for the bottom layer

$$\rho_1 y + \rho_2(1 - y) = 1 \tag{7.9}$$

$$\frac{d}{d\tau}(\rho_1 y) = -\dot{m}_p(y) \tag{7.10}$$

The latter equation has a sink term corresponding to the removal of air from the bottom layer due to its entrainment into the rising fire plume. The form of this equation is very standard (see, e.g. Karlsson and Quintiere 2022 and Novozhilov 2012). If needed, the equation describing conservation of mass for the upper layer may be obtained from (7.9) and (7.10).

Finally, equations reflecting conservation of energy are due. Again, there is a choice regarding which of the two independent equations to consider. The pair may consist of the equation for the whole compartment and for one of the two layers, or for both the layers separately. One pair of equations can always be transformed to another. In the present case, the equations below are written separately for the bottom and the upper layers

$$\frac{d}{d\tau}(\rho_1\theta_1 y) = \dot{Q}_0 - \dot{m}_p(y)\dot{\theta}_p(y) \tag{7.11}$$

$$\frac{d}{d\tau}(\rho_2(1 - y)\theta_2) = \dot{m}_p(y)\dot{\theta}_p(y) \tag{7.12}$$

In view of the identity (7.6), the last two equations may be rewritten as

$$\frac{d}{d\tau}(\rho_1\theta_1 y) = 0 \tag{7.13}$$

$$\frac{d}{d\tau}(\rho_2(1-y)\theta_2) = \dot{Q}_0 \tag{7.14}$$

Initial conditions for Eqs (7.7–7.10), (7.13) and (7.14) are

$$P(0) = 1; y(0) = 1; \rho_1(0) = 1; \theta_1(0) = 1 \tag{7.15}$$

Speaking formally, the set of initial conditions is not complete. Initial properties of the upper layer are undetermined as this layer initially has zero thickness, and as such, they are missing from (7.15).

Importantly, however, the initial conditions (7.15) determine the solution uniquely, as will be seen from the analysis below.

In order to obtain the required solution for the set of Eqs (7.7–7.10), (7.13) and (7.14), the following steps are necessary.

Equation (7.13) upon integration and taking into consideration initial conditions leads to the (expected, from the point of view of physics) conclusion that the temperature of the bottom layer is preserved $\theta_1 = 0$.

Equations (7.7–7.9) and (7.14) allow the upper layer density and temperature to be excluded, which leaves (together with (7.10)) the set of just two equations for the density of the bottom layer and the smoke interface position. The density equation is integrated first to give solution (7.17). The remaining Eq. (7.10) for the smoke layer interface position, upon substitution of solution (7.17), turns out to be integrable by separation of variables. The result is solution (7.21). The remaining unknowns are obtained in straightforward manner from Eqs (7.7)–(7.9).

The final form of the solution of the set of Eqs (7.7–7.10), (7.13) and (7.14) is

$$p = 1 + \dot{Q}_0\tau \tag{7.16}$$

$$\rho_1 = 1 + \dot{Q}_0\tau \tag{7.17}$$

$$\rho_2 = \frac{\left(1-(1+\dot{Q}_0\tau)\beta^{-3/2}\dot{Q}_0\left[\left(1+\beta^{-1}\dot{Q}_0^{2/3}\right)(1+\dot{Q}_0\tau)^{2/3}-1\right]^{-3/2}\right)}{\left(1-\beta^{-3/2}\dot{Q}_0\left[\left(1+\beta^{-1}\dot{Q}_0^{2/3}\right)(1+\dot{Q}_0\tau)^{2/3}-1\right]^{-3/2}\right)} \tag{7.18}$$

$$\theta_1 = 0 \tag{7.19}$$

$$\theta_2 = \frac{\dot{Q}_0\tau}{\left(1-(1+\dot{Q}_0\tau)\beta^{-3/2}\dot{Q}_0\left[\left(1+\beta^{-1}\dot{Q}_0^{2/3}\right)(1+\dot{Q}_0\tau)^{2/3}-1\right]^{-3/2}\right)} \tag{7.20}$$

$$y = \beta^{-3/2}\dot{Q}_0\left[\left(1+\beta^{-1}\dot{Q}_0^{2/3}\right)(1+\dot{Q}_0\tau)^{2/3}-1\right]^{-3/2} \tag{7.21}$$

Smoke filling dynamics, according to solution (7.16–7.21), is illustrated in Figures 7.8 and 7.9, where the smoke interface position and the temperature of the upper layer are plotted against non-dimensional time.

As mentioned earlier, natural initial conditions for the density ρ_2 and the temperature θ_2 do not exist. However, it may be shown by formal asymptotic analysis that these variables have meaningful finite limits at $\tau \to 0$. This is also evident from the plot of the solution presented in Figure 7.9, combined with consideration of Eq. (7.8).

Further analysis of thermal explosion (flashover), along the lines of the model developed in Chapter 6 (Section 6.3), requires a suitable model for radiative feedback from the smoke layer. Radiation from the smoke layer is driven predominantly by soot.

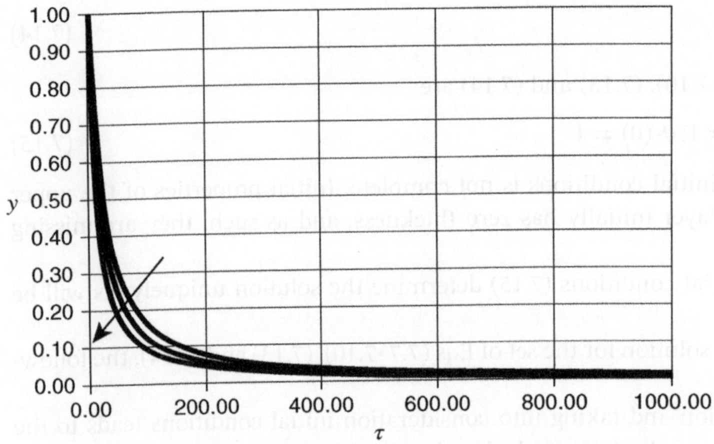

Figure 7.8 Dynamics of the smoke layer interface in non-dimensional variables. The arrow points towards an increase of the non-dimensional HRR for each of the curves: $\dot{Q}_0 = 4.55 \cdot 10^{-3}$; $\dot{Q}_0 = 9.10 \cdot 10^{-3}$; $\dot{Q}_0 = 1.82 \cdot 10^{-2}$. *Source:* Novozhilov (2017) / MDPI / CC BY 4.0.

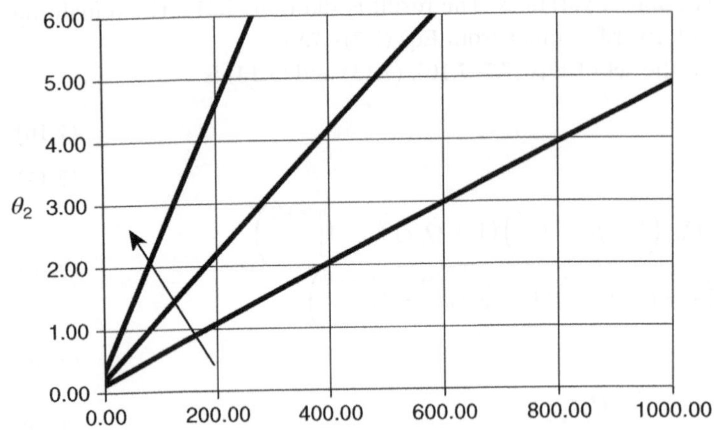

Figure 7.9 Dynamics of the smoke (upper layer) temperature in non-dimensional variables. The arrow points towards an increase of the non-dimensional HRR for each of the curves: $\dot{Q}_0 = 4.55 \cdot 10^{-3}$; $\dot{Q}_0 = 9.10 \cdot 10^{-3}$; $\dot{Q}_0 = 1.82 \cdot 10^{-2}$. *Source:* Novozhilov (2017) / MDPI / CC BY 4.0.

Soot emission is controlled by the absorption coefficient, which can be approximated empirically as (Novozhilov *et al.* 1996)

$$\tilde{k}_s = 1264 f_v \tilde{T} \tag{7.22}$$

where f_v is the soot volume fraction. Note that the numerical coefficient in (7.22) is dimensional. The soot volume fraction is calculated as follows:

$$f_v(\tilde{t}) = \frac{s\tilde{m}_f\tilde{t}}{\tilde{\rho}_s\tilde{S}(\tilde{H} - \tilde{z})} \tag{7.23}$$

and in non-dimensional form

$$f_v(\tau) = \frac{s\dot{m}_f\tau}{\rho_s(1-y)} \tag{7.24}$$

where s is the soot conversion factor (soot yield) of the fuel, $\dot{m}_f$ is the non-dimensional burning rate of the initial fire and $\rho_s = \tilde{\rho}_s/\tilde{\rho}_0$ is the non-dimensional soot density.

Emissivity of the smoke (upper) layer can be written, following Drysdale (2011) as

$$\varepsilon_2 = 1 - \exp\left(-\tilde{k}_s(\tilde{H} - \tilde{S})\right) \tag{7.25}$$

Taking into account relations (7.22–7.24), relation (7.25) may be transformed into

$$\varepsilon_2(\tau) = 1 - \exp\left(-E\frac{s\dot{m}_f\tau}{\rho_s}(\theta_2(\tau) + 1)\right) \tag{7.26}$$

where $E = 1264\tilde{H}\tilde{T}_0$.

Further development is very similar to that of Chapter 6, Section 6.3.

The radiative flux received by the biosolid material is written as

$$\tilde{\dot{q}}''_{rad} = -\frac{\tilde{\sigma}\left(\tilde{T}_g^4 - \tilde{T}_s^4\right)}{\left(\frac{1}{\varepsilon_1} + \frac{1}{\varepsilon_2} - 1\right)} \tag{7.27}$$

Neglecting, in the first approximation, the effect of fuel temperature rise on radiative heat exchange, the last equation is rewritten as

$$\tilde{\dot{q}}''_{rad} = -\frac{\tilde{\sigma}\left(\tilde{T}_g^4 - \tilde{T}_0^4\right)}{\left(\frac{1}{\varepsilon_1} + \frac{1}{\varepsilon_2} - 1\right)} \tag{7.28}$$

Further, the latter heat flux is equal to that transferred into the fuel by conduction ($\tilde{x}$ is Cartesian coordinate in the reacting material, orthogonal to its surface)

$$-\tilde{\lambda}_s\frac{\partial\tilde{T}}{\partial\tilde{x}}(0,\tilde{t}) = \tilde{\dot{q}}''_{rad}(\tilde{t}) \tag{7.29}$$

As in Chapter 6, an integral equivalent of HTE (Landau and Lifshitz 1987) may be applied to express surface temperature

$$\tilde{T}_s(\tilde{t}) - \tilde{T}_0 = -\frac{1}{\sqrt{\tilde{\lambda}_s\tilde{\rho}_s\tilde{c}_s}}\frac{1}{\sqrt{\pi}}\int_0^{\tilde{t}}\frac{\tilde{\dot{q}}''_{rad}(\tilde{\varsigma})}{\sqrt{\tilde{t}-\tilde{\varsigma}}}d\tilde{\varsigma} \tag{7.30}$$

which in non-dimensional form (with $\xi = \frac{\tilde{x}}{\tilde{H}}$ and change of limits of integration) reduces to

$$\theta_s(\tau) = \frac{\sqrt{\tau}}{\sqrt{\pi}}\int_0^1\frac{\frac{\partial\theta}{\partial\xi}(0,\varsigma\tau)}{\sqrt{1-\varsigma}}d\varsigma \tag{7.31}$$

Finally, combining Eqs (7.28), (7.29) and (7.31), the following equation for the evolution of material surface temperature is obtained:

$$\theta_s(\tau) = \frac{Bi_R}{\sqrt{\pi}}\sqrt{\tau}\int_0^1\frac{\left[(\theta_2(\varsigma\tau) + 1)^4 - 1\right]}{\left(\frac{1}{\varepsilon_1(\varsigma\tau)} + \frac{1}{\varepsilon_2} - 1\right)}\frac{d\varsigma}{\sqrt{1-\varsigma}} \tag{7.32}$$

where $Bi_R = \dfrac{\tilde{\sigma}\tilde{T}_0^3\sqrt{\tilde{\kappa}}\sqrt{\tilde{S}}}{\tilde{\lambda}_s\sqrt[4]{\tilde{g}\tilde{H}}^{3/4}}$.

As in Chapter 6, following the well-established (Novozhilov 2010; Drysdale 2011) criterion of critical fuel surface temperature, ignition outside of the initial fire area (full fuel load involvement) is assumed to occur once such an *a priori* assigned surface temperature threshold is achieved. In the present section, the value of non-dimensional critical (ignition) temperature $\theta_s^{cr} = 1.5$, which corresponds to the dimensional ignition temperature of approximately 750 °C, is used. This is generally consistent with reported values for the pilot-free ignition of cellulose-containing materials.

Equation (7.32), combined with the ignition temperature criterion, allows flashover conditions and times to be predicted for various values of controlling parameters. The effect of the most important of these parameters on ignition time is considered below.

The non-dimensional parameters that essentially control the system dynamics are $\dot{m}_f$ ($\dot{Q}_0$ is a function of the latter), E, Bi_R and s.

Non-dimensional ignition time as a function of the non-dimensional burning rate is presented in Figure 7.10. It is evident that, as expected, the ignition time becomes shorter as the mass burning rate increases. Notably, the rate of decrease drops as intensity (mass burning rate) of initial fire increases.

Further, the influence of radiative heat exchange conditions is illustrated in Figure 7.11. These are controlled by the parameter Bi_R (the radiation Biot number). The latter is essentially the ratio between the rate of radiative heat exchange at the fuel surface and the rate of heat conduction into the body of solid fuel. The curves in Figure 7.11 are presented for different values of the parameter E, which allows the sensitivity to this parameter to be assessed. Parameter E is proportional to the clearance height. That explains why ignition time drops as parameter E increases. Reduced clearance height leads obviously to quicker smoke filling of the compartment and to the more sharp growth of average smoke temperature.

Another important parameter quantifying hazard of particular fuel in flashover scenario is its soot yield s. This parameter characterizes fuel soot propensity and, to a large extent, controls the intensity of radiation emitted by smoke layer.

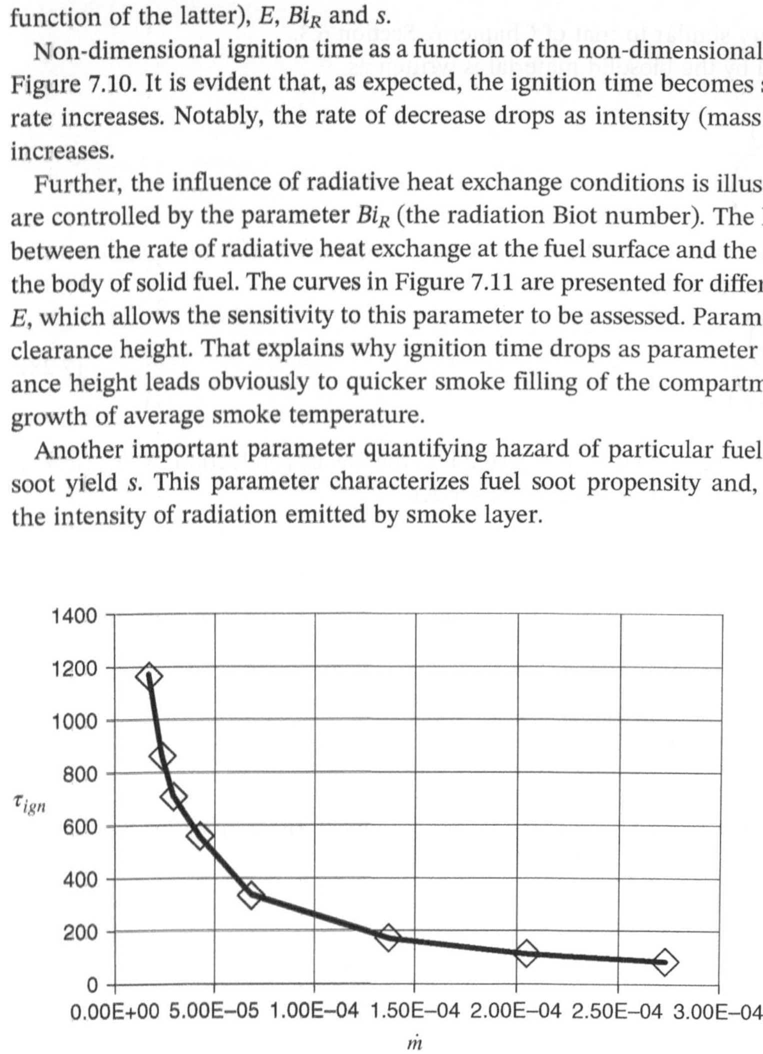

Figure 7.10 Non-dimensional ignition time versus non-dimensional mass burning rate. *Source:* Novozhilov (2017) / MDPI / CC BY 4.0.

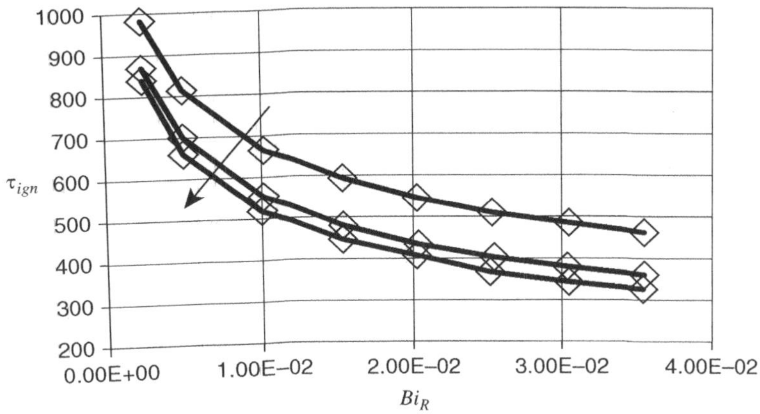

Figure 7.11 Influence of the radiation Biot number on ignition time. Both variables are non-dimensional. The arrow points towards an increase of the parameter E for each of the curves: $E = 3.79 \cdot 10^5$; $E = 1.14 \cdot 10^6$; $E = 1.90 \cdot 10^6$. *Source:* Novozhilov (2017) / MDPI / CC BY 4.0.

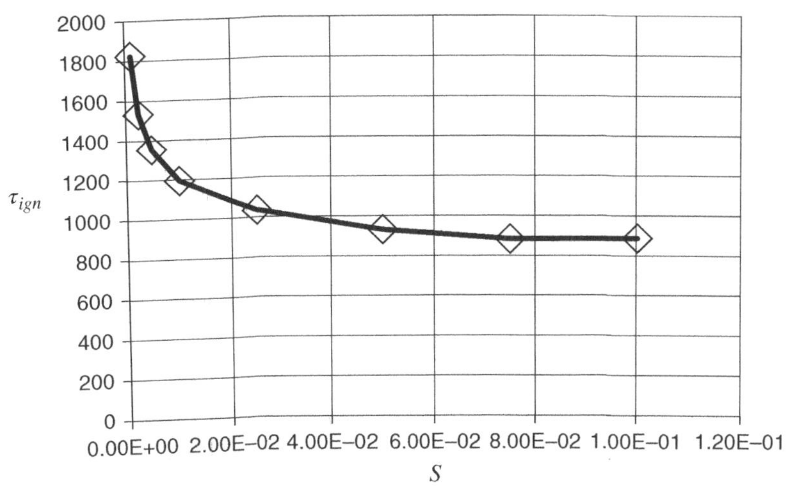

Figure 7.12 Dependence of the ignition time on the fuel soot yield. Both variables are non-dimensional. *Source:* Novozhilov (2017) / MDPI / CC BY 4.0.

Dependence of the ignition time on the fuel soot propensity is presented in Figure 7.12. It is clear that while ignition time drops sharply with increased fuel soot propensity in the range of soot yields $0 < s < 2 \cdot 10^{-2}$, for larger values of the parameter the dependence levels off.

It is also instructive to consider the key dependences in dimensional variables, in order to get a feeling of real scales involved in the process.

The plot of non-dimensional ignition time versus non-dimensional burning rate (Figure 7.10) may be converted into the dependence of dimensional ignition time on initial fire HRR $\tilde{\dot{Q}}_0$. This dependence is shown in Figure 7.13. The plot demonstrates the same trend as in Figure 7.10. However, dimensional dependence makes it easier to reach some important practical conclusions.

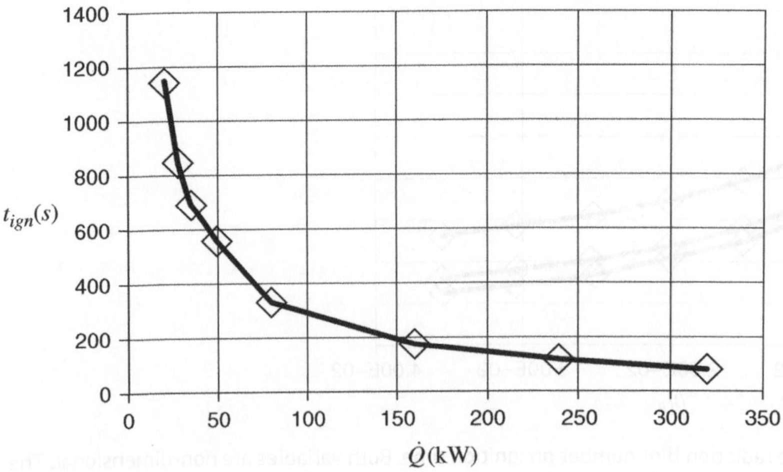

Figure 7.13 Dimensional ignition time as a function of the dimensional initial fire HRR. $\tilde{H}$ = 3 m; $\tilde{S}$ = 16 m^2. *Source:* Novozhilov (2017) / MDPI / CC BY 4.0.

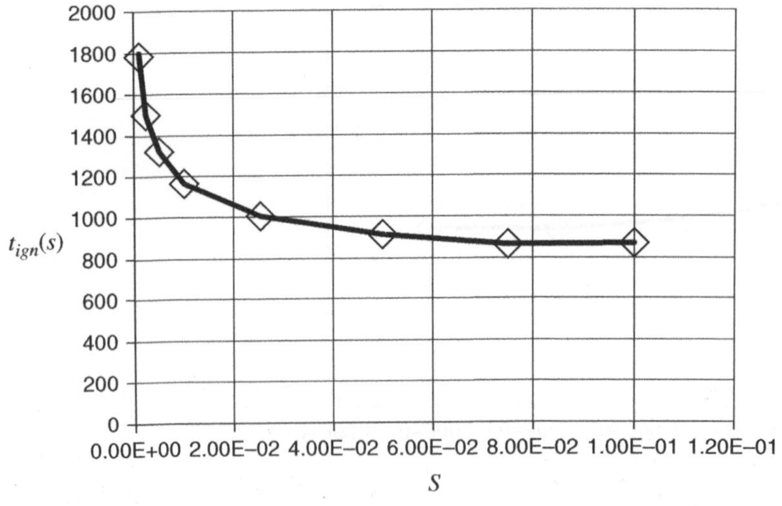

Figure 7.14 Dependence of the dimensional ignition time on the non-dimensional fuel soot yield. $\tilde{H}$ = 3 m; $\tilde{S}$ = 16 m^2. *Source:* Novozhilov (2017) / MDPI / CC BY 4.0.

For example, it is evident that even weak accidental initial fires (say $\tilde{\dot{Q}}_0 \sim 25 - 50$ kW) still result in very short (compared to time scales of storage or transportation) flashover times (within 20 min).

The dependence of dimensional ignition time on soot yield, presented in Figure 7.14, also qualitatively follows the trend of its counterpart, Figure 7.12.

One of the most important factors dictating design requirements for safe storage or transportation of biosolids is the influence of the compartment geometrical scales and configuration on likelihood and time scale of flashover. To this extent, dimensional flashover time is shown as a function of height clearance in Figure 7.15. As expected, higher clearance slows down the smoke filling process.

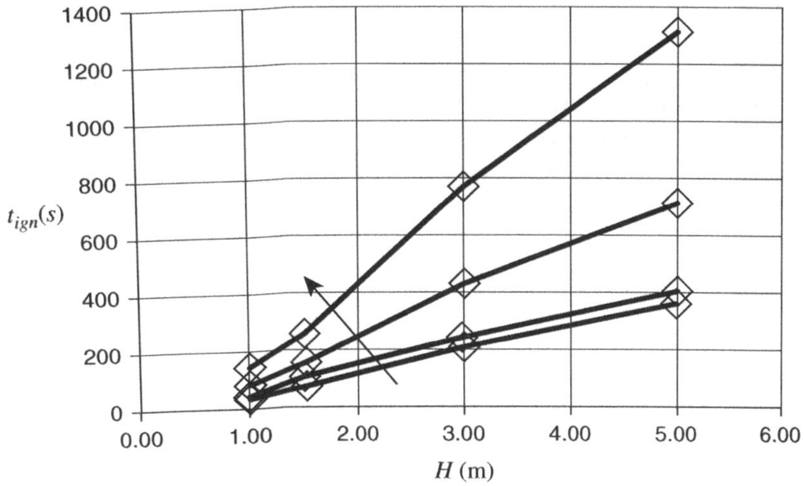

Figure 7.15 Influence of the height clearance on the dimensional ignition time. The arrow points towards an increase of the fuel area $\tilde{S}$ for each of the curves: $\tilde{S}$ = 12 m^2; $\tilde{S}$ = 24 m^2; $\tilde{S}$ = 48 m^2; $\tilde{S}$ = 100 m^2. *Source:* Novozhilov (2017) / MDPI / CC BY 4.0.

Consequently, ignition (flashover) time increases. Besides, the influence of fuel surface area $\tilde{S}$ is also shown in the same figure. Compartment volume is directly proportional to this parameter. Therefore, at a fixed height clearance, increased fuel surface area also leads to delay in flashover.

In summary, the study by Novozhilov (2017) demonstrates that transition from initially localized biosolid fire to flashover occurs (in a closed compartment) under a wide range of conditions. As such, biomass solid fuels, accumulated in large quantities, represent a significant fire hazard. That hazard must clearly be addressed when developing and implementing technologies for biosolid fuel storage and transportation.

The same conclusion is confirmed by many other studies, which considered specific biomass solid fuels.

One of the typical biosolid fuels is compost. Being stored in large piles, this substance is prone to biological heating at low (compared to combustion) temperatures. The primary mechanism of such heating is microorganism activity. Such microorganisms may include aerobic mould-fungi and bacteria. Although heating rate provided by this mechanism is initially weak, it is persistent over long periods and may eventually (after substantial induction period) lead to self-ignition of the bulk of solid material and consequently to dangerous fires.

Details of relevant kinetics are rather complicated and not considered here. It is sufficient to say that biologically driven heating does not follow the conventional Arrhenius law but exhibits more complicated form of reaction rate dependence on temperature, where inhibitory effects should also be taken into account (Roels 1983; Luangwilai *et al.* 2013). Biologically induced heat release rate in a compost pile peaks at about 50 °C while starting to decrease soon above this temperature until at about 80 °C, microorganisms die or become inactive (Rynk 2000).

Detailed analysis, both theoretical and numerical, of self-heating and self-ignition in compost piles has been provided in a series of studies by Nelson and co-authors (Sidhu *et al.* 2007; Luangwilai *et al.* 2010, 2012, 2013, 2018). Both biological and oxidative mechanisms of self-heating were accounted for. The studies included one- and two-dimensional formulations.

Along with ambient temperature, the other parameters, such as moisture content and external airflow, affect long-term behaviour of self-heating compost pile.

Generally, moisture content of the pile has to be within certain range to sustain biological energy generation (Luangwilai *et al.* 2012). Too dry or too wet piles damp heat generation by microorganisms. External airflow also has significant effect on thermal stability of compost piles (Luangwilai *et al.* 2010). It has twofold effect, enriching medium with oxygen and thus promoting oxidation-driven energy release while, on the other hand, cooling the pile via convective heat transfer mechanism.

Very sophisticated, three-dimensional computational fluid dynamics (CFD) models (Zambra *et al.* 2012) produce essentially the same trends. Compost pile in an open atmosphere environment is an extremely complex chemical and physical system when compared to simple confined gaseous mixtures considered first by Semenov. It is remarkable, in this regard, to observe that the behaviour of some key parameters of such a system follows qualitatively predictions of the basic unsteady Semenov model. As an example, the maximum temperature inside the pile, predicted by Zambra *et al.* (2012), is presented in Figure 7.16. Temperature time histories are presented for three cases, differentiated by the pile height and its volume-to-area ratio. The temperature time histories corresponding to cases 1 and 3 demonstrate a typical asymptotic approach to steady-state equilibrium. Case 2 manifests thermal explosion and pile self-ignition.

Striking similarity of these curves to predictions of most simple, zero-dimensional, unsteady Semenov model (Figure 2.9, Chapter 2) emphasizes, once again, the universal nature of the Semenov thermal explosion theory and the power of its fundamental approach.

A more complicated behaviour may be observed if a spotted explosion is considered. Aganetti *et al.* (2016) present a model, very similar in nature to those developed by Nelson and co-authors and Zambra *et al.* (2012). The model is applicable to a broad range of reacting materials, for example, biosolids, coal, food grains and compost. Kinetics of biologically driven energy release is modelled in exactly the same way as by Luangwilai *et al.* (2013). This is essentially two-dimensional

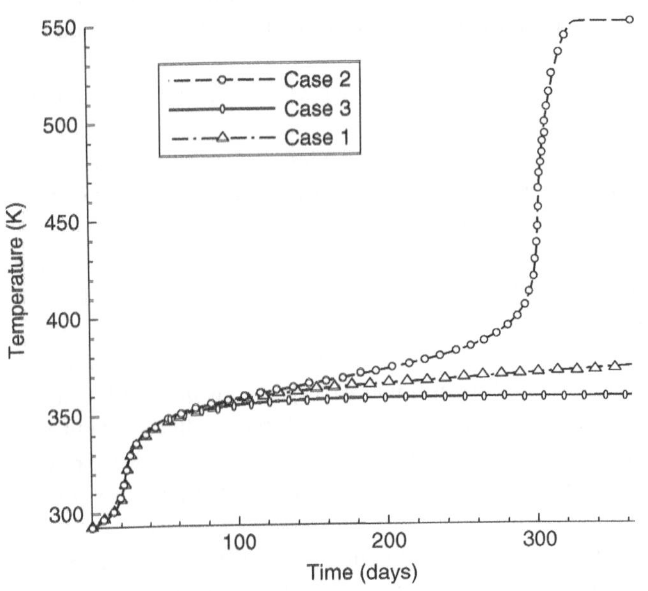

Figure 7.16 Maximum temperature inside the compost file for three cases considered. *Source:* Reproduced with permission from Zambra *et al.* (2012) / Elsevier.

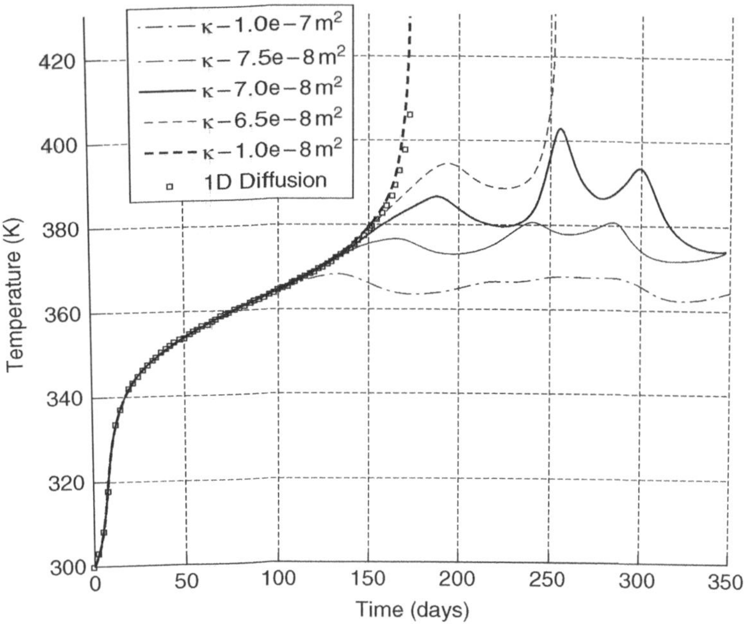

Figure 7.17 Temperature evolution of a hot spot inside a stockpile of biosolid fuel with different permeabilities. *Source:* Reproduced with permission from Aganetti *et al.* (2016) / Elsevier.

CFD model, which, for some cases considered, has been reduced (with reasonable approximation) to one-dimensional. Figure 7.17 shows the temperature evolution of a hot spot immersed into the stockpile of biosolid fuel, at different values of pile permeability. The figure shows both sub-critical (high permeability allowing for pile cooling by convection) and super-critical behaviour (low permeability resulting in reduced rate of energy dissipation and prevalence of energy generation). Thermal explosion occurs for permeabilities below $\sim 10^{-8}\,m^2$.

Transition from sub-critical to supercritical conditions is accompanied by oscillations, which occur while oxidative reactions start to pick up. The exact cause of these oscillations is not revealed by Aganetti *et al.* (2016). It is plausible to suggest that oscillations arise as a result of non-linear interactions between processes of energy generation and flow within the pile subjected to specific boundary conditions. In this regard, since boundary conditions are thought to play a major role, observed oscillation patterns are again reminiscent of predictions of a much more simple, basic Semenov model. The latter predicts oscillating behaviour exerted by temperature oscillations at the system boundary (Novozhilov 2016, 2018).

There are other types of biosolid materials prone to self-heating and subsequent ignition.

For bagasse piles (bagasse are pulpy fibrous remains after crushing sugarcane or sorghum stalks to extract their juice), such phenomena have been investigated by Gray *et al.* (1984), Dixon (1988), Sexton *et al.* (2001). Similar to other biosolids (compost, RDF, RPF), wet bagasse experiences noticeable self-heating in the range of temperatures 50 – 70 °C, which is likely to be driven by the same mechanisms as for the above fuels. Due to this fact, development of self-heating and self-ignition in bagasse piles may be expected to be similar, in its major features, to those of compost, RDF and RPF.

Qualitatively similar behaviour is observed in biofuels made of wood products. Such are large-scale wood piles (Ferrero *et al.* 2009) and yard trimmings (Buggeln and Rynk 2002). Authors of both studies note the fundamental contribution of aerobic processes to the overall time history of energy release leading to thermal explosion. Jones and Raj (1988) experimented with small (10^3 cm^3) cubic volume filled with Eucalyptus saligna leaves and observed that the rise of ambient temperature of the order of $2 - 3\,°C$, around the value of $\tilde{T}_0 \sim 427\,°C$ led to self-ignition. A similar experiment with wood shavings, in a 64 cm^3 cubic volume, confirmed autoignition temperature of about $\tilde{T}_0 \sim 500\,°C$.

Sisson *et al.* (1992) proposed a quite general reaction-diffusion model for self-heating of damp cellulosic materials and hypothesized that inclusion of moisture effects is necessary to explain large discrepancies between safe pile sizes predicted by classical Frank-Kamenetskii theory and really observed ones. In a subsequent paper (Sisson *et al.* 1993), they investigated some of the steady states of their model. The results were obtained in both spatially uniform (Semenov), as well as in spatially distributed (Frank-Kamenetskii) formulations. However, the exhaustive picture of the model's steady-state behaviour was not obtained.

Another typical biomass solid fuel receiving attention recently is biochar, the form of charcoal created by controlled pyrolysis at very high temperatures, over extended periods.

Phounglamcheik *et al.* (2022) investigated self-heating of biochar using a simplified one-dimensional model. They discovered typical transitions to thermal explosion upon increasing ambient temperature, sample size and Zeldovich number. An interesting observation of the study was the effect of biochar bed void fraction on temperature evolution in the sample. It appeared that the temperature at the centre of the container containing biochar decreased with the reduced value of the void fraction. This was explained by the reduced amount of oxygen in the container, available for oxidation. The above observation allowed the authors to put forward certain recommendations for the design of pyrolysis processes used to produce biochar.

7.4 Thermal Explosion of Electric Batteries

Modern electric batteries represent technology that has become increasingly popular for energy storage applications in both industrial and domestic domains. With all their advantages, electric batteries are prone to accidental heating and thermal runaway, which carry substantial risks.

This technology is quite interesting in the context of thermal explosion development as batteries are substantially different, as reacting systems, from any other media previously studied by thermal explosion theory.

In view of high practical importance, the studies of electric batteries self-ignition have been developing extremely rapidly, and the present section does not pretend to be an exhausting overview of the subject, especially with respect to the most recent publications. It is also worthwhile to note that the methods of mitigation of electric batteries self-ignition are left for discussion in Chapter 9.

There are many types of electric batteries technologies (Cavanagh *et al.* 2015), for example, lithium-ion, lead-acid, advanced lead-acid, nickel-cadmium, zinc bromide flow and vanadium redox flow.

With respect to thermal runaway phenomenon, lithium-ion batteries (LiB) are by far the most studied (Ghiji *et al.* 2020).

At the moment, lithium-ion cells are commercially available in various, namely button, cylindrical, pouch and prismatic, designs. Single lithium-ion cell is a primary electrical energy storage device, consisting of cathode, anode and separator in a closed volume filled with electrolyte. During cell discharge phase, lithium ions move from anode to cathode (and become intercalated between layers of cathode crystals) and during the charge phase in the opposite direction. Upon initial charge, lithium ions react with the solvent of the electrolyte and form on the anode specific layer called solid–electrolyte interface (SEI). SEI is permeable for lithium ions but not for electrolyte (Tarascon and Armand 2001). Its stability is a deciding factor in the cell safety (Ghiji *et al.* 2020).

In order to achieve parameters of the energy storage/supply required in applications, cells are typically aggregated into battery systems. This is done by connecting cells either in parallel, to increase capacity, or sequentially, to ramp up voltage. Design hierarchy starts with a single cell and then develops bottom-up to include the levels of battery (stack of cells connected in parallel), module (many batteries connected in series), battery pack (several modules or many batteries) and, finally, battery system (several battery packs connected in parallel) (Otto *et al.* 2012).

Thermal explosion of a single lithium-ion cell is typically triggered by one of the following 'external' events:

1) *Electrical abuse (over- charging/discharging)* (Lisbona and Snee 2011; Li *et al.* 2014; Yuan *et al.* 2015; Ye *et al.* 2016). Achieving any voltage outside the manufacturer-specified charge window may lead to lithium plating or dendrite formation on the anode. Over time, this may cause piercing of the separator, short circuit between electrodes and subsequently thermal explosion.
2) *Thermal abuse (overheating)* (Lee *et al.* 2000; Spotnitz and Franklin 2003; Yang *et al.* 2005; Guo *et al.* 2017). Internal cell temperatures in the range of 90 – 120 °C start exothermic reaction of SEI layer decomposition. Further, at temperatures above about 200 °C, additional heat is released due to thermal decomposition of hydrocarbon-based electrolyte.
3) *Mechanical abuse (penetration, pinch, bend, etc.).* (Lamb and Orendorff 2014; Ruiz *et al.* 2018). Such events may happen, for example, during crash of electric vehicle. Mechanical abuse may cause short circuit between the electrodes, resulting in localised heating and onset of thermal explosion.
4) *Internal short circuit (ISC)* (Feng *et al.* 2018). An ISC occurs due to the malfunction of the separator allowing contact between the electrodes via electrolyte. This may occur due to any of the abuse conditions (1–3) or due to manufacturing fault.

Whatever is the triggering event (1–4) (or their combination), leading to initial energy release, the ultimate reason behind lithium-ion cell thermal explosion lies in its internal design and the range of materials used for its fabrication. Their combination allows a number of exothermic reactions to occur in the course of internal temperature rise within the cell (Spotnitz and Franklin 2003; Ghiji *et al.* 2020). Such are, for example, exothermic reactions of SEI decomposition, which can cause loss of protection of anode. This, in turn, may lead to exothermic reaction between fluorinated binder within the anode and lithiated carbon.

An important circumstance, which makes lithium-ion cell susceptible to thermal runaway, is the fact that both the separator and the electrolyte are fabricated from combustible materials. Separators are typically made of microporous polyolefin films, such as polyethylene (PE), polypropylene (PP) or laminates of polyethylene and polypropylene (Arora and Zhang 2004). Electrolytes in lithium-ion cells are usually flammable carbonate-based organic solvents, such as ethylene carbonate (EC), dimethyl carbonate (DMC), diethyl carbonate (DEC) and ethyl methyl carbonate (EMC) and/or propylene carbonate (PC) (Xu 2004). Additives may include lithium hexafluorophosphate ($LiFP_6$), lithium hexafluoroarsenate monohydrate ($LiAsF_6$), lithium perchlorate ($LiClO_4$) and lithium tetrafluoroborate ($LiBF_4$) (Wang *et al.* 2006; Ghiji *et al.* 2020).

Such a composition makes it possible for electrolyte to react exothermically with intercalated lithium. Additionally, electrolyte may react with oxygen, produced as a result of decomposition of cathode.

Escalation of exothermic electrochemical reaction processes, which are above normal operating temperature and are very complex and many of which occur simultaneously (Ghiji *et al.* 2020), leads eventually to the establishment of positive feedback loop typical for thermal explosion.

It is clear that there are multiple paths to this feedback loop, depending on which of the conditions (1–4) have led to initial energy release (Ghiji *et al.* 2020).

Selected real data on thermal explosion development in lithium-ion cells are presented in Figures 7.18 and 7.19. Figure 7.18 shows criticality with respect to charging current for fixed

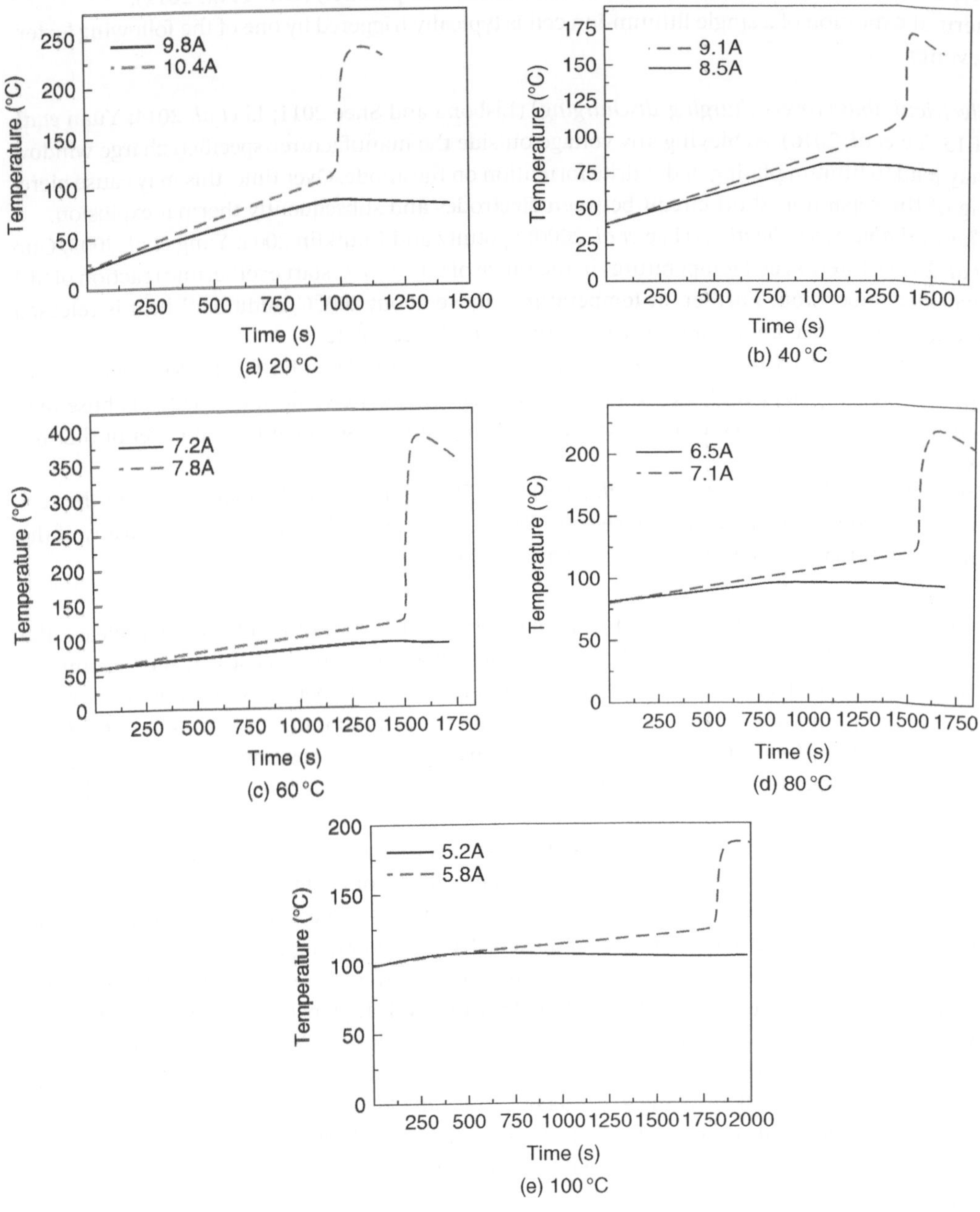

Figure 7.18 Temperature time histories for lithium-ion battery for different ambient temperatures and charging currents. *Source:* Reproduced with permission from Guo *et al.* (2017) / Elsevier.

26/9/2025 2:24

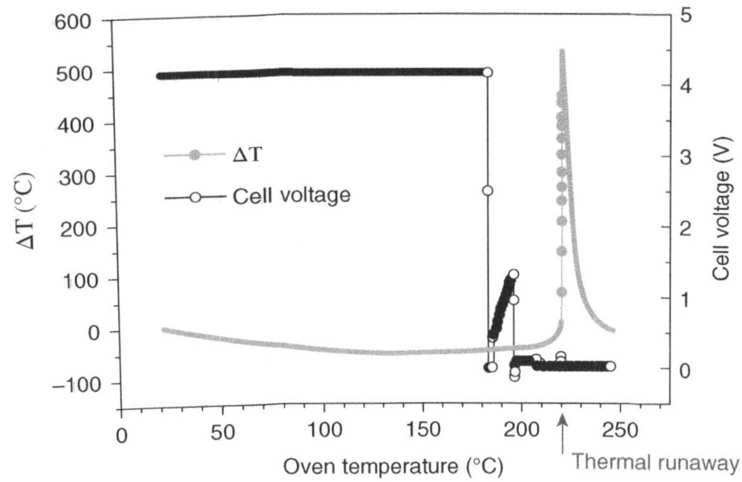

Figure 7.19 Time history of differential temperature for Samsung 18650 laptop cell under external heating conditions. Heating rate approximately 3.67 K/min. *Source:* Larsson *et al.* (2016) / MDPI / CC BY 4.0.

specified environment temperatures (and fixed all the other parameters that may affect critical conditions). This figure proves that charging current is one of the parameters controlling the onset of thermal runaway (i.e. one of the components in the set of critical conditions). Figure 7.18 allows estimation of critical charging current at each of considered levels of ambient temperature (with the relative error below 6%). It may be deduced easily from the presented data that critical charging current decreases almost linearly with increasing ambient temperature. The study by Guo *et al.* (2017) also contains data on critical conditions with respect to other parameters: ambient temperature and heat dissipation rate.

Figure 7.19 demonstrates time history of differential temperature for Samsung 18650 lithium-ion cell, subjected to oven heating from 0 °C to approximately 220 °C in about one hour. Differential temperature is the temperature difference between the average cell surface temperature and the oven temperature. Thermal explosion is evident at the external temperature of ~220 °C (Larsson *et al.* 2016). It is interesting to note that well before thermal explosion melting of the separator takes place leading to cell voltage breakdown.

Thermal explosion was also observed in experiments carried out by Lee *et al.* (2000).

Thermal explosion leads to pressure build-up within the cell, which may lead to cell swelling, rupture and gas venting. Since vented gas jet typically contains not previously reacted flammable components, jet fire (diffusion jet flame) often follows thermal runaway in lithium-ion cell. Such a fire is driven, to a large extent, by oxidation of yet unburned portions of electrolyte and separator. Combustion of the latter two components accounts for over 75% of the total HRR in jet fire (Ghiji *et al.* 2020).

Thermal explosion is especially high hazard in those applications where LiBs are aggregated into large battery systems, such as in electric vehicles (EV) and hybrid electric vehicles (HEV) (Al-Hallaj and Selman 2002; Larsson *et al.* 2016; Ruiz *et al.* 2018; Feng *et al.* 2018), aircrafts and submarines (Buckingham *et al.* 2008). While thermal explosion starts typically in a single cell (due to basic probabilistic considerations), cell-to-cell propagation of thermal runaway is a dangerous scenario, which indeed occurs in reality. Spinner *et al.* (2015) observed such propagation in cell packages of hexagonal configurations, starting from one of the cells being overcharged and subsequently overheated. They pointed out that the main reason for failure propagation within the pack was lack of cooling due to dense packing of the cells. This emphasizes the increased risk created by cell aggregation into Battery Packs and Systems since these must typically have dense packing, due to both energy release density requirements and space limitations. Lamb *et al.* (2015) studied thermal explosion propagation in the modules of cells of different types and with different electrical configurations. They found that electrical configuration is an additional factor that affects thermal explosion propagation. Parallel cell connection provides additional, in comparison to series configuration, heat conduction paths between different cells in the module and thus increases heating rate of the cells during induction period. This leads to faster propagation of thermal explosion through the module. However, this effect becomes important only for less dense packing where direct contact area of adjacent cells is limited (e.g. battery consisting of cylindrical cells). For dense packing (such as in poach cell battery), direct contact area between cells is significant, and heat transfer through this area dominates.

High risks associated with the exploitation of LiBs prompted the development of safety testing methods and standards (Ruiz *et al.* 2018). Test methods essentially seek to recreate the 'external' events (1–4) triggering thermal explosion and identified above and amount to various abuse techniques grouped into mechanical (shock, drop, penetration, etc.), electrical (overcharge/overdischarge, ISC), environmental (thermal stability, thermal shock, overheating, etc.) and chemical (emissions, flammability) clusters.

Continuous use of tests is warranted in view of apparent challenges faced by modelling approach. Internal structure of electric cells is complicated, and this makes successful development of analytical models (similar to Semenov or Frank-Kamenetskii models for gases) very unlikely. Further, potential models have to include both the thermo-chemical and electro-chemical phenomena, which presents an additional difficulty.

Nevertheless, some progress with modelling of thermal behaviour of lithium-ion cells has been made using field (CFD-type) models. These attempts fall essentially into the two categories:

1) The studies aimed to predict temperature time histories and/or distributions within electric cells.
2) The studies aimed to predict not only temperature field inside the cell but also the occurrence of thermal explosion.

Within the first group of investigations, Chen and Evans (1996) considered prismatic stack of lithium-ion cells. They solved three-dimensional unsteady HTE with energy generation source term modelled as a function of the cell's electrical and thermodynamic properties. Predicted temperatures at several locations within the stack agreed well with experimental data during charge and discharge processes. Al-Hallaj *et al.* (1999) used simplified, one-dimensional model, which produced nevertheless good agreement with the temperature data on Sony 18650 cell. Good predictions of temperatures within lithium-ion cells have also been reported by Sato (2001) and Funabashi *et al.* (2002). Further discussion of thermal modelling of lithium-ion cells and batteries is provided in the review by Wang *et al.* (2012).

As far as the second group of studies is concerned, already mentioned simulations of Chen and Evans (1996) indicated that thermal explosion is possible in the case of localised ISC. Spotnitz and Franklin (2003) presented geometrically simple one-dimensional model considering cell as a slab consisting of layers of different materials. On the other hand, they have been able to include rather comprehensive electrochemical kinetics. Using their model, the authors simulated temperature distributions within lithium-ion cells during various abuse tests: oven (overheating), short-circuit, overcharging, nail (mechanical penetration and resulting localised heating). They observed thermal explosion development, upon reaching critical conditions, in all the above tests. However, key combination of exothermic reactions leading to thermal runaway was different in each scenario. Unfortunately, no comparison with experimental data on critical conditions was provided.

Finally, among the attempts to model thermal explosion, theoretical analysis of Wang *et al.* (2010) may be pointed out. The authors considered the problem from the point of view of Catastrophe theory (Arnold 1984). They considered zero-dimensional (similar to Semenov) energy balance equation taking into account, however, additional effects: Joule heating and thermal radiation heat losses. Further, approximating Arrhenius function by a quadratic polynomial, they manage to write the equation in potential form. The exact form of this potential function describes the swallowtail catastrophe (Arnold 1984). Based on this finding, the authors were able to point out thermal explosion regions in the space of control parameters. A simple practical example was considered, and comparison with experimental data was encouraging.

It should be pointed out, however, that polynomial approximation of Arrhenius function requires, in the context of presented catastrophe analysis, special justification. Small perturbations of the original system may, generally speaking, lead to catastrophe of a different type. Besides, the degree of approximating polynomial was chosen rather arbitrary, and it is not clear whether the presented analysis is qualitatively (at least) invariant with respect to variation of the latter degree.

8

Control Problem in the Theory of Thermal Explosion

8.1 Problem Formulation

The aim of the present chapter is to examine the possibilities of actively influencing the process of thermal explosion development. The ultimate goal of such control has to be the prevention of thermal explosion. It is this goal that is fundamentally important from both theoretical and practical points of view.

Despite such importance, research in this direction is essentially missing, apart from one study by Gelman *et al.* (1992) and its digest in the monograph by Khudyaev (2003). The specific control goal, considered by Gelman *et al.* (1992), was executing the chemical reaction at the explosion limit, that is avoiding thermal explosion, but at a sufficiently fast rate. The study used numerical techniques exclusively.

It follows from the above that control is understood in this chapter as an *active control*, meaning that the system is initially at supercritical conditions, and actions to transform it to subcritical conditions are being made during the system evolution. Moreover, the quantitative parameters of such actions may be determined by the current and/or past states of the system.

Informally, the idea of the thermal explosion control is reflected in Figure 8.1. Here, typical reacting mixture temperature time histories are shown, including those corresponding to the thermal explosion conditions (i.e. growing infinitely with time), as well as those corresponding to the thermal quasi-equilibrium.

In terms of the definition of thermal explosion adopted in the present book, transition to subcritical conditions must be enforced before the time instant when the maximum of the second derivative of the EGD function is achieved.

This chapter provides a purely theoretical analysis of the problem formulated above. This means that in the following analysis, the system parameters are assumed to be variable (within the limits preserving their physical significance, e.g. positiveness) in the way required for successful control. Practical techniques to vary parameters in certain ways are discussed in Chapter 9.

The simplest model, that is Semenov model with no kinetic (reactant consumption) effects is considered for the present analysis.

Methods developed in this chapter may be in principle generalized to include reactant consumption. This generalization is not considered, however, in this text.

Control function, or control is understood in this book in a generalized sense, referring not only (in line with its conventional meaning in control theory) to the law of change of known functions in a differential equation, but also to the law of change of boundary conditions.

Thermal Explosion: Theory and Application, First Edition. Vasily B. Novozhilov.

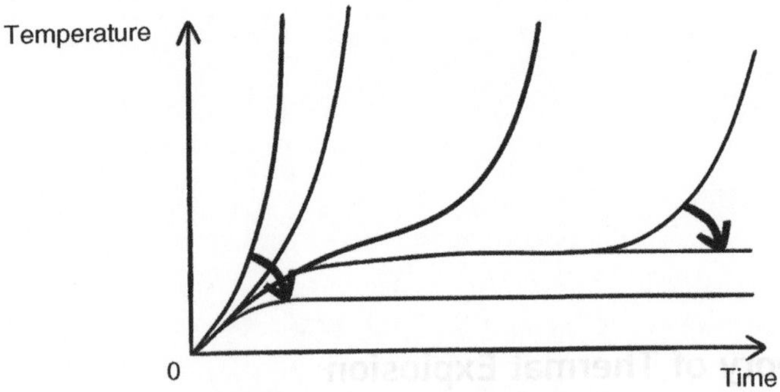

Figure 8.1 Schematic of thermal explosion control. Curved arrows show transition of the system evolution from thermal explosion trajectories to trajectories corresponding to subcritical conditions.

The aim of the control, in the context of this book, is to prevent thermal explosion. The problem is considered in the Semenov formulation (Eqs. (2.5) and (2.6)). This means that control (for a particular chemical reaction and ambient temperature, i.e. fixed Arrhenius number) may only be achieved by variation of physical parameters: either the Semenov number (which is a constant parameter in the original Semenov model, Chapter 2), or the ambient conditions (constant ambient temperature in the original model). Note that the latter enters the model implicitly, via the definition of the nondimensional temperature (Eq. (2.2)). There are no other parameters admitting variation in the Semenov formulation.

Three possible types of control are considered in this chapter. These are: instantaneous, autonomous, and non-autonomous controls. In more precise ways, these controls are defined and analysed in the relevant sections later.

There are also other types of control that are generally possible. These include, for example, autonomous or non-autonomous controls with delay(s) (where the control function is shifted in its argument(s) with respect to solution).

Such types of control are not considered in this book.

8.2 Instantaneous Control

Under the application of instantaneous control, control functions are chosen as step functions, that is, they change instantly from one fixed value to another.

As explained above, the two control methods are considered here: (i) instantaneous change in the ambient temperature, and (ii) instantaneous change in the Semenov number.

Although the latter change may generally be achieved by varying several parameters (see definition of the Semenov number (2.2)), in most natural situations of given vessel and chemical reaction, this change may be conveniently comprehended as the change of the heat transfer coefficient only. Both control methods are illustrated in Figure 8.2.

For simplicity of illustration the initial condition in this figure is assumed as $\theta(0) = 0$, although, in fact, the point depicting the non-negative initial temperature may be thought to lie anywhere to the left of the vertical line AB (Figure 8.2).

Any initial condition for the system in Figure 8.2 is supercritical, that is, for any such condition the system undergoes the thermal explosion as the EGD function (curve 1) has no real roots.

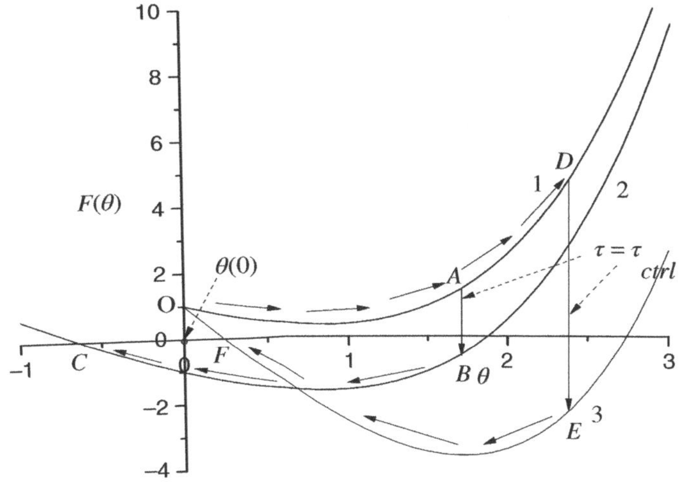

Figure 8.2 Thermal explosion control with instantaneously changed ambient temperature or Semenov number. $Ar = 0.01$. 1 – $F(\theta)$ Eq. (8.1). $Se = 0.43$; 2 – $F(\theta)$ Eq. (8.2). $Se = 0.43$, $\tilde{T}_{ctrl} = \tilde{T}_0 - \left(\frac{\tilde{R}\tilde{T}_0^2}{\tilde{E}}\right)\theta_{ctrl}$, $\theta_{ctrl} = 0.86$; 3 – $F(\theta)$ Eq. (8.1). $Se = 0.19$. Vertical arrows correspond to instantaneous change of parameters at particular control time instances $\tau = \tau_{ctrl}$.

The evolution of the system, until the control application, is governed by the Semenov Eqs (2.5) and (2.6).

$$\frac{d\theta}{d\tau} = F(\theta); 0 \le \tau < \tau_{ctrl}$$

$$F(\theta) = \exp\left(\frac{\theta}{1 + Ar\theta}\right) - \frac{1}{Se}\theta \tag{8.1}$$

Here τ_{ctrl} is time of control application.

Trajectory governed by Eq. (8.1) is shown as curve 1 in Figure 8.2. Point $O(0, 1)$ is the initial point of the trajectory. For easier comprehension of results trajectories of the system, which strictly speaking belong to the θ-axis, are being projected here onto the corresponding EGD curves. It is in this sense that the term 'system trajectory' is being used in the descriptions of Figures 8.2 and 8.3.

Consider first application of the control in the form of an instantaneous change (reduction) of ambient temperature.

If the latter is abruptly dropped at $\tau = \tau_{ctrl}$ to a certain value $0 < \tilde{T}_{ctrl} < \tilde{T}_0$, then the subsequent evolution of the system is described by the following equation

$$\frac{d\theta}{d\tau} = F(\theta); \tau_{ctrl} \le \tau < \infty$$

$$F(\theta) = \exp\left(\frac{\theta}{1 + Ar\theta}\right) - \frac{1}{Se}(\theta + \theta_{ctrl}) \tag{8.2}$$

where an additional energy sink

$$\theta_{ctrl} = \frac{(\tilde{T}_0 - \tilde{T}_{ctrl})}{\left(\frac{\tilde{R}\tilde{T}_0^2}{\tilde{E}}\right)} \tag{8.3}$$

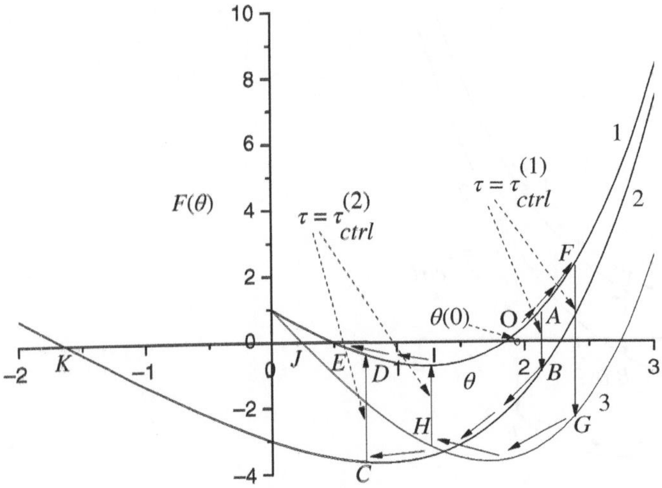

Figure 8.3 Thermal explosion impulse control with instantaneously changed ambient temperature or Semenov number. $Ar = 0.01$. 1 – $F(\theta)$ Eq. (8.1). $Se = 0.43$; 2 – $F(\theta)$ Eq. (8.2). $Se = 0.43$, $\tilde{T}_{ctrl} = \tilde{T}_0 - \left(\frac{\tilde{R}\tilde{T}_0^2}{\tilde{E}}\right)\theta_{ctrl}$, $\theta_{ctrl} = 1.72$; 3 – $F(\theta)$ Eq. (8.1). $Se = 0.19$. Vertical arrows correspond to instantaneous change of parameters at particular control time instances $\tau = \tau_{ctrl}^{(1,2)}$.

is proportional to the temperature difference between the original ambient temperature $\tilde{T}_0$ and the temperature after application of control $\tilde{T}_{ctrl}$.

The EGD function $F(\theta)$ in Eq. (8.2) may be called a control EGD function, as opposed to the original EGD function in Eq. (8.1).

Such control application is shown at the point A, Figure 8.2, and results in the trajectory of the system dropping instantaneously to point B on curve 2. Subsequently, the system is attracted along curve 2 to the stable stationary point C, making the aim of control achieved.

It is important to note that the end point of the trajectory switch B must be strictly below the abscissa axis (e.g. satisfying the condition $F(B) = -\varepsilon$, where ε is a small number). Moving the trajectory to the point that lies exactly on the axis $F = 0$ would mean that the system will be in an unstable condition, and a thermal explosion may still develop. The other necessary condition is that the time of control $\tau = \tau_{ctrl}$ must be such that the desired point below the axis is achievable for some $\tilde{T}_{ctrl}$, satisfying the inequalities $\tilde{T}_0 > \tilde{T}_{ctrl} > \tilde{T}_{abs} = 0°K$.

In summary, the system undergoing control by instantaneous change of the ambient temperature, follows the trajectory *OABC* in Figure 8.2, thus avoiding the thermal explosion that will happen should no control action be taken at point A.

Consider now the second control method, which is the instantaneous change of the Semenov number (i.e. the heat transfer coefficient). As in the previous example, assume that the system evolves along curve 1, Figure 8.2, starting from point O. The control action is taken at point D, and represents the change of the Semenov number. The trajectory of the system drops, accordingly, to point E, and further system evolution, for the times $\tau_{ctrl} \leq \tau < \infty$, is governed by the same Eq. (8.1), but with the reduced Semenov number. This evolution follows curve 3 in Figure 8.2 and the system eventually reaches the stable stationary point F.

In summary, the system undergoing control by instant change of the heat transfer coefficient (or, equivalently, the Semenov number), follows the trajectory *ODEF* in Figure 8.2. Thermal explosion, which will happen should no control action be taken at point D, is avoided.

A modified version of the above two methods may be developed in the case when the original EGD function has two distinctive roots (Figure 8.3).

Similar to Figure 8.2, the trajectory governed by Eq. (8.1) is shown as curve 1. The initial state of the system, however, is different. It has to be supercritical, for the control action to make sense. Therefore, the point depicting the initial condition $\theta(0)$ is to the right of the largest root of the EGD function. Point O just above is the initial point of the trajectory, projected onto the EGD function.

Figure 8.3 shows the control actions taken at the times $\tau = \tau_{ctrl}^{(1)}$ by changing either the ambient temperature (trajectory moving along line AB to curve 2), or the heat transfer coefficient (trajectory moving along line FG to curve 3).

In principle, after these instantaneous changes the system may be left to evolve along the either curve 2 or curve 3, until the trajectories end at the stable stationary points K (trajectory $OABK$) and J (trajectory $OFGJ$), respectively, avoiding thermal explosion. Such control actions would be completely identical to the ones shown in Figure 8.2 at points A (trajectory $OABC$) and D (trajectory $ODEF$), respectively.

In the present case (two separate roots of the EGD function), however, the other type of control is possible by returning the ambient temperature and the Semenov number to their original values at points C and H, respectively. Geometrically, this means that the trajectories return back to original curve 1 (trajectories $OABCDE$ and $OFGHIE$, respectively), bypassing the largest (unstable) stationary point of the system.

Control of such type may be called an impulse control as control functions return to their original values, having effectively rectangular impulse (⊔-)-shaped profiles.

The case with a single root of the EGD function (i.e. $Se = Se_{cr}$) may be considered analogously. Returning to the control by the instantaneous change of the ambient temperature, one may define the important parameters, namely, the critical control time and the critical control temperature of the thermal explosion. These are defined as follows.

The theoretical minimum ambient temperature is that of absolute zero $\tilde{T}_0^{min} = 0°\ \mathrm{K} \approx -273°\ \mathrm{C}$. Therefore, the theoretical maximum value of the additional energy sink (8.3) is

$$\theta_{ctrl}^{max} = \theta_{ctrl}\Big|_{\tilde{T}_{ctrl} = \tilde{T}_0^{min} = 0} = \frac{\tilde{T}_0}{\left(\dfrac{\tilde{R}\tilde{T}_0^2}{\tilde{E}}\right)} = \frac{1}{Ar} \tag{8.4}$$

This value defines the maximum theoretically achievable magnitude of the control action.

The explosion critical control temperature is defined as such temperature θ_{ctrl}^{cr} that at any time when the current temperature of the system satisfies condition $\theta(\tau) < \theta_{ctrl}^{cr}$ control action with additional sink (8.4) prevents thermal explosion.

The control action with the value of the additional sink (8.4) is fully analogous to the transition AB of the system trajectory from curve 1 (original EGD function) to curve 2 (EGD augmented by additional sink term) in Figure 8.2.

It is easy to see that the critical control temperature can be found from the following equation

$$\exp\left(\frac{\theta_{ctrl}^{cr}}{1 + Ar\theta_{ctrl}^{cr}}\right) - \frac{1}{Se}\left(\theta_{ctrl}^{cr} + \theta_{ctrl}^{max}\right) = 0 \tag{8.5}$$

or, equivalently

$$\exp\left(\frac{\theta_{ctrl}^{cr}}{1 + Ar\theta_{ctrl}^{cr}}\right) - \frac{1}{Se}\theta_{ctrl}^{cr} - \frac{1}{SeAr} = 0 \tag{8.6}$$

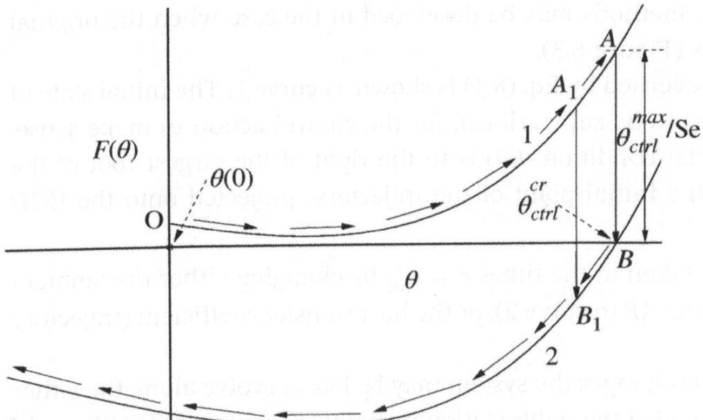

Figure 8.4 Definition of explosion critical control temperature. 1 – $F(\theta)$ Eq. (8.1); 2 – $F(\theta)$ Eq. (8.2). $\theta_{ctrl} = \theta_{ctrl}^{\max}$.

Geometrically, the condition described by Eqs. (8.5) and (8.6) means that the point B (trajectory reverse point after control action AB) lies right on the θ-axis (Figure 8.4).

Such a point is unstable; therefore, the actual control action must be taken before the temperature θ_{ctrl}^{cr} is reached. This explains the strict inequality in the above definition of the critical control temperature. An example of a successful control action (transition A_1B_1), where the control EGD function is strictly less than zero at reverse point B_1, is also shown in Figure 8.4.

The critical control temperature may be readily estimated by the numerical solution of the algebraic Eq. (8.6). For example, for $Ar = 0.01$ and $Se = 1.0$, the critical control temperature is $\theta_{ctrl}^{cr} \approx 4.85$, that is several characteristic temperature intervals over the initial temperature. This estimation broadly agrees with the quantitative necessary condition for thermal explosion (inequality (1.8)), which stipulates that the possibility of thermal explosion is determined by the reaction system dynamics over the temperature scale of the order of a characteristic temperature interval.

Also note that the above estimated value is well below the temperature at which the energy generation (Arrhenius) function has a maximum of its second derivative (recall definition of thermal explosion, Chapter 1), thus confirming that the definition of the critical control temperature makes sense.

In more detail (for different system parameters), the critical control temperatures are shown in Figure 8.5. Here, the considered Semenov numbers are in the supercritical range. The critical control temperature drops, at a fixed Arrhenius number, with an increase in the Semenov number. This evident trend in Figure 8.5 is easily predictable, taking into account that the EGD function, at a fixed temperature, increases with an increase in the Semenov number (formula (8.1)).

Corresponding to the critical control temperature, the critical control time τ_{ctrl}^{cr} may be defined as

$$\tau_{ctrl}^{cr} = \int_{\theta(0)}^{\theta_{ctrl}^{cr}} \left[\exp\left(\frac{\theta}{1 + Ar\theta} \right) - \frac{1}{Se}\theta \right]^{-1} d\theta \tag{8.7}$$

that is the time required for the system with the initial condition $\theta(0) = \theta_0$ to reach the critical control temperature.

The dependence of the critical control time on the Arrhenius and Semenov numbers is presented in Figure 8.6 for the trivial initial condition $\theta(0) = 0$. Therefore, the considered values of the

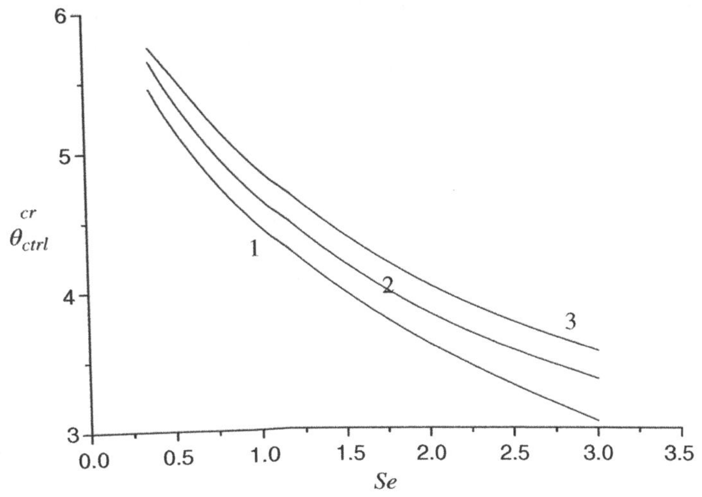

Figure 8.5 Dependence of critical control time on Arrhenius and Semenov numbers. 1 – $Ar = 0.02$; 2 – $Ar = 0.015$; 3 – $Ar = 0.012$.

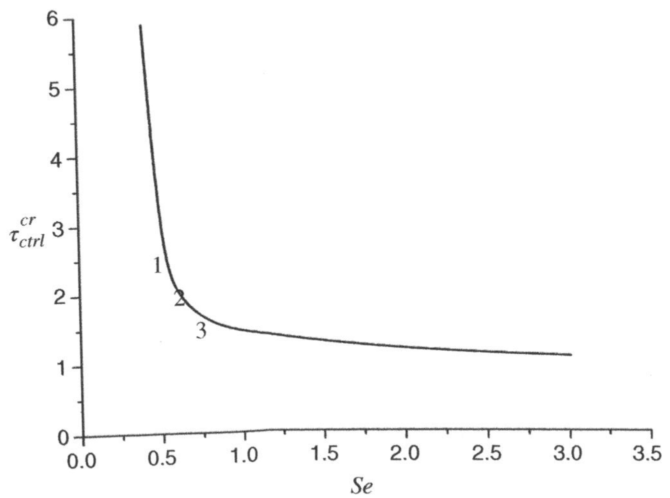

Figure 8.6 Dependence of critical control time on Arrhenius and Semenov numbers. 1 – $Ar = 0.02$; 2 – $Ar = 0.015$; 3 – $Ar = 0.012$ The curves are indistinguishable in the presented scale.

Semenov numbers must be supercritical, as indeed is the case in Figures 8.5 and 8.6. It should be noted the upper limit in the integral (8.7), which is the critical control temperature, is a function of Ar and Se itself (Eqs. (8.5) and (8.6)).

Figure 8.6 demonstrates the set of curves similar to that of Figure 8.5, in the sense that they are plotted for the same values of Arrhenius numbers and over the same range of Semenov numbers.

Remarkably, the curves for different values of Ar in Figure 8.6 collapse (in the presented scale) onto a single dependence. This is explained by a relatively small difference in the critical control temperature between the curves $1-3$ in Figure 8.5, combined with the steep rise of the EGD functions over the corresponding ranges of temperature change.

The other distinctive feature of the plot in Figure 8.6 is the existence of the two very different regions: the one in which the dependence $\tau_{ctrl}^{cr}(Se)$ drops steeply (Semenov numbers of up to $Se \approx 0.75$), and the other one in which the dependence quickly levels off ($Se > \approx 0.75$).

8.3 Smooth Control

Under smooth control, the governing parameters would change continuously, according to the prescribed laws, as the solution for the system temperature evolves.

The aim of this section is to construct such controls that would keep the system temperature, over an infinite time interval, under a prescribed temperature threshold. The latter must be set below the temperature corresponding to the maximum of the second derivative of the Energy Generation function (see the definition of thermal explosion, Chapter 1). Besides, an analytical treatment of the problem is sought, as much as possible.

The general idea of construction is the following.

A particular law of change of the Semenov number is used as the means of control, that is, the control function for the system will have a form of the EGD function (2.6) with a variable Semenov number.

Without the loss of generality, the Frank-Kamenetskii transformation (i.e. formal setting $Ar = 0$ in the EGD function) is used throughout this section. All the developments below, however, proceed with very minor changes for the general form ($Ar \neq 0$) of the EGD function, given by Eq. (2.6).

Majorizing control function is constructed first in such a form that it bounds the real control function from above, and the system temperature governed by the majorizing control function always stays below the prescribed threshold. In addition, the law of variation of the Semenov number may be chosen in such a way that Eqs (2.5) and (2.6) with right-hand side (RHS) replaced by this majorizing function are solvable analytically.

Assuming that the law of the Semenov number change is identical in both the majorizing and real control functions, the comparison theorem of the theory of ODE will ensure that the temperature of the real system is also bounded by the same threshold.

The analytical solution for majorizing control function will provide an explicit estimation of the behaviour of system temperature under control.

The major step in constructing the majorizing control function is illustrated in Figure 8.7. Here θ^* is the temperature at which control is enforced, and $\theta_1 < \theta_{F_{max}^{(2)}}$ is prescribed temperature limit (globally bounding temperature).

The linear majorizing function $\dot{G}^{maj}(\theta)$ is sought for the Energy Generation function $\dot{G}(\theta) = \exp(\theta)$ on the interval $[\theta^*, \theta_1]$. This linear function (line 2, Figure 8.7) is easily defined explicitly; the trivial exact formulae are presented later in this section.

Once the function $\dot{G}^{maj}(\theta)$ is defined, the following EGD functions may be identified:

$F^{maj}(\theta)$ – majorizing function for the original EGD function $F(\theta)$, defined by $F^{maj}(\theta) \equiv \dot{G}^{maj}(\theta) - \frac{1}{Se_0}\theta$. Obviously, similar to $\dot{G}^{maj}(\theta)$, it has a linear segment for $\theta^* \leq \theta \leq \theta_1$ (line 2, Figures 8.8 and 8.9), but with a different slope. What is important, the majorizing property of the linear segment is preserved upon transition from the energy generation to the EGD function (curves 1 and 2, Figures 8.8 and 8.9);

$F_{ctrl}(\theta)$ – EGD (control) function $F(\theta)$ with Semenov number replaced by the prescribed control function (curves 4 or 5, Figures 8.8 and 8.9). This is the control function governing the real system temperature under the control action;

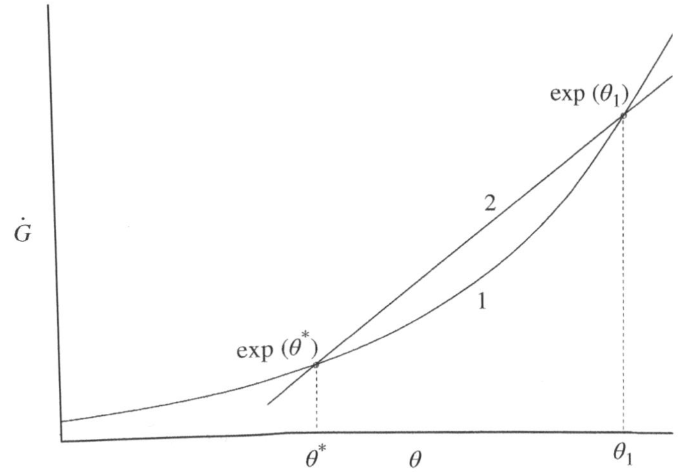

Figure 8.7 Linear majorizing function for the energy generation (Frank-Kamenetskii-transformed) function. 1 – original energy generation function $\dot{G}(\theta) = \exp(\theta)$; 2 – linear majorizing function $\dot{G}(\theta) \equiv \dot{G}^{maj}(\theta)$ for original $\dot{G}(\theta)$ on the interval $[\theta^*, \theta_1]$. Temperatures: $\theta_{ctrl} = \theta^*$ – start of control; $\theta = \theta_1$ – end of control.

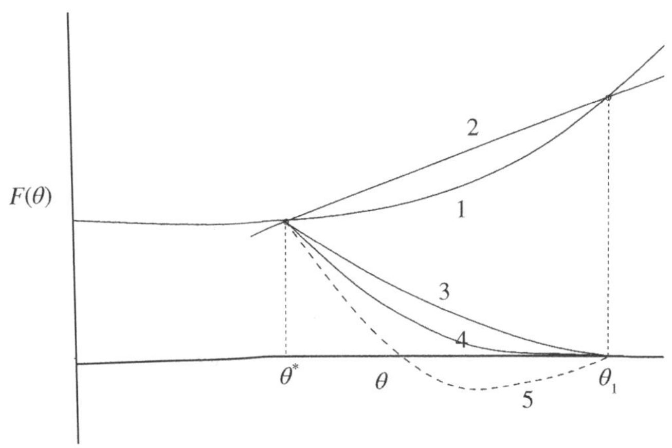

Figure 8.8 EGD functions under control action. 1 – original EGD function $F(\theta) = \exp(\theta) - \dfrac{1}{Se_0}\theta$; 2 – linear majorizing function $F^{maj}(\theta) \equiv \dot{G}^{maj}(\theta) - \dfrac{1}{Se_0}\theta$ for original $F(\theta)$ on the interval $[\theta^*, \theta_1]$; 3 – majorizing function upon application of control $F^{maj}_{ctrl}(\theta) \equiv \dot{G}^{maj}(\theta) - \dfrac{1}{Se(\theta,\tau)}\theta$; 4 – original EGD function under control $F_{ctrl}(\theta) = \exp(\theta) - \dfrac{1}{Se(\theta,\tau)}\theta$; 5 – one of possible alternative ways of behaviour of original EGD function under control $F_{ctrl}(\theta) = \exp(\theta) - \dfrac{1}{Se(\theta,\tau)}\theta$. Temperatures: $\theta_{ctrl} = \theta^*$ – start of control; $\theta = \theta_1$ – end of control. Curves under control action are shown only up to the end of control point.

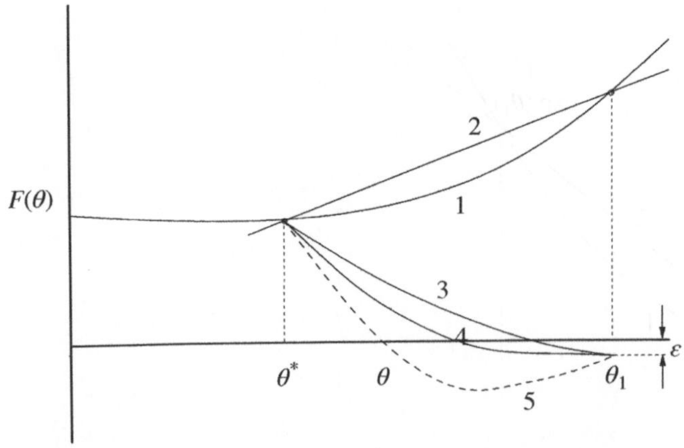

Figure 8.9 EGD functions under control action. Stable control. 1 – original EGD function $F(\theta) = \exp(\theta) - \frac{1}{Se_0}\theta$; 2 – linear majorizing function $F^{maj}(\theta) \equiv \dot{G}^{maj}(\theta) - \frac{1}{Se_0}\theta$ for original $F(\theta)$ on the interval $[\theta^*, \theta_1]$; 3 – majorizing function upon application of control $F^{maj}_{ctrl}(\theta) \equiv \dot{G}^{maj}(\theta) - \frac{1}{Se(\theta,\tau)}\theta$; 4 – original EGD function under control $F_{ctrl}(\theta) = \exp(\theta) - \frac{1}{Se(\theta,\tau)}\theta$; 5 – one of possible alternative ways of behaviour of original EGD function under control $F_{ctrl}(\theta) = \exp(\theta) - \frac{1}{Se(\theta,\tau)}\theta$. Temperatures: $\theta_{ctrl} = \theta^*$ – start of control; $\theta = \theta_1$ – end of control. Curves under control action are shown only up to the end of control point.

$F^{maj}_{ctrl}(\theta)$ – majorizing (control) function for $F_{ctrl}(\theta)$ with the identical law of change of the Semenov number (curve 3, Figures 8.8 and 8.9). This is control function governing the imaginable (overestimated) system temperature under the control action.

Behaviour of control functions depends on the chosen law of change of the Semenov parameter. Figures 8.8 and 8.9 illustrate (qualitatively) the choice needed for successful control. Indeed, the majorizing control function $F^{maj}_{ctrl}(\theta)$ (curve 3) turns into zero at or before the threshold temperature θ_1, thus the system temperature governed by this functions (as a RHS of equation analogous to (2.5) and (2.6)) cannot exceed this limit. Since the control function $F_{ctrl}(\theta)$ is bound by the majorizing control function $F_{ctrl}(\theta) \le F^{maj}_{ctrl}(\theta)$, then thanks to the basic comparison theorem of the theory of ODE the real system temperature, governed by $F_{ctrl}(\theta)$, would not exceed the same limit either. The aim of control would be achieved. The significance of majorizing control function $F^{maj}_{ctrl}(\theta)$ is that, as will be shown later, it delivers in certain cases an analytical solution that serves as an estimate (upper bound) for behaviour of real system temperature under the control action.

The principal difference between Figures 8.8 and 8.9 is that the strategy illustrated in Figure 8.9 (majorizing control function turns into slightly negative value, i.e. it is being 'submerged' below the abscissa axis, at $\theta = \theta_1$) ensures stable control. If it simply turns into zero (Figure 8.8), then random fluctuations at that point may, generally speaking, force the system temperature to exceed the limit $\theta = \theta_1$.

If the control provided by majorizing control function is stable (Figure 8.9), the same will be true for the real solution, governed by control function $F_{ctrl}(\theta)$.

The general form of the prescribed law of the Semenov number change is $Se = Se(\theta, \tau)$, that is, it may depend both on the solution and, explicitly, on the independent variable (time).

Control where the Semenov parameter depends explicitly on solution only $Se = Se(\theta)$ will be called an autonomous control, as the governing equation (of the form of (2.5) and (2.6)) will have in this case autonomous form. The opposite case of the joint explicit dependence $Se = Se(\theta, \tau)$ will be called non-autonomous control as it requires analysis of the non-autonomous equation.

Figures 8.8 and 8.9 are drawn in the autonomous variable, but the illustrations are applicable to both autonomous and non-autonomous controls as long as the solution of the ODE with control functions, analogous to (2.5) and (2.6), is monotonic. This is true up to the point where the RHS of the equation turns into zero. Beyond this point, the solution will not increase any further (i.e. trajectory actually ends up at the point of intersection of curve 4 or 5 with the abscissa axis, at the value less than θ_1), and no consideration or illustration is really necessary.

It should be noted that profiles of control functions (curves 3, 4 and 5) in Figures 8.8 and 8.9 are only indicative. Only the essential features, which are the functions being positive at $\theta = \theta^*$ and slightly negative (Figure 8.9) or zero (Figure 8.8) at $\theta = \theta_1$ are shown. In-between these two points, the control functions may behave somewhat differently, depending on the particular law of Semenov number variation $Se = Se(\theta, \tau)$ that has been chosen for the control. It is clear, however, that for the implementation of the control strategy shown in Figures 8.8 and 8.9 Semenov parameter must decrease between the initial and final control points, that is the requirement $Se(\theta^*, \tau(\theta^*)) > Se(\theta_1, \tau(\theta_1)) > 0$ must be fulfilled. In such a case, the difference between the heat generation and dissipation rates decreases as well (consider the analogue of EGD function (2.6) with variable Semenov number). In particular, the latter requirement is met by monotonically decreasing the Semenov parameter profile. Such type of profile has been implicitly assumed when drawing Figures 8.8 and 8.9.

Also, control functions are drawn only up to the point $\theta = \theta_1$ (end of control) in these figures. Turn now to precise quantitative formulae.

The linear majorizing segment, between $\theta = \theta^*$ and $\theta = \theta_1$, for the Energy Generation function $\dot{G}(\theta) = \exp(\theta)$ is defined as follows (Figure 8.7)

$$\dot{G}^{maj}\big|_{\theta^* \le \theta \le \theta_1} = l_1\theta + l_2 \tag{8.8}$$

where the parameters l_1, l_2 are defined from the obvious conditions

$$\begin{aligned} \dot{G}^{maj}(\theta^*) &= \dot{G}(\theta^*) = \exp(\theta^*) \\ \dot{G}^{maj}(\theta_1) &= \dot{G}(\theta_1) = \exp(\theta_1) \end{aligned} \tag{8.9}$$

to be

$$\begin{aligned} l_1 &= \frac{[\exp(\theta_1) - \exp(\theta^*)]}{[\theta_1 - \theta^*]} \\ l_2 &= \exp(\theta^*) - \frac{[\exp(\theta_1) - \exp(\theta^*)]}{[\theta_1 - \theta^*]}\theta^* \end{aligned} \tag{8.10}$$

The majorizing function for the Energy Generation function $\dot{G}(\theta) = \exp(\theta)$ is therefore

$$\dot{G}^{maj}(\theta) = \begin{cases} \dot{G}(\theta) = \exp(\theta), & 0 \le \theta \le \theta^* \\ l_1\theta + l_2, & \theta^* \le \theta \le \theta_1 \end{cases} \tag{8.11}$$

Consequently, there exists majorizing function for the original EGD function $F(\theta) = \exp(\theta) - \frac{1}{Se_0}\theta$, namely $F^{maj}(\theta) \equiv \dot{G}^{maj}(\theta) - \frac{1}{Se_0}\theta$, given by

$$F^{maj}(\theta) = \begin{cases} \exp(\theta) - \dfrac{1}{Se_0}\theta, & 0 \le \theta \le \theta^* \\ l_1\theta + l_2 - \dfrac{1}{Se_0}\theta, & \theta^* \le \theta \le \theta_1 \end{cases} \tag{8.12}$$

It also possesses the linear segment on $\theta \in [\theta^*, \theta_1]$ (Figures 8.8 and 8.9), and coincides with the original EGD function at points $\theta = \theta^*$ and $\theta = \theta_1$.

In the most general case, the smooth control function has the form

$$F_{ctrl}(\theta,\tau) = \dot{G}(\theta) - \frac{1}{Se(\theta,\tau)}\theta = \begin{cases} \dot{G} - \dfrac{1}{Se_0}\theta, & 0 \le \theta \le \theta^* \\ \dot{G} - \dfrac{1}{Se(\theta,\tau)}\theta, & \theta^* \le \theta \le \theta_1 \end{cases} = \begin{cases} \exp(\theta) - \dfrac{1}{Se_0}\theta, & 0 \le \theta \le \theta^* \\ \exp(\theta) - \dfrac{1}{Se(\theta,\tau)}\theta, & \theta^* \le \theta \le \theta_1 \end{cases} \tag{8.13}$$

where the precise law of change of the Semenov number

$$Se = \begin{cases} Se_0, & 0 \le \theta \le \theta^* \\ Se(\theta,\tau), & \theta^* \le \theta \le \theta_1 \end{cases} \tag{8.14}$$

$Se = Se(\theta, \tau) > 0$ on the interval $[\theta^*, \theta_1]$ needs to be specified. The control function is identical to the original EGD function up to the temperature $\theta = \theta^*$; the actual control is engaged in the temperature interval $\theta \in [\theta^*, \theta_1]$.

The majorizing control function is defined as

$$F^{maj}_{ctrl}(\theta,\tau) \equiv \dot{G}^{maj}(\theta) - \frac{1}{Se(\theta,\tau)}\theta = \begin{cases} \exp(\theta) - \dfrac{1}{Se(\theta,\tau)}\theta, & 0 \le \theta \le \theta^* \\ l_1\theta + l_2 - \dfrac{1}{Se(\theta,\tau)}\theta, & \theta^* \le \theta \le \theta_1 \end{cases} \tag{8.15}$$

The ODEs governing temperature evolution, under control, in the real and imaginable systems are, respectively,

$$\frac{d\theta}{d\tau} = F_{ctrl}(\theta,\tau); \tau \in [0, \tau(\theta_1)] \tag{8.16}$$

and

$$\frac{d\theta}{d\tau} = F^{maj}_{ctrl}(\theta,\tau); \tau \in [0, \tau(\theta_1)] \tag{8.17}$$

By virtue of the construction of majorizing functions,

$$l_1\theta + l_2 \ge \exp(\theta), \quad \theta^* \le \theta \le \theta_1 \tag{8.18}$$

therefore

$$F_{ctrl} \le F^{maj}_{ctrl}, \quad 0 \le \theta \le \theta_1 \tag{8.19}$$

The comparison theorem of the theory of ODE (Arnold 1992) states that the solution of Eq. (8.16) is bounded by the solution of Eq. (8.17) for all $\tau \in [0, \tau(\theta_1)]$.

Assuming that a suitable choice of controlling parameter variation has been chosen (to conform with general control concept, Figure 8.9), the solution controlled by majorizing control function will always stay below the limit $\theta = \theta_1$. Therefore, the same will also be true for the controlled temperature of the real system.

If the solution of the latter 'majorizing' equation (with the RHS given by majorizing control function) is obtained in analytical form, then an explicit estimation of the system temperature behaviour under control action may be provided.

In the following two subsections both autonomous and non-autonomous controls are considered. An example of the specific form of the Semenov number profiles $Se = Se(\theta)$, or $Se = Se(\tau)$ will be presented. A mixed case of non-autonomous control, with the dependence $Se = Se(\theta, \tau)$, as well as controls with delay, is not considered.

8.3.1 Smooth Autonomous Control

For simplicity, without the loss of generality, monotonic functions $Se = Se(\theta)$ are considered. Then, for successful control, the Semenov number must be a decreasing function of temperature.

In this subsection the law of the Semenov number change $Se = Se(\theta)$ is considered. All the ODEs considered below are, therefore, autonomous.

Temperature evolves, according to the Semenov model, with majorizing control function as (Chapter 2)

$$\frac{d\theta}{d\tau} = F_{ctrl}^{maj} \equiv \dot{G}^{maj}(\theta) - \frac{1}{Se(\theta)}\theta; \theta(0) = 0; 0 \leq \tau < \infty \tag{8.20}$$

with the solution (separating variables)

$$\int_0^\theta \left[\frac{1}{\dot{G}^{maj} - \dfrac{1}{Se(\theta)}\theta} \right] d\theta = \tau; 0 \leq \tau < \infty \tag{8.21}$$

and, for the real system, as

$$\frac{d\theta}{d\tau} = \exp(\theta) - \frac{1}{Se(\theta)}\theta; \theta(0) = 0; 0 \leq \tau < \infty \tag{8.22}$$

with the solution (separating variables)

$$\int_0^\theta \left[\frac{1}{\exp(\theta) - \dfrac{1}{Se(\theta)}\theta} \right] d\theta = \tau; 0 \leq \tau < \infty \tag{8.23}$$

As an example, the following form of monotonically decreasing, over the control interval, function $Se = Se(\theta)$ is considered

$$Se(\theta) = \begin{cases} Se_0; & 0 \leq \theta \leq \theta^* \\ \dfrac{1}{C_1\theta + C_2}; & \theta^* \leq \theta \leq \theta_1 \end{cases} \tag{8.24}$$

where $C_1 > 0$. The choice of a positive value of the parameter C_1 is possible for the following reasons explained below.

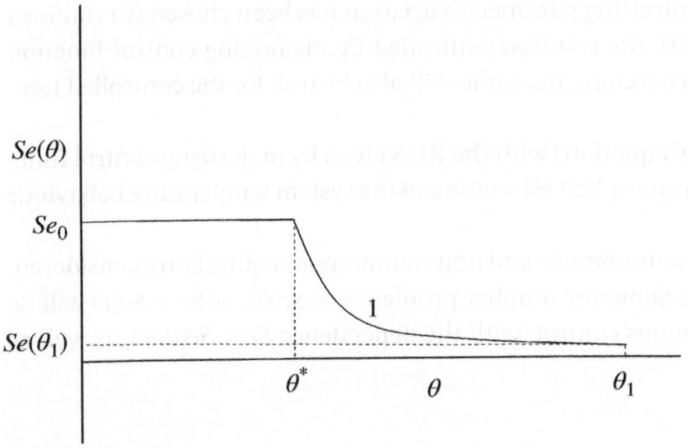

Figure 8.10 Sketch of Semenov number law of change for autonomous control. 1 – profile (8.24).

A sketch of the profile (8.24) is shown in Figure 8.10. Parameters C_1 and C_2 are determined from the requirements

$$Se(\theta^*) = Se_0 \tag{8.25}$$

and ('submerging' of majorizing control function)

$$F_{ctrl}^{maj}(\theta_1) = -\varepsilon \tag{8.26}$$

Physically meaningful values of the Semenov numbers are positive. Therefore, the following restriction on the values C_1,C_2 must also be fulfilled

$$Se(\theta) = \frac{1}{C_1\theta + C_2} > 0; \theta^* \leq \theta \leq \theta_1 \tag{8.27}$$

The requirements (8.25) and (8.26) transform, respectively, to the following equations for the parameters C_1 and C_2

$$C_1\theta^* + C_2 = \frac{1}{Se_0}$$
$$l_1\theta_1 - (C_1\theta_1 + C_2)\theta_1 + l_2 = -\varepsilon \tag{8.28}$$

The solution of these equations is easily obtained

$$C_1 = \frac{\left(l_1 + \dfrac{l_2 + \varepsilon}{\theta_1}\right) - \dfrac{1}{Se_0}}{(\theta_1 - \theta^*)}$$

$$C_2 = \frac{\dfrac{1}{Se_0}\theta_1 - \left(l_1 + \dfrac{l_2 + \varepsilon}{\theta_1}\right)\theta^*}{(\theta_1 - \theta^*)} \tag{8.29}$$

In the view of the first of Eq. (8.28), for fulfilment of the requirement (8.27), it is sufficient to have $C_1 > 0$.

Since $\theta_1 > \theta^*$ by definition (Figure 8.7), $l_1 > 0$ (Eq. (8.10)). It is clear that parameter l_2 is also positive for sufficiently small θ^* and θ_1 (Figure 8.7). It follows then, in the view of Eq. (8.29), that constant C_1 can be made positive by choosing the start and end of control temperatures θ^* and θ_1 to be sufficiently close to the initial temperature.

For $0 \leq \theta \leq \theta^*$, the majorizing control function coincides with the original EGD function $F_{ctrl}^{maj}(\theta) \equiv F(\theta)$, therefore, the solution for the system temperature has the form

$$\int_0^\theta \left[\frac{1}{\exp(\theta) - \frac{1}{Se(\theta)}\theta} \right] d\theta = \tau; 0 \leq \tau \leq \tau^* \tag{8.30}$$

where

$$\tau^* = \int_0^{\theta^*} \left[\frac{1}{\exp(\theta) - \frac{1}{Se(\theta)}\theta} \right] d\theta \tag{8.31}$$

is the time when the temperature reaches point $\theta = \theta^*$. This is also the time of the start of the control action, that is, $\tau_{ctrl} \equiv \tau^*$.

Beyond this point, for $\tau^* \leq \tau < \infty$, analytical solution may be obtained for the Eq. (8.20), with the initial condition replaced by $\theta(\tau^*) = \theta^*$, where the RHS is given by the function (following Eqs. (8.15) and (8.24))

$$\begin{aligned} F_{ctrl}^{maj} &= l_1\theta + l_2 - \frac{1}{Se(\theta)}\theta \\ &= -C_1\theta^2 + (l_1 - C_2)\theta + l_2;\ \theta^* \leq \theta \leq \theta_1 \end{aligned} \tag{8.32}$$

The system temperature time history will be majorized by this analytical expression. Since the latter, in turn, is bounded by construction of majorizing control function by the threshold θ_1, the real temperature profile will also be bounded at $\tau \to \infty$ by the same value. Thus the aim of the control action will be achieved.

Solution of the form (8.21)

$$\int_{\theta^*}^\theta \left[\frac{1}{l_1\theta + l_2 - \frac{1}{Se(\theta)}\theta} \right] d\theta = \tau - \tau^*; \tau^* \leq \tau < \infty \tag{8.33}$$

is transformed into the analytical form in the following way.

The integral on the left-hand side (LHS) takes the form

$$\int_{\theta^*}^\theta \left[\frac{1}{l_1\theta + l_2 - \frac{1}{Se(\theta)}\theta} \right] d\theta = \int_{\theta^*}^\theta \left[\frac{1}{-C_1\theta^2 + (l_1 - C_2)\theta + l_2} \right] d\theta \tag{8.34}$$

and is expressed analytically (Prudnikov, Brychkov and Marichev 1986), depending on the value of the discriminant

$$D = (l_1 - C_2)^2 + 4C_1 l_2 \tag{8.35}$$

Solution (8.33) is written then (in the regions where respective denominators are different from zero) in the following implicit analytical form

$$
\begin{aligned}
&-\frac{1}{C_1\left(\theta^{(1)}-\theta^{(2)}\right)}\left(\ln\left|\frac{\theta-\theta^{(1)}}{\theta-\theta^{(2)}}\right|-\ln\left|\frac{\theta^*-\theta^{(1)}}{\theta^*-\theta^{(2)}}\right|\right)=\tau-\tau^*;D>0\\
&-2\left[\frac{1}{(-2C_1\theta+(l_1-C_2))}-\frac{1}{(-2C_1\theta^*+(l_1-C_2))}\right]=\tau-\tau^*;D=0\\
&\frac{2}{\sqrt{-D}}\left[Arctg\left(\frac{-2C_1\theta+(l_1-C_2)}{\sqrt{-D}}\right)-Arctg\left(\frac{-2C_1\theta^*+(l_1-C_2)}{\sqrt{-D}}\right)\right]=\tau-\tau^*;D<0
\end{aligned}
\tag{8.36}
$$

Here $\theta^{(1)} < \theta^{(2)}$ are the real roots of the polynomial $F_{ctrl}^{maj}(\theta) = -C_1\theta^2 + (l_1 - C_2)\theta + l_2$

$$
\theta^{(1)} = \frac{(l_1 - C_2) - \sqrt{D}}{2C_1};\theta^{(2)} = \frac{(l_1 - C_2) + \sqrt{D}}{2C_1}
\tag{8.37}
$$

As discussed above, the easiest way to ensure the physical significance of the Semenov number is to choose parameters C_1 and l_2 to be positive. In this case the discriminant (8.35) will be positive as well $D > 0$. In fact, the choice $D > 0$ in (8.36) is dictated just by the assumed shape of majorizing control function, that is, curve 3 in Figure 8.9.

Indeed, majorizing control function assumes a positive value at $\theta = \theta^*$, and negative one at $\theta = \theta_1$ (Figure 8.9). For a quadratic polynomial, this is only possible when its discriminant is positive. That is, condition $D > 0$ is forced, in fact, by the requirements (8.25) and (8.26). (Condition (8.25) means that the majorizing control function at $\theta = \theta^*$ is equal to the original EGD function, which is assumed to correspond to supercritical conditions. Thus, the value $F_{ctrl}^{maj}(\theta^*)$ is positive.)

For the particular example considered, where the leading coefficient of the quadratic polynomial $F_{ctrl}^{maj}(\theta)$ is negative (8.32), curve 3 in Figure 8.9 will in fact be convex. The root of $F_{ctrl}^{maj}(\theta)$ (curve 3, Figure 8.9) lying between $\theta = \theta^*$ and $\theta = \theta_1$ will be, therefore, its largest root $\theta^{(2)}$ (8.37).

Solution (8.36), for the case $D > 0$, may be converted into an explicit form on the interval $[\theta^*, \theta^{(2)}]$. Taking into account that on this interval

$$
\theta - \theta^{(1)} > 0; \theta - \theta^{(2)} < 0; \theta^* - \theta^{(1)} > 0; \theta^* - \theta^{(2)} < 0
\tag{8.38}
$$

the solution (8.36) for $D > 0$ is readily transformed to

$$
\ln\left\{\frac{\left(\theta^{(2)}-\theta^*\right)}{\left(\theta^*-\theta^{(1)}\right)}\frac{\left(\theta-\theta^{(1)}\right)}{\left(\theta^{(2)}-\theta\right)}\right\}=C_1\left(\theta^{(2)}-\theta^{(1)}\right)(\tau-\tau^*);\theta\in\left[\theta^*,\theta^{(2)}\right]
\tag{8.39}
$$

and, with obvious further transformations, to the final explicit form

$$
\theta=\frac{\left\{\theta^{(1)}+\frac{\left(\theta^*-\theta^{(1)}\right)}{\left(\theta^{(2)}-\theta^*\right)}\exp\left[C_1\left(\theta^{(2)}-\theta^{(1)}\right)(\tau-\tau^*)\right]\theta^{(2)}\right\}}{\left\{1+\frac{\left(\theta^*-\theta^{(1)}\right)}{\left(\theta^{(2)}-\theta^*\right)}\exp\left[C_1\left(\theta^{(2)}-\theta^{(1)}\right)(\tau-\tau^*)\right]\right\}};\theta\in\left[\theta^*,\theta^{(2)}\right]
\tag{8.40}
$$

From formula (8.40) it is easy to see that the solution $\theta(\tau)$ approaches the root $\theta^{(2)}$ from the left at an infinitely large time

$$
\lim_{\tau\to\infty}\theta(\tau)=\theta^{(2)}
\tag{8.41}
$$

Due to the fact that majorizing control function, that is the quadratic polynomial (8.32), changes sign from positive to negative between $\theta = \theta^*$ and $\theta = \theta_1$ (conditions (8.25) and (8.26), Figure 8.9)

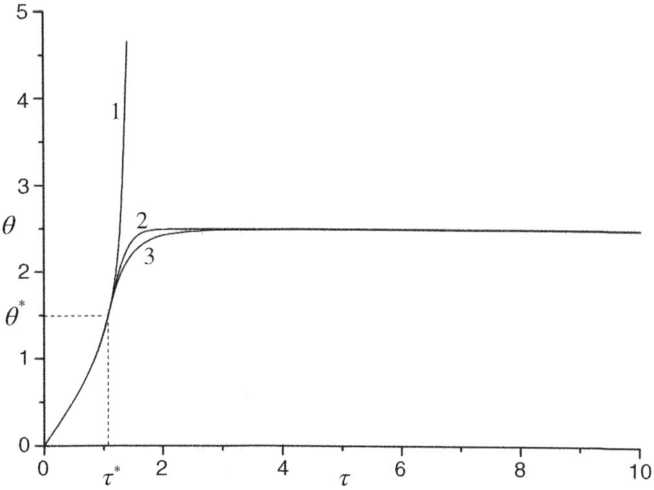

Figure 8.11 Example of smooth autonomous control. $\theta^* = 1.5$; $\theta_1 = 2.5$. 1 – uncontrolled thermal explosion, $Se = 1.0$. Curve similar to curve 2, Figure 2.9, but with Frank-Kamenetskii transformation being used ($Ar = 0$); 2 – analytical control solution (8.40) with majorizing control function. Semenov number variation (8.24); 3 – numerical control solution without majorization. Semenov number variation (8.24).

and its leading coefficient is negative its largest root $\theta^{(2)} < \theta_1$. Therefore, the solution (8.40), governed by majorizing control function, is bounded by θ_1, and the same, by virtue of the discussed above comparison theorem, is true for the temperature of the real system under control. If θ_1 is chosen below the temperature of the maximum of the second derivative of the EGD function, then the thermal explosion would be prevented by the control action of the present example.

An example of smooth autonomous control is presented in Figure 8.11. Curve 1 represents development of uncontrolled thermal explosion, for parameters identical to curve 2, Figure 2.9. Control action is taken at point $\theta^* = 1.5$, prescribed temperature limit is set as $\theta_1 = 2.5$. These input fix the parameters (8.10) of the majorizing function (8.11) as $l_1 \approx 7.70$, $l_2 \approx -7.07$. Further, the parameters (8.29) are estimated as $C_1 \approx 3.88 > 0$, $C_2 \approx -4.82$.

The discriminant (8.35) appears to be positive $D \approx 47.02 > 0$, therefore there are two real roots, given by (8.37), $\theta^{(1)} \approx 0.73$ and $\theta^{(2)} \approx 2.50$. The solution with majorizing control function is available in the explicit analytical form (8.40).

Figure 8.11 confirms that under performed controlled action the solution based on majorizing control function (curve 2) is bounded by $\theta^{(2)}$, as intended by the very construction of the majorizing function. Correspondingly, thanks to the comparison theorem, the actual solution is also bounded, upon control action, by the same finite value. This proves success of the performed control action.

8.3.2 Smooth Non-autonomous Control

In this subsection, the Semenov number is assumed to be a function of time only $Se = Se(\tau)$, with particular law of its change specified. This is the simplest form of the non-autonomous control function, compared to the general non-autonomous form $Se = Se(\theta, \tau)$. Consequently, the counterparts of ODEs (8.20) and (8.22) in subsection 8.3.1 become non-autonomous.

The temperature in the Semenov model with majorizing control function evolves now as

$$\frac{d\theta}{d\tau} = F_{ctrl}^{maj} \equiv \dot{G}^{maj}(\theta) - \frac{1}{Se(\tau)}\theta; \theta(0) = 0; 0 \le \tau < \infty \tag{8.42}$$

and, for the real system, as

$$\frac{d\theta}{d\tau} = \exp(\theta) - \frac{1}{Se(\tau)}\theta; \theta(0) = 0; 0 \leq \tau < \infty \tag{8.43}$$

In contrast to Section 8.3.1, the variables cannot be separated in neither of these equations.

However, with the majorizing energy generation function $\dot{G}^{maj}(\theta)$ defined by (8.11), Eq. (8.42) becomes linear (with variable coefficients) on the interval $[\theta^*, \theta_1]$, and its solution may be written in explicit integral form.

The development of this solution follows the standard procedure.

The Eq. (8.42) is written in the form

$$\frac{d\theta}{d\tau} = \left(l_1 - \frac{1}{Se(\tau)}\right)\theta + l_2; \tau^* \leq \tau < \infty \tag{8.44}$$

with the initial condition

$$\theta(\tau^*) = \theta^* \tag{8.45}$$

where the time τ^* corresponds to the temperature at the start of control $\tau^* = \tau(\theta^*)$. Solution $\theta(\tau)$ is monotonically increasing up to this point, so that temperature and time are uniquely related and the former inverse function exists.

First, the homogeneous equation

$$\frac{d\theta}{d\tau} = \left(l_1 - \frac{1}{Se(\tau)}\right)\theta; \tau^* \leq \tau < \infty \tag{8.46}$$

is solved.

Separating the variables, the general solution of Eq. (8.46) is readily written as

$$\theta = C\exp\left(l_1\tau - \int_{\tau^*}^{\tau} \frac{1}{Se(s)} ds\right); \tau^* \leq \tau < \infty \tag{8.47}$$

where $C \in \mathbb{R}$ is arbitrary constant.

A particular solution of the non-homogeneous equation is found by the variation of the constant method, that is, by setting $C = C(\tau)$ in (8.47) and substituting solution in this form back into the non-homogeneous Eq. (8.44).

The following ODE is then obtained for the function $C = C(\tau)$

$$\frac{dC}{d\tau} = l_2 \exp\left(\int_{\tau^*}^{\tau} \frac{1}{Se(s)} ds - l_1\tau\right); \tau^* \leq \tau < \infty \tag{8.48}$$

Its solution is

$$C(\tau) = l_2 \int_{\tau^*}^{\tau} \exp\left\{\int_{\tau^*}^{s} \frac{1}{Se(\xi)} d\xi - l_1 s\right\} ds + C_1; \tau^* \leq \tau < \infty \tag{8.49}$$

where $C_1 \in \mathbb{R}$ is undetermined so far constant.

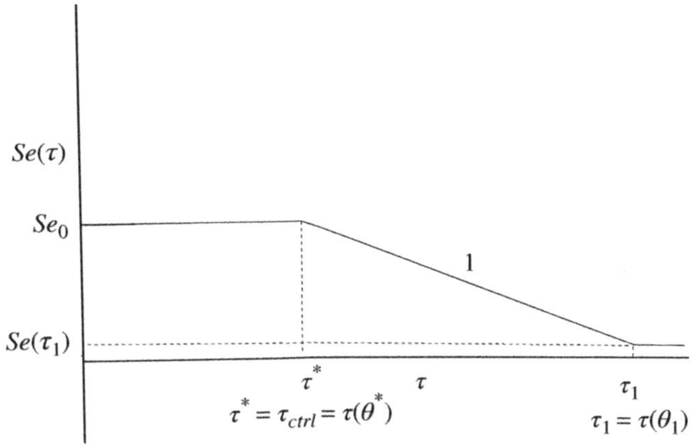

Figure 8.12 Sketch of Semenov number law of change for non-autonomous control. 1 – Profile (8.51)

The solution of the non-homogeneous equation is the sum of the general solution (8.47) and the particular solution. The constant C_1 is fixed by the initial condition $\theta(\tau^*) = \theta^*$.

The final result is

$$\theta(\tau) = \left\{ l_2 \int_{\tau^*}^{\tau} \exp\left\{ \int_{\tau^*}^{s} \frac{1}{Se(\xi)} d\xi - l_1 s \right\} ds + \theta^* \cdot \exp(-l_1\tau^*) \right\} \exp\left\{ l_1\tau - \int_{\tau^*}^{\tau} \frac{1}{Se(s)} ds \right\} ; \tau^* \leq \tau < \infty \tag{8.50}$$

Consider now three particular examples of non-autonomous control functions.

The first one employs the linear law of the Semenov number change on the interval $[\theta^*, \theta_1]$ (or, equivalently, $[\tau^*, \tau_1]$), shown in Figure 8.12.

$$Se = \begin{cases} Se_0; \tau \in [0, \tau^*] \\ Se_0 - \varphi(\tau - \tau^*); \tau \in [\tau^*, \tau_1]; \quad \varphi > 0 \\ Se_0 - \varphi(\tau_1 - \tau^*); \tau \in [\tau_1, \infty) \end{cases} \tag{8.51}$$

End of control time τ_1 is supposed to correspond to the limiting (which is also end of control) temperature θ_1 on the solution that develops under control. This value is not known *a priori*, and the choice of its particular value in Eq. (8.51) needs further discussion. This is provided later in this section.

According to the general control concept, the 'submerging' condition (8.26) must be fulfilled. This becomes

$$\left(l_1 - \frac{1}{Se(\tau_1)} \right) \theta_1 + l_2 = -\varepsilon \tag{8.52}$$

or

$$Se(\tau_1) = \frac{\theta_1}{l_1\theta_1 + l_2 + \varepsilon} \tag{8.53}$$

Condition (8.52) explains why control parameter φ in Eq. (8.51) must be chosen to be positive. Indeed,

$$\left(l_1 - \frac{1}{Se_0}\right)\theta^* + l_2 = F(\theta^*) > 0 \tag{8.54}$$

by the choice of the original EGD function, and fulfilment of (8.52) requires $Se(\tau_1) < Se_0$. For the adopted Semenov number profile (8.51), this is equivalent to the requirement $\varphi > 0$. The choice of the positive control parameter φ will be discussed later.

In addition, to maintain its physical significance, the Semenov number must always remain positive under the control action, therefore, the restriction

$$0 < \frac{\theta_1}{l_1\theta_1 + l_2 + \varepsilon} < Se_0 \tag{8.55}$$

has to be satisfied by the choice of parameters l_1, l_2 and θ_1.

Conditions (8.51) and (8.53) mean that parameter φ and end of control time τ_1 are related by

$$\varphi = \frac{1}{(\tau_1 - \tau^*)}\left[Se_0 - \frac{\theta_1}{l_1\theta_1 + l_2 + \varepsilon}\right] \tag{8.56}$$

In view of the right of the inequalities (8.55) parameter φ calculated by Eq. (8.56) is always positive.

The Semenov number profile (8.51) can be substituted into Eq. (8.50) to obtain the explicit form of the solution on the interval $\tau \in [\tau^*, \tau_1]$.

Using a number of intermediate results

$$\int_{\tau^*}^{s} \frac{1}{Se(\xi)} d\xi = \ln\left(\frac{Se_0}{Se_0 - \varphi(s - \tau^*)}\right)^{\frac{1}{\varphi}}; s \in [\tau^*, \tau_1] \tag{8.57}$$

$$\exp\left\{l_1\tau - \int_{\tau^*}^{\tau} \frac{1}{Se(s)} ds\right\} = \left[1 - \frac{\varphi}{Se_0}(\tau - \tau^*)\right]^{\frac{1}{\varphi}} \exp(l_1\tau); \tau \in [\tau^*, \tau_1] \tag{8.58}$$

$$\int_{\tau^*}^{\tau} \exp\left\{\int_{\tau^*}^{s} \frac{1}{Se(\xi)} d\xi - l_1 s\right\} ds = \frac{1}{l_1} \exp\left[-l_1\left(\frac{Se_0}{\varphi} + \tau^*\right)\right]\left(\frac{\varphi}{l_1 Se_0}\right)^{-\frac{1}{\varphi}} \times \int_{l_1\left(\frac{Se_0}{\varphi} - (\tau - \tau^*)\right)}^{l_1\frac{Se_0}{\varphi}} \eta^{-\frac{1}{\varphi}} \exp(\eta) d\eta; \tau \in [\tau^*, \tau_1] \tag{8.59}$$

solution (8.50) is written finally in the form

$$\theta(\tau) = \frac{l_2}{l_1} \exp\left[-l_1\left(\frac{Se_0}{\varphi} + \tau^*\right)\right]\left(\frac{\varphi}{l_1 Se_0}\right)^{-\frac{1}{\varphi}} \int_{l_1\left(\frac{Se_0}{\varphi} - (\tau - \tau^*)\right)}^{l_1\frac{Se_0}{\varphi}} \eta^{-\frac{1}{\varphi}} \exp(\eta) d\eta + \theta^*\left[1 - \frac{\varphi}{Se_0}(\tau - \tau^*)\right]^{\frac{1}{\varphi}} \exp(l_1(\tau - \tau^*)); \tau \in [\tau^*, \tau_1] \tag{8.60}$$

It should be noted that in the case of the specific Semenov number dependence (8.51), the general solution (8.50) may be written in the form (8.60) only up to point $\tau = \tau_1$. For $\tau > \tau_1$ Semenov number remains constant (formula (8.51)) and the intermediate results (8.57–8.59) will be expressed differently. However, due to 'submerging' restriction $F_{ctrl}^{maj}(\theta_1) < 0$ (Eqs. (8.26), or (8.52)), the solution governed by majorizing control function remains bounded by $\theta = \theta_1$ at all times $0 \leq \tau < \infty$. By virtue of the comparison theorem, real temperature profile, governed by the control function, will also be bounded at $\tau \to \infty$ by the same value.

The above conclusions are only true if $\theta(\tau_1) = \theta_1$ for the solution governed by majorizing control function. As mentioned earlier in the present subsection, this requirement cannot be fulfilled *a'priori* and requires special treatment.

The condition may be satisfied in the following way.

One may choose, as a first approximation, $\tau_1 = \tau(\theta_1)$ using solution for original EGD function. The latter is supposed to be chosen to correspond to the supercritical conditions, so the value $\theta = \theta_1 > \theta^*$ is achieved at some point in time. This value also fixes some positive first approximation value for the slope φ, due to formula (8.56).

Consider now the solution (8.60) with the latter value of φ. At time $\tau = \tau_1$ (defined above) such solution provides some temperature $\theta(\tau_1) \neq \theta_1$.

Solution (8.60) is a continuous function of $\varphi > 0$. Also, φ is continuous function of τ_1, due to formula (8.56). Therefore, the solution (8.60) is a continuous function of τ_1. Due to this continuity, values of τ_1 close to τ^* will result in solutions with $\theta(\tau_1)$ close to θ^*. On the other hand, as $\tau_1 \to \infty$ $\varphi \to 0$, that is, profile (8.51) flattens out and approaches asymptotically the original EGD profile $Se(\tau) \equiv Se_0$. As the original EGD corresponds, by definition of the control problem, to supercritical conditions, $\theta(\tau_1) \to \infty$ as $\tau_1 \to \infty$. In other words $\theta(\tau_1)$ is a continuous function on $\tau_1 \in [\tau^*, \infty)$ assuming values both smaller and larger than θ_1. Therefore, there exists such value τ_1 that $\theta(\tau_1) = \theta_1$. This value of τ_1 must be used to obtain the value of the slope $\varphi > 0$ from (8.56), and then obtain majorizing control solution from Eq. (8.60). It must be emphasized that the requirement $\theta(\tau_1) = \theta_1$ may only be satisfied by an iterative procedure involving repetitive numerical evaluation of the solution (8.60) with different values of $\tau_1 \in [\tau^*, \infty)$ (and with corresponding to them, according to (8.56), different values of the parameter φ).

Solution (8.60) is bounded by construction, due to the 'submerging' condition (8.26), by the value $\theta = \theta_1$. In the view of the comparison theorem, the same is true for the temperature history of the real system under the control action.

The second example is delivered by the convex exponential profile

$$Se(\tau) = \begin{cases} Se_0; \tau \in [0, \tau^*] \\ C_1 - \exp(\varphi(\tau - C_2)); \tau \in [\tau^*, \tau_1] \\ C_1 - \exp(\varphi(\tau_1 - C_2)); \tau \in [\tau_1, \infty) \end{cases} \tag{8.61}$$

sketched in Figure 8.13.

The profile (8.61) has three parameters. Two of them, the constants C_1, C_2, are fixed by the requirements at the start and the end of the control. The Semenov number at the end of control $Se(\tau_1)$ is fixed exactly as in the previous example (formula (8.53)) with the positiveness requirement (the left of two inequalities (8.55)).

Therefore,

$$Se(\tau^* = \tau_{ctrl}) = Se_0$$

$$Se(\tau = \tau_1) = \frac{\theta_1}{l_1\theta_1 + l_2 + \varepsilon} \tag{8.62}$$

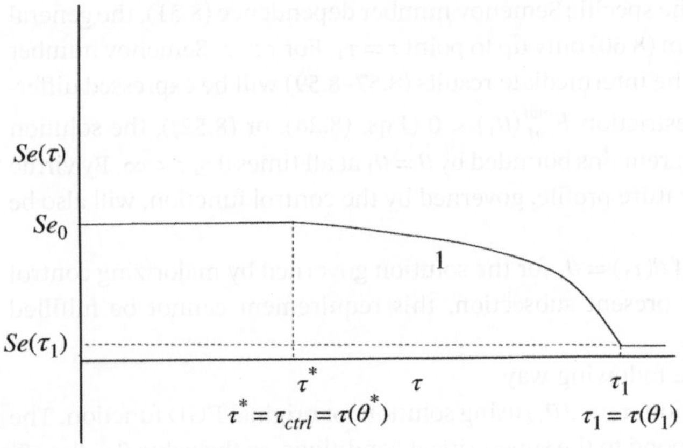

Figure 8.13 Sketch of Semenov number law of change for non-autonomous control. 1 – profile (8.61).

Parameter φ provides, in contrast to the previous example, an extra degree of freedom and may be chosen to assume any positive value. Under such conditions, profile (8.61) always (in fact, for arbitrary C_1, C_2, not necessarily restricted by (8.62)) specifies monotonically decreasing between $\tau = \tau^*$ and $\tau = \tau_1$ function, thus fulfilling the requirement of the right of the inequalities (8.55).

Eqs. (8.62) translate into the following set of equalities:

$$\begin{cases} C_1 = Se_0 + \exp(\varphi(\tau^* - C_2)) \\ C_1 = Se(\tau_1) + \exp(\varphi(\tau_1 - C_2)) \end{cases} \tag{8.63}$$

from where

$$\exp(-\varphi C_2)[\exp(\varphi\tau_1) - \exp(\varphi\tau^*)] = Se_0 - Se(\tau_1) \tag{8.64}$$

and further

$$\exp(\varphi C_2) = \frac{[\exp(\varphi\tau_1) - \exp(\varphi\tau^*)]}{(Se_0 - Se(\tau_1))} \tag{8.65}$$

From the last equation

$$C_2 = \frac{1}{\varphi}\ln\left\{\frac{[\exp(\varphi\tau_1) - \exp(\varphi\tau^*)]}{(Se_0 - Se(\tau_1))}\right\} \tag{8.66}$$

Now C_1 can be found from the first of Eqs. (8.63), with the help of (8.64)

$$C_1 = Se_0 + \frac{(Se_0 - Se(\tau_1))\exp(\varphi\tau^*)}{[\exp(\varphi\tau_1) - \exp(\varphi\tau^*)]} \tag{8.67}$$

The law of the Semenov number change (8.61) with the determined constants C_1, C_2 ((8.67) and (8.66)) can now be substituted into the formula for the general solution (8.50). The result reads

$$\begin{aligned}\theta(\tau) = &\left\{ l_2 \int_{\tau^*}^{\tau} \exp\left\{ \int_{\tau^*}^{s} \frac{d\xi}{[C_1 - \exp(\varphi(\xi - C_2))]} - l_1 s \right\} ds \right. \\ &\left. + \theta^* \cdot \exp(-l_1\tau^*) \right\} \exp\left\{ l_1\tau - \int_{\tau^*}^{\tau} \frac{ds}{[C_1 - \exp(\varphi(s - C_2))]} \right\}; \tau \in [\tau^*, \tau_1]\end{aligned} \tag{8.68}$$

In order to achieve control aim solution (8.68) must be bounded by $\theta = \theta_1$. This occurs upon forcing the correspondence $\theta(\tau_1) = \theta_1$, due to exactly the same arguments as in the previous example. The latter forcing can be done by exactly the same reasoning, and by the same type of numerical iterative procedure as in the previous example.

The result will be bounded behaviour of temperature time history under the majorizing control function and, correspondingly, identically bounded behaviour of the temperature time history for the real reacting system at all times $0 \le \tau < \infty$.

The final example of the non-autonomous control features profile which is also exponential, but concave in this case

$$Se(\tau) = \begin{cases} Se_0; \tau \in [0, \tau^*] \\ C_1 \exp(-C_2\tau); \tau \in [\tau^*, \tau_1] \\ C_1 \exp(-C_2\tau_1); \tau \in [\tau_1, \infty) \end{cases} \tag{8.69}$$

The sketch of this profile is presented in Figure 8.14. Development proceeds in exactly the same way as in the two previous examples.

The value $Se(\tau_1)$ is fixed by the requirement (8.53), and the inequalities (8.55) still need to be fulfilled by appropriately choosing parameters l_1, l_2 and θ_1.

The set of equations

$$\begin{cases} C_1 \exp(-C_2\tau^*) = Se_0 \\ C_1 \exp(-C_2\tau_1) = Se(\tau_1) \end{cases} \tag{8.70}$$

determines the constants C_1, C_2 uniquely.

In particular,

$$\exp(C_2(\tau_1 - \tau^*)) = \frac{Se_0}{Se(\tau_1)} \tag{8.71}$$

from where

$$C_2 = \frac{1}{(\tau_1 - \tau^*)} \ln\left[\frac{Se_0}{Se(\tau_1)}\right] \tag{8.72}$$

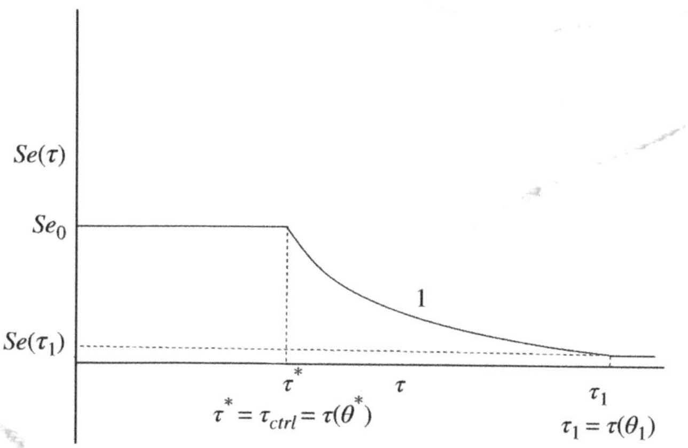

Figure 8.14 Sketch of Semenov number law of change for non-autonomous control. 1 – profile (8.69).

Then C_1 is easily found

$$C_1 = Se_0 \left[\frac{Se_0}{Se(\tau_1)} \right]^{\frac{\tau^*}{(\tau_1 - \tau^*)}} \tag{8.73}$$

Being substituted into the formula for the general solution (8.50), the law of time history of the Semenov number change (8.69) generates, with the help of the following intermediate results

$$\int_{\tau^*}^{s} \frac{1}{Se(\xi)} d\xi = \frac{1}{C_1 C_2} [\exp(C_2 s) - \exp(C_2 \tau^*)] \tag{8.74}$$

$$\int_{\tau^*}^{\tau} \frac{1}{Se(s)} ds = \frac{1}{C_1 C_2} [\exp(C_2 \tau) - \exp(C_2 \tau^*)] \tag{8.75}$$

the final explicit analytical formula for the solution

$$\begin{aligned} \theta(\tau) = \Bigg\{ & l_2 \exp\left[-\frac{1}{C_1 C_2} \exp(C_2 \tau^*) \right] \int_{\tau^*}^{\tau} \exp\left\{ \frac{1}{C_1 C_2} \exp(C_2 s) - l_1 s \right\} ds \\ & + \theta^* \exp(-l_1 \tau^*) \Bigg\} \exp\left\{ l_1 \tau - \frac{1}{C_1 C_2} [\exp(C_2 \tau) - \exp(C_2 \tau^*)] \right\}; \tau \in [\tau^*, \tau_1] \end{aligned} \tag{8.76}$$

The condition $\theta(\tau_1) = \theta_1$ is forced in exactly the same way as in the previous two examples, resulting in the bounded, by $\theta = \theta_1$ at all times, solutions under both majorizing and real control functions.

An example of smooth non-autonomous control may be constructed in the way similar to that illustrated in Figure 8.11.

9

Thermal Explosion Prevention

9.1 Concept of Thermal Management. Passive and Active Methods

As has been shown in Chapter 8, thermal explosion may be prevented by changing condition in which chemical reaction occurs, in particular by varying Semenov number of the reactive system. This emphasizes, from both theoretical and practical points of view, the importance of careful investigation and use of control possibilities arising in course of unfolding of a particular physical and/or chemical process. The concept of controlling evolution of any thermodynamic system, to ensure its safe, robust and optimum (in some sense) operation is known as thermal management.

Common examples of thermal management are restrictions of operating temperatures and energy release densities in various electronic components and devices (computers, cell phones, electrical domestic appliances, etc.).

Advanced, very sophisticated methods of design, control and optimization of thermodynamic systems have been developed by Bejan, based on the proposed so-called constructal law (Bejan *et al.* 1996; Bejan 2000). This approach allows a number of important problems to be solved. For example, it suggests optimum pathways for heat dissipation in electronic devices, the problem already mentioned above. Other examples include emergence of tree-like structures as optimum design for convective mass transfer between point source/sink and given volume. Such considerations explain structures of human respiratory system and river drainage basins. This understanding of natural systems may greatly assist in design required for engineering applications having analogous optimization goals and constraints.

Similar methods for optimum design may be applied to problems of heat transfer.

Nevertheless, optimization approaches based on Bejan's constructal law have not yet been exploited in the problems related to thermal runaway.

This may be partially explained by the fact that the problem of managing energy dissipation aggravates dramatically in the latter case, compared to conventional electronic devices, due to a different law governing heat generation: highly non-linear and looped into positive feedback mechanism.

Generally, thermal management methods may be classified into the two groups: passive and active.

Taking thermal explosion phenomenon specifically, passive thermal management methods are precautions aimed at eliminating the very possibility of explosion development. They are implemented prior to potential start of thermal runaway, and are typically related to overall system design.

Thermal Explosion: Theory and Application, First Edition. Vasily B. Novozhilov.

Active thermal management methods aim at suppressing thermal runaway during its induction period. Such methods require some kind of intervention, resulting in a change, instantaneous or dynamical, of parameters of the system. As examples of active thermal management, some cases of thermal explosion control by change of the system parameters have been discussed in Chapter 8.

The difference between the two groups may be understood easily taking as an illustration example, familiar in general to everyone, of house fire protection.

In this case examples of passive protection methods are the use of non- or hardly- combustible materials, compartmentation (to limit smoke spread), provisions for fire exits and stairwells. All these measures are implemented at design and construction stages, well ahead of potential fire break-out.

Active methods of fire management (control) assumes the form of operation of water sprinklers, direct intervention by fire brigade, etc. These measures are implemented during fire development.

Specific passive and active methods are discussed below in Sections 9.2–9.5.

Passive methods are discussed below in Section 9.2.

Some active protection methods, related to dynamic change of the system's Semenov number, have been discussed already in Section 8.3, Chapter 8. Additional methods are discussed in Sections 9.3–9.5. The last of these Sections is devoted specifically to prevention of fire flashover.

9.2 Passive Methods

Passive prevention methods, as outlined above, refer to measures taken in advance in order to eliminate the very possibility of thermal explosion development. As such, the aim of application of such methods is to provide long-term safe storage of substances susceptible to exothermic chemical transformations.

The most straightforward of such methods are keeping mixture compositions outside of flammability limits, or (if within flammability limits) below the minimum autoignition temperature.

Producing mixture compositions that are outside flammability limits (at intended storage pressure and temperature) provides a very conservative protection. Indeed, even within flammability limits and below the minimum autoignition temperature an external source of energy (i.e. spark ignition) is required to propagate flame through the mixture (Beyler 1995).

If stored mixture has to be within its flammability limits, then it must be kept below the minimum (across all the flammable compositions) autoignition temperature. This provides an adequate passive protection against spontaneous ignition.

Examples of flammability limits and minimum autoignition temperatures, for various chemicals, are provided in Tables 9.1 and 9.2.

Beyler (1995) may be consulted for good overview of basic theory and practical implications of flammability limits.

Generally, as a measure of passive protection, Semenov or Frank-Kamenestkii (depending on assumed degree of idealization) parameter of the reactive mixture must be kept below its critical value (i.e. the reactive system must possess a certain degree of sub-criticality). This consideration offers certain choices in concrete passive protection arrangements.

For example, attention should be paid to geometry of the vessel, that is being used for storage. As evident from Table 2.1, spherical vessel provides, among the three symmetrical shapes, the highest critical value of Frank-Kamenetskii parameter. This shape allows, therefore, the use of the larger characteristic size of the vessel for the same degree of system sub-criticality.

Table 9.1 Lower and upper flammability limits (LFL and UFL) of selected gases and vapours in air at $\tilde{T} = 298$ K and $\tilde{P} = 101$ kPa.

Substance	LFL (%) volume	UFL (%) volume
Acetylene	2.5	100
Diethyl ether	1.9	36
Ethane	3.0	12.4
Ethylene	2.7	36
n-Heptane	1.05	6.7
n-Hexane	1.2	7.4
Isobutane	1.8	8.4
Methane	5.0	15.0
Propane	2.1	9.5
Propylene	2.4	11

Source: Adapted from Beyler (1995).

Table 9.2 Minimum autoignition temperature of selected gases and vapours in air at $\tilde{T} = 298$ K and $\tilde{P} = 101$ kPa.

Substance	Minimum autoignition temperature, °K
Acetylene	578
Diethyl ether	433
Ethane	788
Ethylene	763
n-Heptane	488
n-Hexane	498
Isobutane	733
Methane	813
Propane	723
Propylene	733

Source: Adapted from Beyler (1995).

Similarly, for a given degree of system sub-criticality, increased value of the heat transfer coefficient provides an advantage of the use of larger volume of the vessel (see definition of Semenov number (2.2)).

In a more general case of Newton's heat exchange with surroundings (boundary condition of the second type, Section 3.1, Chapter 3) critical conditions depend on Biot number (formulae (3.9), (3.10), (3.13) and (3.14)). The ratio of critical Frank-Kamenetskii parameters at given Bi and at $Bi \to \infty$ (the latter corresponds to classical Frank-Kamenetskii problem with the boundary condition of the first type) is sharply growing function of Bi (formulae (3.13) and (3.14); Figure 3.1).

Therefore, increasing Biot number, by means of increasing heat transfer coefficient, allows in this case also the use of larger size of the vessel at the same degree of sub-criticality.

Advantage of higher values of heat transfer coefficient in case of passive protection suggests, quite naturally, the use of permanent forced convective cooling of reactive system. Basic engineering correlations for forced convection in external flow around bodies of simple shapes (e.g. plate, cylinder, sphere) are well known (Incropera and DeWitt 2002; Özişik 1985). These correlations may be used for calculations in support of development of specific passive protection designs.

Forced convective cooling may also be used in more complicated arrangements involving not just a single vessel, but a number of bodies susceptible to thermal explosion. This type of thermal management design is applicable, for example, to protection of packs of electric batteries. Schematic sketch of such arrangement is presented in Figure 9.1.

Forced convective cooling may be further enhanced by application of two-phase flow, in most straightforward form by loading flow with water mist. If thermal explosion starts to develop, then heating up, as well as partial or complete evaporation of fine droplets provide an additional intake of energy by cooling stream, extracting it at the same time from reacting system.

Practical calculation methods and correlations for all stages of droplet heat-up and evaporation are widely available (Lefebvre 1989; Incropera and DeWitt 2002; Özişik 1985).

Various very specific passive protection methods are applicable to Lithium-ion Batteries (Lee *et al.* 1999; Wang *et al.* 2012).

Such passive protection measures include:

a) coating of cathode materials. Surface of cathode materials may be coated with oxides such as MgO, Al_2O_3, SiO_2, TiO_2, ZnO, SnO_2, ZrO_2. Such material modification leads to a number of benefits (Wang *et al.* 2012): direct contact between cathode and electrolyte solution is prevented, phase transition suppressed, structural stability is improved and disorder of cations in crystal sites reduced. As a result, heat generation during charging cycling is decreased, thus providing passive explosion protection.
b) anode surface treatment. Since thermal decomposition of SEI is the most easily triggered exothermic reaction in lithium-ion cells, SEI thermal stability is a deciding factor in battery safety. This stability may be improved by mild oxidation, deposition of metals and metal oxides, coating with polymers and other carbon-containing substances. These modifications improve surface stability of graphitic carbon anodes by removing some reactive sites and/or defects from the surface, forming an oxide layer and covering active edge structures on the surface. In this way direct contact of graphite with electrolyte solution is prevented, its surface reactivity with electrolyte, decomposition of electrolyte and co-intercalation of solvated lithium ions are all reduced.

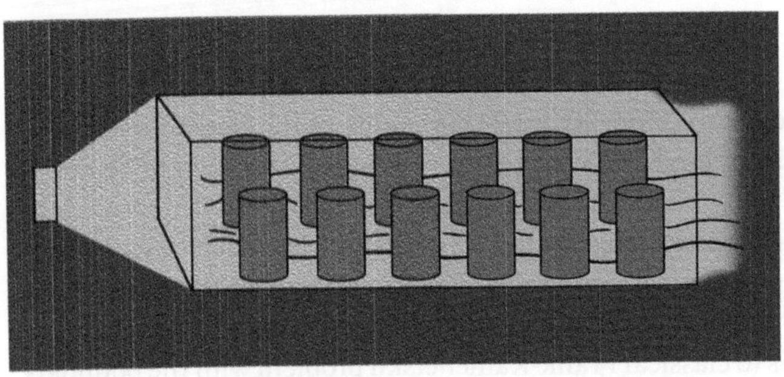

Figure 9.1 Schematic illustration of forced convective cooling flow around an array of cylinders.

c) modification of electrolyte with additives. The types of additives directly related to passive thermal explosion protection are flame retardant additives and overcharge additives.

Potential Flame Retardants (FRs) should not only be efficient in suppressing flammability of liquid electrolytes, but should also be electrochemically stable on both cathode and graphitic anode. FRs applied in Lithium-ion batteries are typically based on organic phosphorous compounds and their halogenated derivatives. These are, for example, trimethyl phosphate (TMP), dimethyl methylphosphonate (DMMP), triphenyl phosphate (TPP), cresyl diphenyl phosphate (CDP), diphenyl octyl phosphate (DOP), alkyl phosphate, hexamethylphosphoramide (HMPA), tributyl phosphate (TBP), and others (Wang *et al.* 2012).

As for non-phosphorus FRs, the use of fluorinated propylene carbonates and methyl nonafluorobutyl ether (MFE) have been attempted. Another electrolyte additive, succinonitrile can reduce the amount of gas emitted at high temperatures, increase the onset temperatures of exothermic reactions and decrease their heat release rates (Wang *et al.* 2012).

Lee *et al.* (1999) reported synthesis of a new FR additive to electrolyte, hexamethoxycyclotriphosphazene which, according to their experiments, is capable of reducing self-heating rate of lithium-ion battery by up to 70%.

Overcharge additives, according to their functionality, are classified into the two groups: redox shuttle additives and shutdown additives.

Redox shuttle is a molecule which can readily transfer electrons between the cell electrodes. It facilitates such movement of charge by undergoing reversible oxidation-reduction (redox) reactions at each of the electrodes. Overcharge is prevented by dissipating excess energy as heat through this electron exchange process.

Shutdown additive, being exposed to high temperatures, undergoes a reaction which causes the electrolyte to solidify or significantly change its properties, effectively stopping the flow of ions and shutting the cell down.

It is important to note that redox shuttle additives protect the cell form overcharge reversibly, while shutdown additives terminate cell operation permanently.

Compounds explored as redox shuttle additives include thianthrene derivatives, polysulfides, 2,5-ditertbutyl-1,4,-dimethoxybenzene (DDB), 4-tertbutyl-1,2-dimethoxybenzene (TDB), 3-chloroanisole (3CA), 2,2,6,6-tetramethylpiperidine-1-oxyl (TEMPO), diphenylamine (DPA), and others (Wang *et al.* 2012). Unfortunately, these substances are active only at high charging voltages, consequently they do not actually respond to heat generation in batteries.

Redox chemical shuttles can only provide limited overcharge protection and cannot prevent catastrophic failure induced by severe overcharging. This means that, in the view of paramount importance of safety, additional shutdown protection mechanisms such as polymerizable additives must be incorporated, even at the cost of terminating the battery's useful life (Wang *et al.* 2012).

The shutdown additives, such as cyclohexylbenzene, biphenyl, xylene and others terminate cell operation permanently.

d) application of safety devices. The aims of safety devices are to release excessive pressure and/or energy accumulating during induction period of thermal runaway.

These goals are achieved by application of safety vents, thermal fuses, Positive Temperature Coefficients (PTS) devices and shutdown separators at cell level, as well as by application of special Battery Management Systems (BMS) and Thermal Management Systems (TMS) at Module/Pack levels, and Fire Detection and Suppression Systems at compartment levels (Wang *et al.* 2012; Ghiji *et al.* 2020).

Pressure relief vents open at elevated internal cell pressures facilitating hot gas evacuation from the cell and preventing cell rupture. Caution, nevertheless, should be exercised in application of these devices as after the relief event ambient air containing oxygen enters the cell with renewed risk of starting exothermic reactions.

Thermal fuses are wires made of alloys that melt once current flow passing through it reaches at a pre-set level. They shut down the battery permanently. However, if the voltage is too high disconnection of thermal fuse may not be sufficient for prevention of thermal explosion (Wang *et al.* 2012). Thermal fuses typically melt at about 30–50°C above the maximum operating temperature of the battery.

PTS elements consist of conductive polymer forming a closed circuit in normal conditions. It is activated at abnormally high currents and high temperatures by melting the polymer. This causes the resistivity of the element to rise sharply, which in turn leads to fall of the cell's current and corresponding decrease of energy release rate.

Shutdown separators consist of layered films with different phase transition temperatures which can close the separator pores and halt lithium ion transport (Ghiji *et al.* 2020).

BMS is the primary safety device at Module and Pack levels. It controls and prevents over-charge, over-discharge, and operates the battery pack at optimum performance level to maximize its life-span. It functionality normally includes data acquisition, battery state determination, as well as electrical, thermal and safety management. TMS works alongside with BMS to maintain optimum operating temperature (typically from 20 to 40° C) for each cell and to smooth temperature gradients within and between modules.

At the level of compartments, hosting Battery Systems, appropriate Fire Detection and Suppression Systems need be installed. These have been reviewed extensively by Ghiji *et al.* (2020).

9.3 Inertization

Reactive mixture inertization is a method which can potentially be used in active mode to prevent thermal explosion.

Consider the system described by Semenov theory which is at initial supercritical condition. It is assumed, for simplicity, that at this initial state mixture consists of one reactant, undergoing first order chemical decomposition, and one inert additive. Amounts of both substances are fixed initially.

Assume that the same additive being added to the system at prescribed mass flow (injection) rate $\tilde{I}(\tilde{t})$. As a result, the mass of the system would increase, and energy, generated by chemical reaction, spent on heating up unit of mass decrease. Therefore, heating rate of the mixture during induction period would decrease as well. It may be expected that an appropriate choice of time injection profile $\tilde{I}(\tilde{t})$, combined with slowing down of chemical reaction due to reactant consumption, would prevent thermal runaway.

Consider corresponding mathematical formulation.

Mass and density of the system increase as

$$\tilde{M}(\tilde{t}) = \tilde{M}_0 + \int_0^{\tilde{t}} \tilde{I}(s)ds \qquad (9.1)$$

$$\tilde{\rho}(\tilde{t}) = \frac{\tilde{M}(\tilde{t})}{\tilde{V}}$$

This means that density time history may be written as

$$\tilde{\rho}(\tilde{t}(\tau)) = \tilde{\rho}_0[1 + \psi(\tilde{t}(\tau))] \tag{9.2}$$

where $\psi(\tau)$ is non-dimensional function

$$\psi(\tau) = \frac{1}{\tilde{\rho}_0 \tilde{V}} \int_0^{\tilde{t}(\tau)} \tilde{I}(s)ds \tag{9.3}$$

Here the definition of non-dimensional time τ is modified (compared to (2.2)) to reflect thermodynamic process occurring at constant volume

$$\tau = \frac{\tilde{Q}\tilde{k}_0\tilde{E}}{\tilde{\rho}_0\tilde{c}\tilde{R}\tilde{T}_0^2}\tilde{t} \tag{9.4}$$

By repeating the derivation of the energy conservation equation for Semenov model (2.4), it is easy to see that for the present case of open system it transforms into

$$\frac{d}{d\tau}[(1 + \psi(\tau))\theta] = \exp\left(\frac{\theta}{1 + Ar\theta}\right) - \frac{1}{Se}\theta \tag{9.5}$$

In a similar way, kinetic Eq. (4.9) (for equation of the first order; $m = 1$) transforms into

$$\frac{d\eta}{d\tau} = \frac{(1 + \psi(\tau))}{Td}\exp\left(\frac{\theta}{1 + Ar\theta}\right)(1 - \eta) \tag{9.6}$$

where Todes number is modified (compared to (4.8)) as

$$Td = \frac{\tilde{Q}\tilde{E}\tilde{C}_0}{\tilde{\rho}_0\tilde{c}\tilde{R}\tilde{T}_0^2} \tag{9.7}$$

Equations (9.5) and (9.6), with the initial conditions

$$\theta(0) = \eta(0) = 0 \tag{9.8}$$

form an analogue of Semenov model for an open system, undergoing continuous (inert) mass injection.

9.4 Cooling Media Injection

In a similar way, thermal explosion prevention by injecting reacting medium (gas) with temperature below ambient one may be considered.

Assuming, as a basic case, reaction order to be zero (no kinetic effects) and injected media (medium) uniform and identical to the one present in the vessel initially, the problem is formulated by modifying the Semenov model (2.4).

If again (as in Section 9.3) $\tilde{I}(\tilde{t})$ is mass flow rate of injected gas, with the constant temperature $\tilde{T}_{in} < \tilde{T}_0$, then an additional non-dimensional volume-averaged enthalpy flux received by the system is

$$\frac{\tilde{I}\tilde{c}(\tilde{T}_{in} - \tilde{T}_0)}{\tilde{V}} \tag{9.9}$$

and Eq. (2.4) for Semenov model is modified as follows

$$\frac{d}{d\tau}[(1+\psi(\tau))\theta] = \exp\left(\frac{\theta}{1+Ar\theta}\right) - \frac{1}{Se}\theta - \wp(\tau) \tag{9.10}$$

receiving, apart from the changed LHS, an extra (generally, unsteady) term on the RHS

$$\wp(\tau) = \frac{\tilde{I}(\tilde{t}(\tau))\tilde{c}(\tilde{T}_0 - \tilde{T}_{in})}{\tilde{Q}\tilde{k}_0\tilde{V}} \tag{9.11}$$

9.5 Prevention of Fire Flashover

Phenomenon of fire flashover, its mechanism and relation to thermal explosion have been discussed in Chapter 6.

As shown qualitatively in Figure 6.2, there may, in principle, exist regimes in which compartment fire is either controlled or suppressed (dashed curves, Figure 6.2) by some active protection measures. In these cases flashover is prevented.

Primary cause of flashover is radiation from hot smoke layer, looped in positive feedback mechanism (Chapter 6). Therefore, conceptually, the strategy of active protection may be removal of excessive heat from smoke layer which develops at early stages of fire, before flashover. Such intervention would reduce hot smoke temperature and thus interrupt positive feedback loop.

Conventional automatic active fire protection measure is application of water sprays, either in the form of standard sprinkler ones or fine water mists (DiNenno *et al.* 1995; Cote 1997).

Possibility of flashover prevention by water spray-induced energy dissipation from smoke layer may be investigated within the framework of the flashover model discussed in Section 6.2, Chapter 6. This model is essentially described by Eq. (6.22).

Critical boundary, according to this model, is presented by the curve $a_2 = \varphi(a_1)$ in Figure 9.2 (This curve repeats the critical curve in Figure 6.4, Chapter 6).

Figure 9.2 shows basic idea of flashover suppression. If initial fire conditions are super-critical (i.e. set of parameters (a_1, a_2) belongs to flashover region, as shown in Figure 9.2), then action of water spray on smoke layer must provide an additional cooling, resulting in increase of the

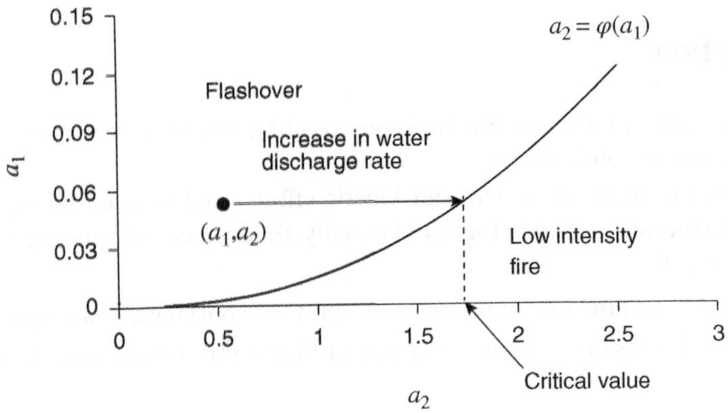

Figure 9.2 Flashover critical boundary (Graham *et al.* 1995, 1999) and the effect of water discharge rate. *Source:* Reproduced with permission from Novozhilov (2001) / Elsevier.

coefficient a_2 (accounting for energy losses in Eq. (6.22)). Upon such an action the point depicting set of control parameters (a_1, a_2) would shift to the right, over the critical boundary, to the region of quasi-steady (low intensity) fire. Flashover would be prevented. Updated value of the coefficient a_2 depends on magnitude of energy transfer rate between spray and hot layer. The latter, in turn, is determined by the regime of heat transfer dictated by spray parameters, such as total mass flow rate, and droplet size and velocity distributions. In particular, position of the coefficient a_2 just on the curve $a_2 = \varphi(a_1)$ (Figure 9.2) describe critical conditions (critical spray parameters) for flashover in the presence of additional heat extraction.

Design of appropriate flashover suppression strategy reduces, therefore, to the analysis of thermal interaction between water spray and smoke layer.

The following analysis (Novozhilov 2001) provides estimations of critical water discharge rates, required to prevent flashover, for various spray parameters.

Consider interaction of water spray with hot smoke layer (Figure 9.3) of thickness $\tilde{\Delta}$ and temperature $\tilde{T}$. The following analysis assumes that the time scale of droplet motion through the layer is smaller than the time scale of smoke temperature change. It should be noted that the scale used below for smoke layer temperature is ambient temperature $\tilde{T}_0$, to be conformant with Chapter 6 and the model (6.22). It is assumed further that absorption of heat by water spray(s) is uniform within the hot layer. This may be achieved (approximately) by employing sufficiently large number of sprinkler/water mist nozzles with overlapping spray patterns, and adjusting discharge cone angles for each of these sprays. In particular, choosing the spray cone with predominantly horizontal injection (e.g. using horizontal spray deflectors) will facilitate strong overlapping and uniform droplet distribution.

Energy absorption rate by water spray may be estimated by calculating the same rate for a single droplet, and integrating the result using the function describing droplet size distribution in the spray. Calculation of energy absorption rate by a single droplet generally requires analysis of both its dynamics and evaporation.

In general, droplet life time in sufficiently hot environment may be reasonably accurately divided into the following two stages (Lefebvre 1989):

1) heat-up to wet-bulb temperature without evaporation, and
2) evaporation at wet-bulb temperature, with constant rate of decrease of surface area of the droplet, until the end of droplet life time.

The above two-stage assumption becomes increasingly accurate (and also web-bulb temperature approaches boiling temperature) as the environment (smoke layer in the present case) gets hotter.

Depending on whether droplet residence time in the smoke layer spans over one, or both stages, and whether the second stage is completed over that time, different regimes of spray/hot layer interaction may occur.

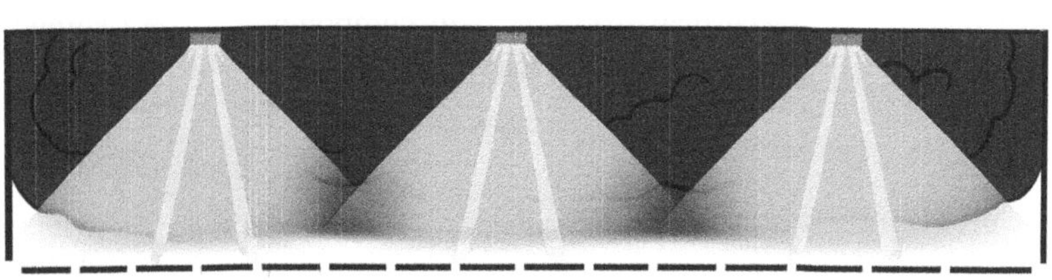

Figure 9.3 Schematic of water sprays interaction with smoke layer.

The following analysis is divided into the separate parts accordingly. Wet-bulb temperature in the expressions below may be approximated by boiling temperature.

a) coarse spray (non-evaporation) limit

In many cases involving conventional sprinklers heat transfer between smoke layer and water spray is predominantly convective (Kung 1977), i.e. droplet experiences only heat-up to wet-bulb temperature without evaporation. Therefore, in this limit water droplet experiences stage (1) only, and its temperature at the exit from smoke layer does not exceed wet-bulb temperature. Such non-evaporation approximation becomes increasingly accurate as droplet size in the spray gets larger and thickness of smoke layer smaller.

Amount of energy absorbed by a single droplet with the mass $\tilde{m}_d$ and specific heat (of water) $\tilde{c}_d$ from smoke layer is given by

$$\tilde{Q}_{drop} = \tilde{m}_d \tilde{c}_d \left(\tilde{T}_d^* - \tilde{T}_0 \right) \tag{9.12}$$

where $\tilde{T}_d^*$ is droplet temperature at the exit from the layer. (It is assumed that droplets are injected into the flow at ambient temperature).

Equation (9.12) shows that only the value of the latter exit temperature is required in order to calculate amount of energy transferred by a single droplet with a given mass. Then total amount of heat dissipated from the smoke layer may be obtained by integration over all the droplets in the spray. ("Spray" here and below refer to either single spray, or a group of sprays issued from different nozzles as in Figure 9.3).

Calculation of the temperature $\tilde{T}_d^*$ requires analysis of droplet dynamics and temperature time history in the flow.

In the appropriately chosen coordinate system (with the origin of the system fixed at the ceiling of the compartment and vertical axis $\tilde{z}$ pointing downwards) equations of droplet dynamics, for velocity components $\tilde{U}$ and $\tilde{V}$ may be written in the following form (Faeth 1983)

$$\begin{aligned} \tilde{m}_d \frac{d\tilde{U}}{d\tilde{t}} &= -\frac{\pi}{8} \tilde{d}^2 \tilde{\rho} C_D \sqrt{\tilde{U}^2 + \tilde{V}^2}\, \tilde{U} \\ \tilde{m}_d \frac{d\tilde{V}}{d\tilde{t}} &= -\frac{\pi}{8} \tilde{d}^2 \tilde{\rho} C_D \sqrt{\tilde{U}^2 + \tilde{V}^2}\, \tilde{V} + \tilde{m}_d \tilde{g} \end{aligned} \tag{9.13}$$

For a typical conventional sprinkler the volumetric mean droplet diameter is of the order of 1 – 2 mm. Taking characteristic droplet size 1 mm and typical droplet velocities at the orifice 10 – 20 m/s, particle Reynolds number may be estimated to be of the order of Re = 700 – 1500. In this range drag coefficient for spherical particles is a weak function of particle Reynolds number (Faeth 1983) and varies between approximately 0.47 and 0.6. Therefore, rendering it constant (for example, $C_D = 0.55$) in Eq. (9.13) is a good approximation.

Even with a constant drag coefficient the Eq. (9.13) are still coupled with each other. However, in the view of the argument presented above, the vertical droplet velocity component may be assumed small compared to the horizontal one, $|\tilde{V}| \ll |\tilde{U}|$. This greatly simplifies the problem by uncoupling equations in the set (9.13)

$$\begin{aligned} \tilde{m}_d \frac{d\tilde{U}}{d\tilde{t}} &= -\frac{\pi}{8} \tilde{d}^2 \tilde{\rho} C_D |\tilde{U}| \tilde{U} \\ \tilde{m}_d \frac{d\tilde{V}}{d\tilde{t}} &= -\frac{\pi}{8} \tilde{d}^2 \tilde{\rho} C_D |\tilde{U}| \tilde{V} + \tilde{m}_d \tilde{g} \end{aligned} \tag{9.14}$$

In order to estimate droplet traverse time through the layer it is sufficient to consider only half of symmetrical discharge cone. Therefore, the restriction $\tilde{U} > 0$ may be introduced, and the set of Eq. (9.14) turns into

$$\begin{aligned}\tilde{m}_d \frac{d\tilde{U}}{d\tilde{t}} &= -\frac{\pi}{8}\tilde{d}^2\tilde{\rho}C_D\tilde{U}^2 \\ \tilde{m}_d \frac{d\tilde{V}}{d\tilde{t}} &= -\frac{\pi}{8}\tilde{d}^2\tilde{\rho}C_D\tilde{U}\tilde{V} + \tilde{m}_d\tilde{g}\end{aligned} \tag{9.15}$$

Equation (9.15) may be written in non-dimensional form by introducing the following scales: $\left|\vec{\tilde{V}}\right|_{\max}$ (maximum velocity magnitude in the spray; $\vec{\tilde{V}} = (\tilde{U}, \tilde{V})$) for velocity components and $\frac{\left|\vec{\tilde{V}}\right|_{\max}}{\tilde{g}}$ for time. Then the equations are written in the following rather simple form

$$\begin{aligned}\frac{dU}{d\tau} &= -\beta\tilde{U}^2 \\ \frac{dV}{d\tau} &= -\beta UV + 1\end{aligned} \tag{9.16}$$

where $\beta = \frac{3}{4}\frac{\tilde{\rho}}{\tilde{\rho}_d}C_D\frac{\left|\vec{\tilde{V}}\right|^2_{\max}}{\tilde{d}\tilde{g}}$.

The initial conditions

$$U(0) = U_0, \;\; V(0) = V_0 \tag{9.17}$$

must be imposed.

Integration of the set of Eq. (9.17) is straightforward. The first equation admits separation of variables (it should be remembered that $U > 0$) and its solution is

$$U = \left(\frac{1}{U_0} + \beta\tau\right)^{-1} \tag{9.18}$$

Substitution of this solution into the second of Eq. (9.16) leads to the linear ODE

$$\frac{dV}{d\tau} = -\beta\left(\frac{1}{U_0} + \beta\tau\right)^{-1} V + 1 \tag{9.19}$$

with the solution

$$V = \frac{V_0 + \tau\left(1 + \frac{\beta}{2}U_0\tau\right)}{(1 + \beta U_0\tau)} \tag{9.20}$$

Vertical distance travelled by the droplet is given by

$$z(\tau) = \int_0^{\tau} V(s)ds \tag{9.21}$$

In particular, at the exit (traverse) time τ^*

$$\Delta = \int_0^{\tau^*} V(s)ds \tag{9.22}$$

Here non-dimensional thickness of the smoke layer is defined as $\Delta = \frac{\tilde{\Delta}}{\tilde{H}}$, where $\tilde{H}$ is compartment height.

The integral (involving the solution (9.20)) in Eq. (9.22) may be calculated analytically. As a result, Eq. (9.22) for the droplet exit time may be written in the explicit form

$$\frac{1}{4}(\tau^*)^2 + \frac{1}{2\beta U_0}\tau^* + \frac{\left(V_0 - \frac{1}{2\beta U_0}\right)}{\beta U_0}\ln(1 + \beta U_0\tau^*) = \Delta \quad (9.23)$$

Next, energy balance equation for droplet (droplet heat-up process) needs be considered. This equation assumes the form (Novozhilov 2001)

$$\tilde{m}_d\tilde{c}_d\frac{d\tilde{T}_d}{d\tilde{t}} = \frac{\tilde{\lambda}_g Nu}{\tilde{d}}\pi\tilde{d}^2\left(\tilde{T} - \tilde{T}_d\right)$$
$$\tilde{T}_d(0) = \tilde{T}_0 \quad (9.24)$$

or (upon scaling)

$$\frac{dT_d}{d\tau} = \beta_1(\theta - T_d)$$
$$T_d(0) = 1 \quad (9.25)$$

with $\beta_1 = 6Nu\frac{\tilde{\lambda}_g}{\tilde{\rho}_d\tilde{c}_d}\frac{\left|\vec{\tilde{V}}\right|_{max}}{\tilde{d}^2\tilde{g}}(\theta - 1)$.

The appropriate correlation for Nusselt number (Faeth 1983)

$$Nu = 2.0 + 0.6\,\mathrm{Re}_d^{1/2}\mathrm{Pr}^{1/3} \quad (9.26)$$

shows that the latter is, in the range of typical droplet velocities, a rather weak function of particle Reynolds number and may be represented by its constant average value with sufficient accuracy. Then parameter β_1 becomes constant.

The solution of equation Eq. (9.25) is

$$T_d - 1 = (1 - \exp(-\beta_1\tau))(\theta - 1) \quad (9.27)$$

and, therefore, at the exit from smoke layer

$$T_d^* - 1 = (1 - \exp(-\beta_1\tau^*))(\theta - 1) \quad (9.28)$$

In order to calculate total energy loss from smoke layer to water spray Eq. (9.12) (for a single droplet) needs be used in conjunction with the mass droplet size distribution function. The latter function $F(s)$ is defined here in such a way that its value at the point s is equal to the fraction of total spray mass contained in all the droplets with non-dimensional diameters $d < s$, where $d = \frac{\tilde{d}}{\tilde{d}_{1/2}}$ and $\tilde{d}_{1/2}$ is mass median diameter of the spray.

Then if $\tilde{\Psi}$ is total mass flow rate of the spray, $\tilde{\Psi}dF(s)$ gives its fraction carried by all the droplets with diameters satisfying $s \le d \le s + ds$. Energy transfer rate induced by the same portion of droplet flow is $\tilde{\Psi}\tilde{c}_d\left(\tilde{T}_d^* - \tilde{T}_0\right)dF(s)$, and energy transfer rate for the whole spray is

$$\tilde{\dot{Q}}_{spr} = \tilde{\Psi}\tilde{c}_d\tilde{T}_0\int_0^\infty \left(T_d^*(s) - 1\right)dF(s) \quad (9.29)$$

or

$$\tilde{\dot{Q}}_{spr} = \tilde{\dot{\Psi}}\tilde{c}_d\tilde{T}_0 \int_0^\infty f(s)(1 - \exp(-\beta_1\tau^*(s)))ds \cdot (\theta - 1) \tag{9.30}$$

using (9.28) and assuming that distribution function has density $f(s)$.

In order to estimate critical spray discharge rate $\tilde{\dot{\Psi}}_{cr}$, required to suppress flashover, modified model (6.22) needs be used. As follows from the derivation of Eq. (6.22) in Section 6.2, Chapter 6, an additional heat loss due to the spray (proportional to non-dimensional temperature difference $\theta - 1$) is obtained, in non-dimensional form, by dividing expression (9.30) by the HRR of initial fire $\tilde{\dot{Q}}_0$.

Therefore

$$\dot{Q}_{spr} = \frac{\tilde{\dot{\Psi}}\tilde{c}_d\tilde{T}_0}{\tilde{\dot{Q}}_0} \int_0^\infty f(s)(1 - \exp(-\beta_1\tau^*(s)))ds \cdot (\theta - 1) \tag{9.31}$$

and it follows immediately that modified model (6.22) must have coefficient a_2 replaced by A_2, where

$$A_2 = a_2 + \frac{\tilde{\dot{\Psi}}\tilde{c}_d\tilde{T}_0}{\tilde{\dot{Q}}_0} \int_0^\infty f(s)(1 - \exp(-\beta_1\tau^*(s)))ds \tag{9.32}$$

Critical water flow rate, required to prevent flashover, for a given value of coefficient a_1 is given (Figure 9.2) by the requirement $A_2 = \varphi(a_1)$, that is

$$\frac{\tilde{\dot{\Psi}}_{cr}\tilde{c}_d\tilde{T}_0}{\tilde{\dot{Q}}_0} \int_0^\infty f(s)(1 - \exp(-\beta_1\tau^*(s)))ds = \varphi(a_1) - a_2 \tag{9.33}$$

from there

$$\tilde{\dot{\Psi}}_{cr} = \frac{\tilde{\dot{Q}}_0(\varphi(a_1) - a_2)}{\tilde{c}_d\tilde{T}_0} \left[\int_0^\infty f(s)(1 - \exp(-\beta_1\tau^*(s)))ds\right]^{-1} \tag{9.34}$$

The parameters $\tilde{\dot{Q}}_0$ and $\varphi(a_1) - a_2$ both measure severity of the fire (its initial HRR and degree of super-criticality to flashover, respectively). Larger critical discharge rate is required with increase of these parameters, exactly as Eq. (9.34) predicts.

Non-dimensional critical discharge rate $\dot{\Psi}_{cr} = \frac{\tilde{c}_d\tilde{T}_0}{\tilde{\dot{Q}}_0}\tilde{\dot{\Psi}}_{cr}$ is given by

$$\dot{\Psi}_{cr} = (\varphi(a_1) - a_2)\left[\int_0^\infty f(s)(1 - \exp(-\beta_1\tau^*(s)))ds\right]^{-1} \tag{9.35}$$

Figure 9.4 shows (in dimensional form) critical flow rates for monodisperse sprays of various diameters and for different depths of smoke layer. For a monodisperse spray probability density function turns into the δ^-function: $f(s) = \delta(d)$ where d is fixed droplet diameter of monodisperse spray.

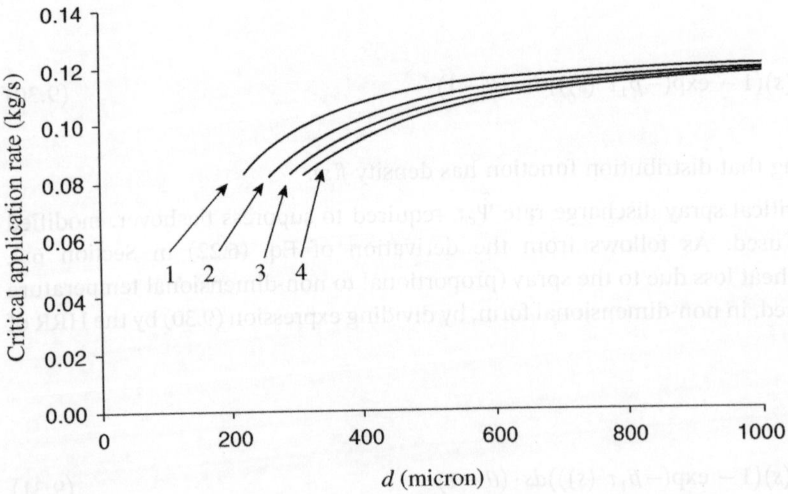

Figure 9.4 Critical water discharge rates for coarse spray (non-evaporation) limit. Different smoke layer thickness. 1 – $\tilde{\Delta}$ = 0.2 m; 2 – $\tilde{\Delta}$ = 0.4 m; 3 – $\tilde{\Delta}$ = 0.6 m; 4 – $\tilde{\Delta}$ = 0.8 m. *Source:* Reproduced with permission from Novozhilov (2001) / Elsevier.

Criterion for applicability of non-evaporation limit (for a smoke layer of given thickness and temperature) follows immediately from Eq. (9.28) by requiring that the droplet exit temperature is equal to wet-bulb temperature

$$T_w - 1 = (1 - \exp(-\beta_1 \tau^*))(\theta - 1) \tag{9.36}$$

Here exit time τ^* is obtained from Eq. (9.23).

Condition (9.36) puts a lower bound on the droplet diameter, droplets with larger diameters travel through smoke layer in a coarse spray regime.

b) fine spray (full evaporation) limit

In the opposite limit of very fine droplets (typically found in water mist sprays) full droplet evaporation within the hot layer occurs. In other words, in this limit droplet pass through both heat-up and evaporation stages, and both stages are fully completed over droplet life time.

Results of CFD simulations (Novozhilov *et al.* 1995–1996) suggest that this is typically the case for droplets with diameters below 100 μm. Assuming total evaporation, Eq. (6.22) is modified as follows

$$\frac{d\theta}{d\tau} = 1 + a_1(\theta^4 - 1) - a_2(\theta - 1) - \Xi \tag{9.37}$$

where an additional heat sink due to spray evaporation is

$$\Xi = \tilde{\eta}\tilde{\Psi}$$
$$\tilde{\eta} = \frac{\left[\tilde{c}_d\left(\tilde{T}_w - \tilde{T}_0\right) + \tilde{H}_{fg}\right]}{\tilde{Q}_0} \tag{9.38}$$

$\tilde{T}_w$ being wet-bulb temperature and $\tilde{H}_{fg}$ latent heat of water vaporization.

Critical conditions may be found from modified form of the set of Eq. (6.27)

$$\begin{aligned} &4\frac{a_1}{a_2}\theta_{cr}^3 = 1 \\ &\frac{a_1}{a_2}\left(\theta_{cr}^4 - 1\right) + 1 + \frac{1}{a_2} - \theta_{cr} = \frac{\Xi}{a_2} \end{aligned} \tag{9.39}$$

from which the following expression for dimensional critical water application rate follows

$$\tilde{\dot{\Psi}}_{cr} = \tilde{\eta}^{-1}\left[a_1\left(\left(\frac{a_2}{4a_1}\right)^{\frac{4}{3}} - 1\right) - a_2\left(\left(\frac{a_2}{4a_1}\right)^{\frac{1}{3}} - 1\right) + 1\right] \tag{9.40}$$

Critical condition (9.40) may be interpreted in the space of three independent parameters $\left(\tilde{\dot{\Psi}}, a_1, a_2\right)$, rather than on the plane as is the case for freely burning fire (Graham *et al.* 1995, 1999). Equation (9.40) determines critical surface in the space of the above parameters shown in Figure 9.5. Flashover area is below the surface (between the critical surface and the plane (a_1, a_2)). For parameters above the surface flashover is impossible. The critical surface intersects with the plane (a_1, a_2) exactly by the critical curve $a_2 = \varphi(a_1)$.

Despite the fact that prediction of critical water flow rate in full evaporation limit is straightforward, establishment of criterion of its applicability still requires analysis of droplet motion and heat-up dynamics.

Analytical integration of droplet motion equations present in this case a formidable challenge, as an assumption of constant drag coefficient becomes grossly invalid. Drag coefficient increases sharply as droplet quickly decelerates and its diameter decreases due to evaporation (and Reynolds particle number reduces accordingly).

Estimation of the limit of validity of full evaporation regime can be made in the following way.

Time span of the first stage, i.e. time it takes for droplet to be heated to wet-bulb temperature, follows directly from Eq. (9.27)

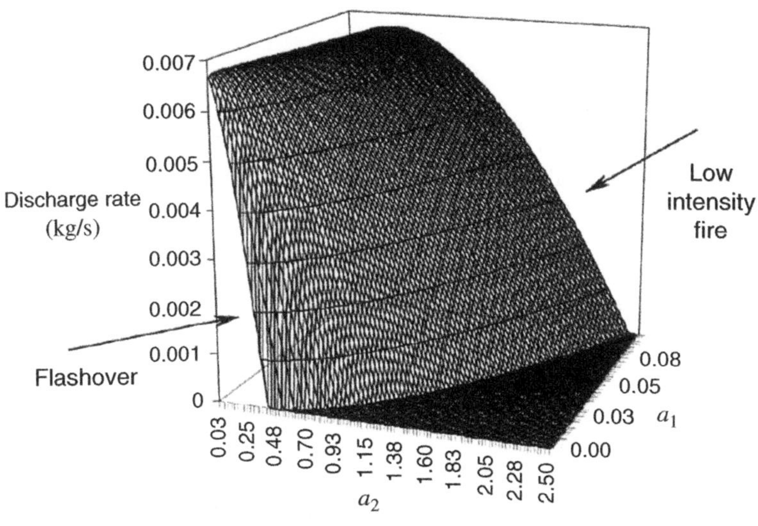

Figure 9.5 Critical surface for flashover for the case of fine spray (full evaporation limit). *Source:* Reproduced with permission from Novozhilov (2001) / Elsevier.

$$\tau_{hp} = -\frac{1}{\beta_1}\ln\left[1 - \frac{(T_w - 1)}{(\theta - 1)}\right] \tag{9.41}$$

Vertical distance travelled by droplet over that time is obtained from Eq. (9.23)

$$l_{hp} = \frac{1}{4}(\tau_{hp})^2 + \frac{1}{2\beta U_0}\tau_{hp} + \frac{\left(V_0 - \frac{1}{2\beta U_0}\right)}{\beta U_0}\ln(1 + \beta U_0 \tau_{hp}) \tag{9.42}$$

As for the full vaporisation stage, the above mentioned "d^2–" law (constant rate of droplet surface area reduction with time; Lefebvre 1989) states that droplet diameter change with time obeys

$$\tilde{d}^2(\tilde{t}) = \tilde{d}_0^2 - \tilde{k}_{ev}\tilde{t} \tag{9.43}$$

where $\tilde{d}$ is current diameter, $\tilde{d}_0$ initial diameter, $\tilde{k}_{ev}$ is the so-called evaporation constant

$$\tilde{k}_{ev} = \frac{8\tilde{\lambda}_g}{\tilde{\rho}_d \tilde{c}_g}\ln\left(1 + \frac{\tilde{c}_g(\tilde{T} - \tilde{T}_w)}{\tilde{H}_{fg}}\right) \tag{9.44}$$

Time of complete evaporation is therefore given by

$$\tilde{t}_{vap} = \frac{\tilde{d}_0^2}{\tilde{k}_{ev}} \tag{9.45}$$

or, in non-dimensional form, using time scale introduced earlier in the present Section, by

$$\tau_{vap} = \frac{\tilde{d}_0^2 \tilde{g}}{\left|\vec{\tilde{V}}\right|_{max} \tilde{k}_{ev}} \tag{9.46}$$

Further, velocity at the end of heat-process, according to Eq. (9.20) is

$$V_{hp} = \frac{V_0 + \tau_{hp}\left(1 + \frac{\beta}{2}U_0\tau_{hp}\right)}{(1 + \beta U_0 \tau_{hp})} \tag{9.47}$$

Assuming that droplet continues to travel with vertical velocity component $V_{vap} = V_{hp}$ during the whole duration of vaporization stage, vertical distance travelled by the droplet over the latter stage is

$$l_{vap} = V_{vap}\tau_{vap} \tag{9.48}$$

Clearly, the above assumption overestimates the distance l_{vap} as the droplet actually decelerates towards the end of its life time, making the criterion derived below an estimate only.

The criterion for full evaporation regime may now be written in the following form

$$l_{hp} + l_{vap} = \Delta \tag{9.49}$$

where the terms on LHS of Eq. (9.49) are given by Eqs. (9.42) and (9.48), respectively.

Condition (9.49) determines the diameter of droplet such that all the droplets with lesser diameters travel through smoke layer fully within fine spray regime. As pointed out, Eq. (9.49) somewhat underestimates the limiting diameter. More accurate estimation may be obtained by CFD analysis, solving equation of motion during evaporation stage numerically.

c) general case

For all the droplets with diameters between those determined by Eqs. (9.49) and (9.36) general case applies. This means that those droplets, as they traverse smoke layer, are being heated up to wet-bulb temperature and partially vaporized.

Critical water application rate in this general case regime may be found approximately by combining analyses for the two limiting cases (a) and (b), and applying assumption of constant vertical velocity component throughout evaporation stage.

Boundaries for the limiting, non-evaporation and full evaporation, regimes are presented in Table 9.3. Droplets with diameters between those listed in the Table experience general case motion/heat transfer regime.

Critical water spray discharge rates for flashover suppression, across all droplet motion/heat transfer regimes, are listed in Table 9.4.

Table 9.3 Limits for non-evaporation and full evaporation regimes of droplet motion/heat transfer in hot layer.

Smoke Layer Thickness, m	Maximum Droplet Diameter for complete evaporation regime, μm	Minimum Droplet Diameter for non-evaporation regime, μm
	(a) $\tilde{T}$ = 500 K	
0.1	6.3	175.0
0.2	31.3	198.1
0.3	37.5	222.5
0.4	43.8	245.0
0.5	50.0	262.6
0.6	51.3	276.4
0.7	52.6	287.7
0.8	53.9	298.9
0.9	55.2	306.3
1.0	56.3	312.5
	(b) $\tilde{T}$ = 700 K	
0.1	50.0	262.7
0.2	56.3	306.4
0.3	61.3	335.0
0.4	62.5	362.5
0.5	68.8	390.8
0.6	70.0	410.2
0.7	75.1	428.7
0.8	75.2	443.6
0.9	75.2	452.5
1.0	75.2	462.9

Source: Adapted from Novozhilov (2001).

Table 9.4 Critical water discharge rates required to prevent flashover as a function of droplet diameter in a monodisperse spray. Fire and model (6.22) parameters: $\tilde{\dot{Q}}_0 = 100$ kW; $\tilde{\Delta} = 0.5$ m; $a_2 = 0.5$. Framed figures indicate general case (heat-up and partial evaporation) regime. Figures above group of framed figures belong to fine spray limit. Figures below group of framed figures belong to coarse spray limit.

Droplet diameter of monodisperse spray, μm	Critical spray Discharge rate, kg/s $a_1 = 0.011$	Critical spray Discharge rate, kg/s $a_1 = 0.018$	Critical spray Discharge rate, kg/s $a_1 = 0.059$
50	0.032	0.038	0.039
100	0.033	0.046	0.055
200	0.036	0.064	0.087
300	0.039	0.082	0.120
400	0.041	0.093	0.135
500	0.046	0.098	0.144
600	0.048	0.100	0.150
700	0.055	0.104	0.154
800	0.055	0.106	0.158
900	0.055	0.108	0.159
1000	0.055	0.110	0.160

Source: Adapted from Novozhilov (2001).

References

Abramov, V.G., Samoilenko, N.G., and Solomonov, V.B. (1978). Thermal explosion in ideal-mixing flow-through reactor with self-catalytic reaction. *Combustion, Explosion, and Shock Waves* 14 (5): 665–669.

Abramowitz, M. and Stegun, I.A. (ed.) (1964). *Handbook of Mathematical Functions*. Washington, USA: National Bureau of Standards.

Adler, J. (1975). Thermal explosion theory for a slab with spatially periodic surface temperature. *Combustion and Flame* 24: 269–271.

Adler, J. and Enig, J.W. (1964). The critical conditions in thermal explosion theory with reactant consumption. *Combustion and Flame* 8 (2): 97–103.

Aganetti, R., Lamorlette, A., Guilbert, E. et al. (2016). Advection and the self-heating of organic porous media. *International Journal of Heat and Mass Transfer* 93: 1150–1158.

Al-Hallaj, S., Maleki, H., Hong, J.S., and Selman, J.R. (1999). Thermal modelling and design considerations of lithium-ion batteries. *Journal of Power Sources* 83: 1–8.

Al-Hallaj, S. and Selman, J.R. (2002). Thermal modeling of secondary lithium batteries for electric vehicle/hybrid electric vehicle applications. *Journal of Power Sources* 110: 341–348.

Aris, R. (1961). *The Optimal Design of Chemical Reactors. A Study in Dynamic Programming*. Academic Press.

Arnold, V.I. (1984). *Catastrophe Theory*. Berlin: Springer-Verlag.

Arnold, V.I. (1988). *Geometrical Methods in the Theory of Ordinary Differential Equations*. New York: Springer.

Arnold, V.I. (1992). *Ordinary Differential Equations*. Berlin: Springer-Verlag.

Arora, P. and Zhang, Z. (2004). Battery separators. *Chemical Reviews* 104: 4419–4462.

Azatyan, V.V. (2021). Role of Semenov's theory of chain reactions in the formation of modern concepts on the processes of combustion, explosion, and detonation of gases. *Russian Journal of Physical Chemistry B* 15 (2): 278–284.

Babrauskas, V. (1979). Full Scale Burning Behaviour of Upholstered Chairs. *NBS technical note No. 1103*. National Bureau of Standards.

Babushok, V.I., Goldshtein, V.M., and Sobolev, V.A. (1990). Critical conditions for thermal explosion with reactant consumption. *Combustion Science and Technology* 70 (1–3): 81–89.

Barenblatt, G.I., Kozyrev, Y.I., Malinin, N.I. et al. (1965). On a vibro-creep of polymer materials. *Journal of Applied Mechanics and Technical Physics* 5: 68–75.

Barenblatt, G.I., Kozyrev, Y.I., Malinin, N.I. et al. (1966). On vibro-creep of polymer materials. *Transactions of the Academy of Sciences of the USSR* 166 (4): 813–816.

Barzykin, V.V. (1973). Thermal explosion under linear heating. *Combustion, Explosion, Shock Waves* 9: 29–42.

Thermal Explosion: Theory and Application, First Edition. Vasily B. Novozhilov.

Barzykin, V.V. (2004). *Thermal Regimes of Exothermic Reactions*. Chernogolovka: ISMAN Press.

Barzykin, V.V., Gontkovskaya, V.T., and Merzhanov, A.G. (1966). On the theory of thermal explosion of self-accelerating reactions. *Combustion, Explosion and Shock Waves* 4: 18–23.

Barzykin, V.V., Gontkovskaya, V.T., Merzhanov, A.G., and Khudyaev, S.I. (1963). On nonsteady theory of thermal explosion. *Journal of Applied Mechanics and Technical Physics* 3: 118–125.

Barzykin, V.V. and Merzhanov, A.G. (1958). Boundary problem in the theory of thermal explosion. *Transactions of the Academy of Sciences of the USSR* 120 (5): 1271–1273.

Barzykin, V.V., Merzhanov, A.G., and Abramov, V.G. (1996). Theory of thermal explosion: from N.N. Semenov to present days. *Chemical Physics* 15 (6): 3–44.

Beard, A.N. (2001). Dependence of flashover on assumed value of the discharge coefficient. *Fire Safety Journal* 36: 25–36.

Beard, A.N. (2003). Dependence of flashover on temperature and aspect ratio of the compartment. *Journal of Fire Sciences* 21: 267–284.

Beard, A.N. (2010). Flashover and boundary properties. *Fire Safety Journal* 45: 116–121.

Beard, A.N., Drysdale, D.D., and Bishop, S.R. (1995). A non-linear model of major fire spread in a tunnel. *Fire Safety Journal* 24: 333–357.

Beard, A.N., Drysdale, D.D., Holborn, P.G., and Bishop, S.R. (1994–95). A model of instability and flashover. *Journal of Applied Fire Science* 4: 3–16.

Bejan, A. (2000). *Shape and Structure, from Engineering to Nature*. Cambridge: Cambridge University Press.

Bejan, A., Tsatsaronis, G., and Moran, M. (1996). *Thermal Design and Optimization*. New York: Wiley.

Benson, S.W. (1960). *The Foundations of Chemical Kinetics*. New York: McGraw Hill.

Berlin, A.A. and Golubkov, G.V. (2021). Nikolai Nikolaevich Semenov, the founder of chemical physics. *Russian Journal of Physical Chemistry B* 15 (2): 199–204.

Beyler, C.L. (1995). Flammability limits of premixed and diffusion flames. In: *SFPE Handbook of Fire Protection Engineering* (ed. P. DiNenno, C.L. Beyler, R.L.P. Custer, et al.), 2-147–2-159. Quincy: National Fire Protection Association.

Bishop, S.R., Holborn, P.G., Beard, A.N., and Drysdale, D.D. (1993). Nonlinear dynamics of flashover in compartment fires. *Fire Safety Journal* 21: 11–45.

Bloshenko, V.N., Khaikin, B.I., and Merzhanov, A.G. (1970). On ignition of particles of metals. *Combustion, Explosion and Shock Waves* 5 (4): 474–488.

Boddington, T. (1963). The growth and decay of hot spots and the relation between structure and stability. *Symposium (International) on Combustion* 9: 287–293.

Boddington, T., Gray, P., and Harvey, D.I. (1971). Thermal theory of spontaneous ignition: criticality in bodies of arbitrary shape. *Philosophical Transactions of the Royal Society of London. Series A, Mathematical and Physical Sciences* 270 (1207): 467–506.

Bogolyubov, N.N. and Mitropolskii, Y.A. (1967). *Asymptotic Methods in the Theory of Nonlinear Oscillations*. New York: Gordon and Breach.

Borman, G.L. and Ragland, K.W. (1998). *Combustion Engineering*. Boston: McGraw-Hill.

Bostanjiyan, S.A., Merzhanov, A.G., and Khudyaev, S.I. (1965). On hydrodynamical thermal explosion. *Transactions of the Academy of Sciences of the USSR* 163 (1): 133–136.

Bradshaw, S., Glasser, D., and Brooks, K. (1991). Self-ignition and convection patterns in an infinite coal layer. *Chemical Engineering Communications* 105 (1): 255–278.

British Broadcasting Corporation (2013). Fresh faults with Boeing Dreamliner planes. https://www.bbc.com/news/business-20950287 (accessed 07 April 2025).

Brooks, K. and Glasser, D. (1986). A simplified model of spontaneous combustion of coal stockpiles. *Fuel* 65 (8): 1035–1041.

Brooks, K., Svanas, N., and Glasser, D. (1988). Evaluating the risk of spontaneous combustion in coal stockpiles. *Fuel* 67 (5): 651–656.

Buckingham, J., Hodge, C. and Hardy, T. (2008). Submarine power and propulsion – application of technology to deliver customer benefit. *Proceedings of the UDT Europe*, Glasgow, Scotland, UK (26–28 June 2008).

Budnick, E.K. and Klein, D.P. (1978). Mobile Home Fire Studies: Summary and Conclusions. NBSIR 78-1530. *National Bureau of Standards.*

Buggeln, R. and Rynk, R. (2002). Self-heating in yard trimmings: conditions leading to spontaneous combustion. *Compost Science and Utilization* 10 (2): 162–182.

Building Research Establishment, Department of the Environment, Fire Research Station (1982). Anatomy of a Fire. http://bufvc.ac.uk/dvdfind/index.php/title/781 (accessed 15 September 2023).

Burkina, R.S. and Knyazeva, A.G. (1991). Effect of autocatalysis on the critical conditions of focal thermal ignition. *Combustion, Explosion and Shock Waves* 27 (2): 143–148.

Carras, J.N. and Young, B.C. (1994). Self-heating of coal and related materials: models, application and test methods. *Progress in Energy and Combustion Science* 20: 1–15.

Cavanagh, K., Ward, J.K., Behrens, S., Bhatt, A.I., Ratnam, E.L, Oliver, E. and Hayward, J. (2015). Electrical Energy Storage: Technology Overview and Applications. CSIRO, Australia, EP154168. *Australian Energy Market Commission.*

Chen, X.D. (1992). On the mathematical modeling of the transient process of spontaneous heating in a moist coal stockpile. *Combustion and Flame* 90: 114–120.

Chen, Y. and Evans, J.W. (1996). Thermal analysis of lithium-ion batteries. *Journal of The Electrochemical Society* 143 (9): 2708–2712.

Chirkov, N.M. (1937). Autocatalytic hydrogen oxidation. *Acta Physicochimica URSS* 6: 1915.

Conesa, J.A. (2020). *Chemical Reactor Design. Mathematical Modeling and Applications.* Weinheim: Wiley-VCH.

Cote, A.E. (ed.) (1997). *Fire Protection Handbook.* Quincy, USA: National Fire Protection Association.

Cox, G. (1995). Compartment fire modelling. In: *Combustion Fundamentals of Fire* (ed. G. Cox), 329–404. London: Academic Press.

De Boer, W., Folman, L.B., Summerbel, R.C., and Boddy, L. (2005). Living in a fungal world: impact of fungi on soil bacterial niche development. *FEMS Microbiology Reviews* 29: 795–811.

DiNenno, P., Beyler, C.L., Custer, R.L.P. et al. (1995). *SFPE Handbook of Fire Protection Engineering.* Quincy, USA: National Fire Protection Association.

Dixon, T. (1988). Spontaneous combustion in bagasse stockpiles. *Proceedings of the Australian Society of Sugar Cane Technologists* 10: 53–61.

Drysdale, D.D. (2011). *An Introduction to Fire Dynamics.* Chichester: Wiley.

Edwards, J.C. (1990). Mathematical Modelling of Spontaneous Heating of a Coalbed. *US Bureau of Mines Report of Investigations 9296*: 15.

Emden, R. (1907). *Gaskugeln.* Leipzig and Berlin: Teubner.

Evans, D.D. (1995). Ceiling jet flows. In: *SFPE Handbook of Fire Protection Engineering* (ed. P.J. DiNenno et al.), 2-32–2-39. Quincy: National Fire Protection Association.

Faeth, G.M. (1983). Evaporation and combustion of sprays. *Progress in Energy and Combustion Science* 9: 1–76.

Fang, J.B. (1975). Measurement of the Behaviour of Incidental Fires in a Compartment. NBSIR 75-679. *National Bureau of Standards.*

Feng, H.-T., Hildyard, C.J., and Scott, S.K. (1992). Thermal runaway in a system with a constant power source. *Combustion and Flame* 88: 433–438.

Feng, X., Ouyang, M., Liu, X. et al. (2018). Thermal runaway mechanism of lithium ion battery for electric vehicles: a review. *Energy Storage Materials* 10: 246–267.

Fengel, D. and Wegener, G. (1984). *Wood Chemistry, Ultrastructure, Reactions.* Berlin: Walter de Gruyter & Co.

Ferrero, F., Lohrer, C., Schmidt, B.M. et al. (2009). A mathematical model to predict the heating-up of large-scale wood piles. *Journal of Loss Prevention in the Process Industries* 22: 439–448.

Fire Prevention and Engineering Bureau of Texas and The National Board of Fire Underwriters (1947). Texas city, Texas, disaster. April 16, 17 1947. http://www.local1259iaff.org/report.htm (accessed 07 April 2025).

Fock, V.A. (1928). On the thermal theory of electric breakdown. *Transactions of the Leningrad Physical-Technical Laboratory* 5: 52–71.

Frank-Kamenetskii, D.A. (1938). Temperature distribution in a reaction vessel and steady-state theory of thermal explosion. *Transactions of the Academy of Sciences of the USSR* 18 (7): 411–412.

Frank-Kamenetskii, D.A. (1939). Temperature distribution in a reaction vessel and steady-state theory of thermal explosion. *Journal of Physical Chemistry* 13 (6): 738–755.

Frank-Kamenetskii, D.A. (2016). *Diffusion and Heat Exchange in Chemical Kinetics*. Princeton: Princeton University Press.

Frenkel, Y.I. (1946). *Kinetic Theory of Liquids*. Oxford: Oxford University Press.

Friedman, M.H. (1963a). Size of "hot spots" in the impact explosion of exothermic materials. *Transactions of Faraday Society* 59 (8): 1865–1873.

Friedman, M.H. (1963b). A correlation of impact sensitivities by means of the hot spot model. *Symposium (International) on Combustion* 9: 294–302.

Friedman, R. (1975). Behaviour of fires in compartments. *Proceedings of the International Symposium on Fire Safety of Combustible Materials*, Edinburgh, United Kingdom. Edinburgh, United Kingdom: Edinburgh University.

Funabashi, A., Kida, Y., Yanagida, K. et al. (2002). Thermal simulation of large-scale lithium secondary batteries using a graphite-coke hybrid carbon negative electrode and $LiNi_{0.7}Co_{0.3}O_2$ positive electrode. *Journal of Power Sources* 104 (2): 248–252.

Gardiner, W.C. Jr. (1972). *Rates and Mechanisms of Chemical Reactions*. Menlo Park: Benjamin.

Gelfand, I.M. (1959). Some problems of the theory of quasi-linear equations. *Russian Mathematical Surveys* 14 (2): 87–158.

Gelman, E.A., Khudyaev, S.I., and Maklakov, S.V. (1992). Application of an averaging method for controlling an exothermic chemical reaction. *Archivum combustionis* 12 (1–4): 53–65.

Ghiji, M., Novozhilov, V., Moinuddin, K. et al. (2020). A review of lithium-ion battery fire suppression. *Energies* 13: 5117. https://doi.org/10.3390/en13195117.

Graham, T.L., Makhviladze, G.M., and Roberts, J.P. (1995). On the theory of flashover development. *Fire Safety Journal* 25: 229–259.

Graham, T.L., Makhviladze, G.M., and Roberts, J.P. (1999). The effects of the thermal inertia of the walls upon flashover development. *Fire Safety Journal* 32: 35–60.

Gray, B.F., Griffiths, J.F., and Hasko, S.M. (1984). Spontaneous ignition hazards in stockpiles of cellulosic materials: criteria for safe storage. *Journal of Chemical Technology and Biotechnology A* 34: 453–463.

Grigoriev, Y.M., Merzhanov, A.G., and Pribytkova, K.V. (1966). Critical conditions of thermal explosion with conductive heat transfer in reaction zone and surrounding media (conjugate problem). *Journal of Applied Mechanics and Technical Physics* 5: 17–24.

Gruntfest, I.J., Young, J.P., and Johnson, N.L. (1964). Temperatures generated by the flow of liquids in pipe. *Journal of Applied Physics* 35 (1): 18–25.

Guo, L.S., Wang, Z.R., Wang, J.H. et al. (2017). Effects of the environmental temperature and heat dissipation condition on the thermal runaway of lithium ion batteries during the charge-discharge process. *Journal of Loss Prevention in the Process Industries* 49: 953–960.

Hägglund, B., Jansson, R. and Onnermark, B. (1974). Fire Development in Residential Rooms after Ignition from Nuclear Explosions. FOA report C 20016-D6 (A3). *Forsvarets Forskningsanstalt*, Stockholm.

Harris, E.J. (1940). The thermal explosion of diethyl peroxide. *Proceedings of the Royal Society A* 175: 254–261.

Health and Safety Executive (1976). The explosion at the Dow chemical factory, King's Lynn. 27th June 1976. https://www.hse.gov.uk/comah/sragtech/casedow76.htm (accessed 07 April 2025).

Herring, J.R. (1983). Spontaneous coal combustion: mechanisms and predictions. *Proceedings of the International Conference on Coal Science*, Pittsburgh, USA (15–19 August 1983). Pittsburgh, USA: International Energy Agency.

Ibitoye, S.E., Mahamood, R.M., Jen, T.-C. et al. (2023). An overview of biomass solid fuels: Biomass sources, processing methods, and morphological and microstructural properties. *Journal of Bioresources and Bioproducts* 8: 333–360.

Incropera, F.P. and DeWitt, D.P. (2002). *Fundamentals of Heat and Mass Transfer*. New York: Wiley.

Inge, L., Semenoff, N., and Walther, A. (1925). Über den durchschlag fester isolatoren. *Zeitschrift für Physik* 32: 273–286.

Istratov, A.G. and Librovich, V.B. (1963). On stability of solutions in a steady-state thermal explosion theory. *Journal of Applied Mathematics and Mechanics* 27 (2): 343–347.

Jones, J.C. and Raj, S.C. (1988). The self-heating and ignition of vegetation debris. *Fuel* 67: 1208–1210.

Kagan, L., Berestycki, H., Joulin, G., and Sivashinsky, G. (1997). The effect of stirring on the limits of thermal explosion. *Combustion Theory and Modelling* 1: 97–111.

Kaganov, S.A. (1965). On the steady-state theory of thermal autoignition. *Journal of Applied Mechanics and Technical Physics* 6: 54–58.

Kaganov, S.A. (1967). The stability of the steady-state solutions in the theory of thermal explosions. *Journal of Applied Mathematics and Mechanics* 31 (6): 1081–1085.

Karlsson, B. and Quintiere, J.G. (2022). *Enclosure Fire Dynamics*. Boca Raton: CRC Press.

Kassoy, D.R. and Linan, A. (1978). The influence of reactant consumption on the critical conditions for homogeneous thermal explosions. *Quarterly Journal of Mechanics and Applied Mathematics* 31 (1): 99–112.

Khan, M.R., Everitt, C.E., and Lui, A.P. (1990). Modeling of oxygen chemisorption kinetics on coal char. *Combustion and Flame* 80: 83–93.

Khudyaev, S.I. (1963). Criterion of solvability of Dirichlet problem for elliptic equations. *Transactions of the Academy of Sciences of the USSR. Mathematics* 148 (1): 44–46.

Khudyaev, S.I. (1964). On boundary value problems for some quasi-linear elliptic equations. *Transactions of the Academy of Sciences of the USSR. Mathematics* 154 (4): 787–790.

Khudyaev, S.I. (1966). On boundary value problems for heat transfer equation with the temperature-dependent sources. In: *Heat and Mass Transfer*, (ed. A.V. Lykov), vol. 6, 149–158. Minsk: Science and Technology.

Khudyaev, S.I. (1995). On a class of integrable equations in problems of combustion and hydrodynamics. *Mathematical Modelling* 7 (1): 35–60.

Khudyaev, S.I. (2003). *Critical Phenomena in Nonlinear Equations*. Moscow: Fizmatlit.

Knyazeva, A.G., Burkina, R.S., and Vilyunov, V.N. (1988). Characteristics of local thermal ignition with various initial temperature distributions. *Combustion, Explosion and Shock Waves* 24 (3): 303–305.

Korobov, V.I. and Ratner, S.B. (1965). Self-heating of polymers under repeated deformation. *Transactions of the Academy of Sciences of the USSR* 161 (4): 824–828.

Koseki, H. (2011). Evaluation of various solid biomass fuels using thermal analysis and gas emission tests. *Energies* 4: 616–627. https://doi.org/10.3390/en4040616.

Krishnaswamy, S., Agarwal, P.K., and Gunn, R.D. (1996). Low-temperature oxidation of coal. 3. Modelling spontaneous combustion in coal stockpiles. *Fuel* 75 (3): 353–362.

Kunakov, M.A., Lidskii, B.V., Lutidze, G.N. et al. (2013). Thermal explosion in a stirred medium. *Russian Journal of Physical Chemistry B* 7 (3): 328–334.

Kung, H.C. (1977). Cooling of room fires by sprinkler spray. *ASME Transactions. Journal of Heat Transfer* 99: 353–359.

Kuo, K.K. (1986). *Principles of Combustion*. New York: Wiley.

Landau, L.D. and Lifshitz, E.M. (1987). *Fluid Mechanics*. Oxford: Butterworth-Heinemann.

Lamb, J. and Orendorff, C.J. (2014). Evaluation of mechanical abuse techniques in lithium ion batteries. *Journal of Power Sources* 247: 189–196.

Lamb, J., Orendorff, C.J., Steele, L.A.M., and Spangler, S.W. (2015). Failure propagation in multi-cell lithium ion batteries. *Journal of Power Sources* 283: 517–523.

Larsson, F., Andersson, P., and Mellander, B.-E. (2016). Lithium-ion battery aspects on fires in electrified vehicles on the basis of experimental abuse tests. *Batteries* 2 (2): 9. https://doi.org/10.3390/batteries2020009.

Lavrentiev, M.A. and Shabat, B.V. (1977). *Problems of Hydrodynamics and their Mathematical Models*. Moscow: Nauka.

Lee, C.W., Venkatachalapathy, R., and Prakash, J. (2000). A novel flame-retardant additive for lithium batteries. *Electrochemical and Solid-State Letters* 3 (2): 63–65.

Lefebvre, A.H. (1989). *Atomization and Sprays*. New York: Taylor and Francis.

Lemke, H. (1913). Über die Differentialgleichungen, welche den Gleichgewichtszustand eines gasförmigen Himmelskörpers bestimmen, dessen Teile gegeneinander nach dem Newtonschen Gesetze gravitieren. *Journal of Mathematics (Berlin)* 142: 118–145.

Li, S., Ma, X., and Yang, C. (2018). Prediction of spontaneous combustion in the coal stockpile based on an improved metabolic grey model. *Process Safety and Environmental Protection* 116: 564–577.

Li, X.R., Koseki, H., and Momota, M. (2006). Evaluation of danger from fermentation-induced spontaneous ignition of wood chips. *Journal of Hazardous Materials* 135: 15–20.

Li, Z., Huang, J., Liaw, B.Y. et al. (2014). A review of lithium deposition in lithium-ion and lithium metal secondary batteries. *Journal of Power Sources* 254: 168–182.

Lidorenko, N.S. and Sidyakin, A.V. (1972). On the possibility of occurrence of thermal explosion in thin metallic powders. *Transactions of the Academy of Sciences of the USSR* 202 (3): 566–569.

Lisbona, D. and Snee, T. (2011). A review of hazards associated with primary lithium and lithium-ion batteries. *Process Safety and Environmental Protection* 89: 434–442.

Luangwilai, T., Sidhu, H.S., and Nelson, M.I. (2012). A two dimensional, reaction-diffusion model of compost piles. *ANZIAM Journal* 53: C34–C52.

Luangwilai, T., Sidhu, H.S., and Nelson, M.I. (2013). Biological self-heating in compost piles: a Semenov formulation. *Chemical Engineering Science* 101: 533–542.

Luangwilai, T., Sidhu, H.S., and Nelson, M.I. (2018). One-dimensional spatial model for self-heating in compost piles: investigating effects of moisture and air flow. *Food and Bioproducts Processing* 108: 18–26.

Luangwilai, T., Sidhu, H.S., Nelson, M.I., and Chen, X.D. (2010). Modelling air flow and ambient temperature effects on the biological self-heating of compost piles. *Asia-Pacific Journal of Chemical Engineering* 5: 609–618.

Makhviladze, G.M. and Myshenkov, V.I. (1977). Thermal mechanism of normal flame propagation speed increase in a pre-discharge electric field. *Journal of Applied Mechanics and Technical Physics* 2: 29–38.

Makshantsev, B.I. and Finkelburg, V.M. (1974). Heat propagation of the luminescence extinguishment wave. *Transactions of the Academy of Sciences of the USSR* 217 (6): 1281–1283.

Maksyutov, S.S., Novozhilov, B.V., and Tsyganov, S.A. (1982). Thermal self-ignition of previously unmixed gases. *Combustion, Explosion and Shock Waves* 18: 407–411.

Merzhanov, A.G. (1966). On critical conditions for thermal explosion of a hot spot. *Combustion and Flame* 10 (4): 341–348.

Merzhanov, A.G. and Abramov, V.G. (1977). Thermal regimes of exothermic processes in continuous stirred tank reactors. *Chemical Engineering Science* 32 (5): 475–481.

Merzhanov, A.G., Barzykin, V.V., and Abramov, V.G. (1996). The theory of thermal explosion: from N.N. Semenov to present day. *Chemical Physics Reports* 15: 793–837.

Merzhanov, A.G., Barzykin, V.V., and Gontkovskaya, V.T. (1963). A problem of local thermal explosion. *Transactions of the Academy of Sciences of the USSR* 148 (2): 380–383.

Merzhanov, A.G., Barzykin, V.V., and Shteinberg, V.T. (1977). Methodological principles in studying chemical reaction kinetics under conditions of programmed heating. *Thermochemica Acta* 21: 301–332.

Merzhanov, A.G. and Dubovitskii, F.I. (1958). Quasi-stationary thermal regime of progression of explosive reactions. *Transactions of the Academy of Sciences of the USSR* 120 (5): 1068–1070.

Merzhanov, A.G. and Dubovitskii, F.I. (1960). Quasi-stationary theory of thermal explosion of self-accelerating reactions. *Journal of Physical Chemistry* 34 (10): 2235–2241.

Merzhanov, A.G. and Dubovitskii, F.I. (1966). Modern state of the theory of thermal explosion. *Advances in Chemistry* 35 (4): 656–683.

Merzhanov, A.G. and Rumanov, E.N. (1987). Nonlinear effects in macroscopic kinetics. *Russian Physical Surveys* 151 (4): 553–593.

Merzhanov, A.G. and Shtessel, E.A. (1971). Thermal explosion in reacting liquid systems in the presence of thermal convection. *Combustion, Explosion, and Shock Waves* 7: 58–65.

Merzhanov, A.G. and Stolin, A.M. (1971). On the thermal theory of flow of viscous liquid. *Transactions of the Academy of Sciences of the USSR* 198 (6): 1291–1294.

Merzhanov, A.G. and Strunina, A.G. (1965). Laws of thermal explosion with constant heating rate. *Combustion, Explosion, Shock Waves* 1: 43–52.

Mischenko, E.F. and Rozov, N.K. (1975). *Differential Equations with Small Parameters and Relaxation.* New York: Plenum Press.

Modest, N.F. (1993). *Radiative Heat Transfer.* New York: McGrow-Hill.

Murasawa, N. and Koseki, H. (2015). Investigation of heat generation from biomass fuels. *Energies* 8: 5143–5158. https://doi.org/10.3390/en8065143.

Nastoyashii, A.F. (1972). The problem of thermonuclear combustion in a bounded region. *Atomic Energy* 32 (1): 43–47.

Nastoyashii, A.F. and Shevchenko, L.P. (1972). Thermonuclear combustion wave in a bounded plasma. *Atomic Energy* 32 (6): 451–455.

Nelson, M.I. and Chen, X.D. (2007). Survey of experimental work on the self-heating and spontaneous combustion of coal. *GSA Reviews in Engineering Geology* 18: 31–83.

Nordon, P. (1979). A model for the self-heating reaction of coal and char. *Fuel* 58 (6): 456–464.

Nordon, P., Young, B.C., and Bainbridge, N.W. (1979). The rate of oxidation of char and coal in relation to their tendency to self-heat. *Fuel* 58 (6): 443–449.

Novozhilov, B.V. (1980). *Chain and Thermal Flame.* Moscow: Znanie.

Novozhilov, V. (2001). Flashover control under fire suppression conditions. *Fire Safety Journal* 36: 641–660.

Novozhilov, V. (2008). Critical conditions for conjugate thermal explosion. *Combustion Theory and Modelling* 12 (3): 433–449.

Novozhilov, V. (2010). Nonlinear dynamical model of compartment fire flashover. *Journal of Engineering Mathematics* 67: 387–400.

Novozhilov, V. (2012). A brief note on the smoke filling equation. *Fire Safety Journal* 47: 16–17.

Novozhilov, V. (2014). Lattice Boltzmann modeling of autoignition in non-uniformly heated mixtures. *Proceedings of the Nineteenth Australasian Fluid Mechanics Conference*, Melbourne, Australia (8–11 December 2014). Perth, Australia: AFMC.

Novozhilov, V. (2016). Thermal explosion in oscillating ambient conditions. *Scientific Reports* 6: 29730. https://doi.org/10.1038/srep29730.

Novozhilov, V. (2017). Fire hazards of some modern solid fuels. *Energies* 10 (1): 113. https://doi.org/10.3390/en10010113.

Novozhilov, V. (2018). Kinetic effects in thermal explosion with oscillating ambient conditions. *Scientific Reports* 8:4030. https://doi.org/10.1038/s41598-018-22341-6.

Novozhilov, V. (2019). Ignition waves in compartment fire flashover. *Proceedings of the 2019 Australian Combustion Symposium*, Adelaide, Australia (4–6 December 2019). Adelaide, Australia: Combustion Institute Australia and New Zealand Section.

Novozhilov, V. and Byrne, C. (2013a). Lattice Boltzmann modelling of thermal explosion in natural convection conditions. *Numerical Heat Transfer, Part A: Applications: An International Journal of Computation and Methodology* 63 (11): 824–839.

Novozhilov, V. and Byrne, C. (2013b). Lattice Boltzmann modelling of convective thermal explosion. *Proceedings of the Ninth Asia-Pacific Conference on Combustion*, Gyeongju, Korea (19–22 May 2013). Gyeongju, Korea: ASPACC.

Novozhilov, B.V. and Novozhilov, V.B. (2020). *Theory of Solid-Propellant Nonsteady Combustion*. Hoboken: Wiley-ASME.

Novozhilov, V., Moghtaderi, B., Fletcher, D.F., and Kent, J.H. (1995–96). Numerical simulation of enclosed gas fire extinguishment by a water spray. *Journal of Applied Fire Science* 5 (2): 135–146.

Novozhilov, V., Moghtaderi, B., Fletcher, D.F., and Kent, J.H. (1996). Computational fluid dynamics modelling of wood combustion. *Fire Safety Journal* 27: 69–84.

Novozhilov, B.V., Samoilenko, N.G., and Manelis, G.B. (2002). Thermal explosion in a stirred medium. *Doklady Physical Chemistry* 385 (1–3): 169–171.

Novozhilov, B.V., Samoilenko, N.G., and Manelis, G.B. (2005). Conditions of the thermal explosion under forced convection of a reacting mixture. *Combustion, Explosion, and Shock Waves* 41 (5): 528–532.

Ölander, A. (1956). Award ceremony speech. Speech presented at a Nobel Prize Award Ceremony in Stockholm, Sweden (11 December 1956). http://www.nobelprize.org/prizes/chemistry/1956/ceremony-speech/ (accessed 04 April 2025).

Oldham, K.B. and Spanier, J. (2002). *The Fractional Calculus*. Mineola: Dover Publications.

Onifade, M. and Genc, B. (2020). A review of research on spontaneous combustion of coal. *International Journal of Mining Science and Technology* 30 (3): 303–311.

Otto, A., Rzepka, S., Mager, T. et al. (2012). Battery management network for fully electrical vehicles featuring smart systems at cell and pack level. In: *Advanced Microsystems for Automotive Applications*, 3–14. Berlin/Heidelberg: Springer.

Ozişik, M.N. (1985). *Heat Transfer-A Basic Approach*. New York: McGrow-Hill.

Parks, J.R. (1961). Criticality criteria for various configurations of a self-heating chemical as functions of activation energy and temperature assembly. *Journal of Chemical Physics* 34 (1): 46–50.

Penner, S.S. (1955). *Introduction to the Study of Chemical Reactions in Flow Systems*. London: Butterworth Scientific Publication.

Philippov, A.A. (2024). Mathematical theory of epidemics and pandemics based on the laws of chain and thermal explosions. *Transactions of the Russian Academy of Sciences* 514 (1): 40–45.

Phounglamcheik, A., Johnson, N., Kienzl, N. et al. (2022). Self-heating of biochar during postproduction storage by O_2 chemisorption at low temperatures. *Energies* 15: 380. https://doi.org/10.3390/en15010380.

Pólya, G. and Szegö, G. (1951). *Isoperimetric Inequalities in Mathematical Physics*. Princeton: Princeton University Press.

Polyanin, A.D. and Zaitsev, V.F. (2002). *Exact Solutions for Ordinary Differential Equations*. Boca Raton: Chapman and Hall/CRC.

Prince, J.C. and Williams, F.A. (2012). Short chemical-kinetic mechanisms for low-temperature ignition of propane and ethane. *Combustion and Flame* 159: 2336–2344.

Prudnikov, A.P., Brychkov, Y.A., and Marichev, O.I. (1986). *Integrals and Series*. New York: Gordon and Breach Science Publishers.

Quintiere, J.G. (2016). *Principles of Fire Behaviour*. Boca Raton: CRC Press.

Ramsden, S.A. and Savic, P. (1964). A radiative detonation model for the development of a laser induced spark in air. *Nature* 203: 1217–1219.

Rizer, Y.P. (1973). *Laser Spark and Propagation of Discharges*. Moscow: Nauka.

Rockett, J.A. (1976). Fire induced gas flow in an enclosure. *Combustion Science and Technology* 12: 165–175.

Roels, J.A. (1983). *Energetics and Kinetics in Biotechnology*. Amsterdam: Elsevier Biomedical Press.

Ruiz, V., Pfrang, A., Kristona, A. et al. (2018). A review of international abuse testing standards and regulations for lithium ion batteries in electric and hybrid electric vehicles. *Renewable and Sustainable Energy Reviews* 81: 1427–1452.

Rychlý, J., Matisova-Rychlá, L., Lazár, M. et al. (2004). Thermal oxidation of cellulose investigated by chemiluminescence: The effect of water at temperatures above 100 °C. *Carbohydrate Polymers* 58: 301–309.

Rynk, R. (2000). Fires at composting facilities. *BioCycle Magazine* 41 (1): 54–58.

Sânchez-Rodriguez, D., Farjas, J., and Roura, P. (2017). The critical conditions for thermal explosion in a system heated at a constant rate. *Combustion and Flame* 186: 211–219.

Sato, N. (2001). Thermal behaviour analysis of lithium-ion batteries for electric and hybrid vehicles. *Journal of Power Sources* 99 (1–2): 70–77.

Schapery, R.A. (1964). Effect of cycling loading on the temperature in viscoelastic media with variable properties. *AIAA Journal* 2 (5): 827–833.

Schmal, D. (1989). Heating of stored coal. In: *Chemistry of Coal Weathering* (ed. C.R. Nelson), 133–215. Amsterdam: Elsevier Science.

Schmidt, L.D. (1945). Changes in coal during storage. In: *Chemistry of Coal Utilization*, vol. I (ed. H.H. Lowry), 627–676. New York: Wiley.

Selezneva, I.K. (1972). Spherically-symmetrical optical discharge as an analogy of diffusion combustion in a mixture of flammable gases. In: *Combustion and Explosion*, (ed. L.N. Stesik), 396–400. Moscow: Nauka.

Semenov, N.N. (1956). Some problems relating to chain reactions and to the theory of combustion. Nobel Lecture. Stockholm, Sweden (11 December 1956). http://www.nobelprize.org/prizes/chemistry/1956/semenov/lecture (accessed 04 April 2025).

Semenov, N.N. (1928a). On theory of the processes of combustion. Communication 1. *Journal of the Russian physical-chemical society* 60 (3): 247–250.

Semenoff, N.N. (1928b). Zur theorie des verbrennungsprozesses. *Zeitschrift für Physik* 48 (8): 571–582.

Semenov, N.N. (2004). *Selected Works*. Moscow: Nauka.

Seplyarskii, B.S. and Afanasiev, B.Y. (1989). Analysis of the unsteady pattern of the heating site ignition. *Combustion, Explosion and Shock Waves* 25: 665–669.

Shouman, A.R. (1999). Solution to the dusty gas explosion problem with reactant consumption. *Combustion and Flame* 119: 189–194.

Shouman, A.R. and El-Sayed, S. (1992). Accounting for reactant consumption in the thermal explosion problem. Part I: mathematical foundation. *Combustion and Flame* 88: 321–344.

Shouman, A.R. and El-Sayed, S. (1997). Accounting for reactant consumption in the thermal explosion problem. Part II: a direct solution with application to the Frank-Kamenetskii problem. *Combustion and Flame* 108: 361–386.

Shouman, A.R. and El-Sayed, S. (1998). Accounting for reactant consumption in the thermal explosion problem. Part III: criticality conditions for the Arrhenius problem. *Combustion and Flame* 113: 212–223.

Shouman, A.R. and El-Sayed, S. (1999). Accounting for reactant consumption in the thermal explosion problem. Part IV: numerical solution of the Arrhenius problem. *Combustion and Flame* 117: 422–428.

Shouman, A.R. and Donaldson, A.B. (1975). The stationary problem of thermal ignition in a reactive slab with unsymmetric boundary temperatures. *Combustion and Flame* 24: 203–210.

Shtessel, E.A., Pribytkova, K.V., and Merzhanov, A.G. (1971). Numerical solution of the problem of a thermal explosion taking account of free convection. *Combustion, Explosion, and Shock Waves* 7: 137–146.

Sexton, M.J., Macaskill, C., and Gray, B.F. (2001). Self-heating and drying in two-dimensional bagasse piles. *Combustion Theory and Modelling* 5 (4): 517–536. https://doi.org/10.1088/1364-7830/5/4/302.

Sisson, R.A., Swift, A., Wake, G.C., and Gray, B.F. (1992). The self-heating of damp cellulosic materials: I. High thermal conductivity and diffusivity. *IMA Journal of Applied Mathematics* 49: 273–291.

Sisson, R.A., Swift, A., Wake, G.C., and Gray, B.F. (1993). The self-heating of damp cellulosic materials: II. On the steady states of the spatially distributed case. *IMA Journal of Applied Mathematics* 50: 285–306.

Sidhu, H.S., Nelson, M.I., and Chen, X.D. (2007). A simple spatial model for self-heating compost piles. *ANZIAM Journal* 48: C135–C150.

Sivashinskii, G.I. (1967). On existence and stability of solutions of steady-state theory of a thermal explosion. *Journal of Applied Mathematics and Mechanics* 31 (1): 137–139.

Spinner, N.S., Field, C.R., Hammond, M.H. et al. (2015). Physical and chemical analysis of lithium-ion battery cell-to-cell failure events inside custom fire chamber. *Journal of Power Sources* 279: 713–721.

Spotnitz, R. and Franklin, J. (2003). Abuse behaviour of high-power, lithium-ion cells. *Journal of Power Sources* 113: 81–100.

Springer, E.L. and Hajny, G.J. (1970). Spontaneous heating in piled wood chips, initial mechanism. *Tappi Journal* 53 (1): 85–86.

Squire, W. (1963). A mathematical analysis of self-ignition. *Combustion and Flame* 7: 1–8.

Strunina, A.G., Gontkovskaya, V.T., and Merzhanov, A.G. (1965). Laws of thermal explosion. *Combustion, Explosion, Shock Waves* 1: 20–22.

Stuart, J.T. (1967). On finite amplitude oscillations in laminar mixing layers. *Journal of Fluid Mechanics* 29 (3): 417–440.

Tanaka, F. and Novozhilov, V. (2019). Predicting fire spread in a tunnel using simple theoretical model. *Proceedings of the Twelfth Asia-Pacific Conference on Combustion*, Fukuoka, Japan (1–5 July, 2019). Fukuoka, Japan: The Combustion Institute and Combustion Society of Japan.

Tarascon, J.-M. and Armand, M. (2001). Issues and challenges facing rechargeable lithium batteries. *Nature* 414: 359–367.

Thomas, P.H. (1961). Effect of reactant consumption on the induction period and critical condition for a thermal explosion. *Proceedings of the Royal Society A* 262: 192–206.

Thomas, P.H. (1965). A comparison of some hot spot theories. *Combustion and Flame* 9 (4): 369–372.

Thomas, P.H. (1995). The growth of fire – ignition to full involvement. In: *Combustion Fundamentals of Fire* (ed. G. Cox), 273–328. London: Academic Press.

Tikhonov, A.N. and Samarskii, A.A. (1963). *Equations of Mathematical Physics*. Oxford: Pergamon Press.

Turns, S.R. (2000). *An Introduction to Combustion*. Boston: McGraw Hill.

Vaganov, D.A., Samoilenko, N.G., and Abramov, V.G. (1978). Periodic regimes of continuous stirred tank reactors. *Chemical Engineering Science* 33 (8): 1133–1140.

Vladimirov, V.S. (1971). *Equations of Mathematical Physics*. New York: Marcel Dekker.

Volpert, A.I. and Khudyaev, S.I. (1975). *Analysis in Classes of Discontinuous Functions and Equations of Mathematical Physics*. Dordrecht: Martinus Nijhoff Publishers.

von Kamke, E. (1977). *Differentialgleichungen. Lösungsmethoden und Lösungen*. Wiesbaden: Vieweg +Teubner Verlag.

Wang, Q., Sun, J., and Chen, C. (2006). Thermal stability of LiPF6/EC + DMC + EMC electrolyte for lithium ion batteries. *Rare Metals* 25: 94–99.

Wang, Q., Ping, P., and Sun, J. (2010). Catastrophe analysis of cylindrical lithium ion battery. *Nonlinear Dynamics* 61: 763–772. https://doi.org/10.1007/s11071-010-9685-7.

Wang, Q., Ping, P., Zhao, X. et al. (2012). Thermal runaway caused fire and explosion of lithium ion battery. *Journal of Power Sources* 208: 210–224.

Waterman, T.E (1966). Determination of Fire Conditions Supporting Room Flashover. Final report IITRI project M6131, DASA 1886. *Defence Atomic Support Agency*, Washington, DC.

Waterman, T.E. (1968). Room flashover – criteria and synthesis. *Fire Technology* 4: 25–31.

Xu, K. (2004). Nonaqueous liquid electrolytes for lithium-based rechargeable batteries. *Chemical Reviews* 104: 4303–4418.

Yang, H., Bang, H., Amine, K., and Prakash, J. (2005). Investigations of the exothermic reactions of natural graphite anode for li-ion batteries during thermal runaway. *Journal of The Electrochemical Society* 152 (1): A73–A79.

Ye, J., Chen, H., Wang, Q. et al. (2016). Thermal behaviour and failure mechanism of lithium ion cells during overcharge under adiabatic conditions. *Applied Energy* 182: 464–474.

Yohe, G.R. (1958). Oxidation of Coal. Illinois State Geological Survey. *US Bureau of Mines Report of Investigations 207*: 51.

Yuan, Q.F., Zhao, F., Wang, W. et al. (2015). Overcharge failure investigation of lithium-ion batteries. *Electrochimica Acta* 178: 682–688.

Zambra, C.E., Moraga, N.O., Rosales, C., and Lictevout, E. (2012). Unsteady 3D heat and mass transfer diffusion coupled with turbulent forced convection for compost piles with chemical and biological reactions. *International Journal of Heat and Mass Transfer* 55: 6695–6704.

Zaturska, M.B. (1974). The interaction of hot spots. *Combustion and Flame* 23 (3): 313–317.

Zaturska, M.B. (1975). Thermal explosion of interacting hot spots. *Combustion and Flame* 25: 25–30.

Zeldovich, Y.B. (1941). On the theory of heat intensity. Exothermic chemical reaction process in a stream. I. *Journal of Technical Physics* 11 (6): 493–500.

Zeldovich, Y.B., Barenblatt, G.I., Librovich, V.B., and Makhviladze, G.M. (1980). *Mathematical Theory of Combustion and Explosion*. Moscow: Nauka.

Zeldovich, Y.B. and Yakovlev, V.I. (1938). Thermal autoignition of nitrous oxide. *Transactions of the Academy of Sciences of the USSR* 19: 699–701.

Zeldovich, Y.B. and Zysin, Y.A. (1941). On the theory of heat intensity. Exothermic chemical reaction process in a stream. II. Accounting for heat loss in the course of reaction. *Journal of Technical Physics* 11 (6): 501–508.

Zinn, J. (1965). Initiation of explosions by hot spots. *Journal of Chemical Physics* 36 (7): 1949.

Ziskin, M.S. (1942). On flammability limit of H_2-O_2 mixture at atmospheric pressure. *Transactions of the Academy of Sciences of the USSR* 34: 279.

Problems

1 Thermal explosion occurs due to extremely sharp non-linear dependence of reaction rate on temperature (Arrhenius kinetics).

The following exercise assists the reader with quantitative appreciation of this extreme sensitivity of reaction rate to increase in temperature.

Consider the gaseous reaction between hydrogen and iodine

$$H_2 + I_2 = 2HI$$

with the reaction activation energy $\tilde{E} = 39.00 Kcalmol^{-1}$.

Estimate, neglecting the temperature dependence of the pre-exponential factor, by how much reaction rate of this reaction increases if the mixture temperature

a) increases by a factor of just about 2 (from 293 K to 600 K);
b) increases by a factor of just about 2.7 (from 293 K to 800 K)

Assume the value of the universal gas constant as

$$\tilde{R} = 8.31\, KJKmol^{-1}K^{-1}$$

2 Estimate, for small values of Arrhenius number, by how much reaction rate of chemical reaction increases if the mixture temperature increases by one characteristic temperature interval.

3 Consider the Semenov model of thermal explosion (Chapter 2).

Show that the intersection point of energy generation (EG) and energy dissipation (ED) rate curves for this model corresponds to stable thermal equilibrium if the ED rate curve has a tangent steeper than the EG rate curve and to unstable thermal equilibrium in the opposite case.

4 Refuse derived fuel (RDF) is being stored in tall cylindrical tanks (silos). Determine the maximum tank volume and its approximate dimensions for safe storage of the fuel (such that potential thermal explosion and subsequent fire is prevented) provided that ambient temperature may reach 35°C, and the height to radius ratio of the silo is 5:1.

Assume that the problem may be considered in the Semenov approximation.

Assume the zero-order reaction.

Assume the following values of parameters:

Reaction pre-exponential factor $\tilde{A} = 0.55 \cdot 10^3 kgm^{-3}s^{-1}$
Activation energy $\tilde{E} = 50 KJmol^{-1}$

Thermal Explosion: Theory and Application, First Edition. Vasily B. Novozhilov.

Heat of reaction $\tilde{Q} = 20 MJkg^{-1}$
Heat transfer coefficient $\tilde{h} = 10^{-2} KWm^{-2}K^{-1}$

5 Prove that the function (3.24) and (3.25) satisfies Eq. (3.22).

6 Derive equations for Frank-Kamenetskii model, taking into account the effects of chemical kinetics (i.e. equations analogous to (4.3) and (4.9) for the Semenov model).
Make the same assumptions as in Section 4.1 of Chapter 4.

7 Derive Eqs. (4.60) for diffusion thermal explosion problem.

8 Obtain the solution (4.64) for Eqs. (4.62) and (4.63) describing diffusion stage of non-premixed thermal explosion.

9 Show that the modification of the kinetic Eq. (4.9), suitable for the flow reactor problem (Chapter 5), is given by the second of Eqs. (5.1).

10 Show that the modification of the standard Semenov model for the case of instantaneous control is given by the problem (8.2) and (8.3).

11 Derive Eqs. (9.5) and (9.6) for inertization model.

12 Derive Eq. (9.10) for cooling media injection model.

Problem Solutions

1 If the temperature dependence of the pre-exponential factor is neglected, then the ratio of reaction rates at two different temperatures is given by the ratio of the respective Boltzmann factors:

$$\frac{w_2}{w_1} = \frac{\exp\left(-\frac{\tilde{E}}{\tilde{R}\tilde{T}_2}\right)}{\exp\left(-\frac{\tilde{E}}{\tilde{R}\tilde{T}_1}\right)} \tag{P1.1}$$

$$\frac{w_2}{w_1} = \exp\left(\frac{\tilde{E}}{\tilde{R}\tilde{T}_1} - \frac{\tilde{E}}{\tilde{R}\tilde{T}_2}\right) = \exp\left(\frac{\tilde{E}}{\tilde{R}}\left(\frac{1}{\tilde{T}_1} - \frac{1}{\tilde{T}_2}\right)\right) = \exp\left(\frac{\tilde{E}}{\tilde{R}}\left(\frac{\tilde{T}_2 - \tilde{T}_1}{\tilde{T}_1\tilde{T}_2}\right)\right) \tag{P1.2}$$

The ratio $\frac{\tilde{E}}{\tilde{R}}$ is estimated as

$$\frac{\tilde{E}}{\tilde{R}} \approx \frac{39.00 \times 4.2 \times 10^3}{8.31} \frac{\text{KJ}}{\text{Kmol}} \frac{\text{KmolK}}{\text{KJ}} \approx 2 \cdot 10^4 \text{K} \tag{P1.3}$$

(a) for $\tilde{T}_1 = 293$ K, $\tilde{T}_2 = 600$ K

$$\frac{w_2}{w_1} = \exp\left(\frac{\tilde{E}}{\tilde{R}}\left(\frac{\tilde{T}_2 - \tilde{T}_1}{\tilde{T}_1\tilde{T}_2}\right)\right) \approx \exp\left(2 \cdot 10^4\left(\frac{600 - 293}{293 \times 600}\right)\right) \approx \exp(2 \cdot 10^4 \times 1.7 \cdot 10^{-3}) \approx 6 \cdot 10^{14} \tag{P1.4}$$

Reaction rate increases by nearly 15 orders of magnitude.
(b) for $\tilde{T}_1 = 293$K, $\tilde{T}_2 = 800$ K

$$\frac{w_2}{w_1} \approx \exp\left(2 \cdot 10^4\left(\frac{800 - 293}{293 \times 800}\right)\right) \approx \exp\left(2 \cdot 10^4 \times 2.2 \cdot 10^{-3}\right) \approx 1.3 \cdot 10^{19} \tag{P1.5}$$

Reaction rate increases by about 19 orders of magnitude.

Thermal Explosion: Theory and Application, First Edition. Vasily B. Novozhilov.

Conclusions:

Reaction rates of typical combustion reactions depend on the temperature in extremely strong, non-linear way.

It is such a dependence of reaction rate on temperature that makes various combustion phenomena such as flame propagation in space, thermal explosion, ignition and others happening.

2 Write down the ratio of reaction rates at the two temperatures separated by characteristic temperature interval: $\tilde{T}_1$ and $\tilde{T}_2 = \tilde{T}_1 + \dfrac{\tilde{R}\tilde{T}_1^2}{\tilde{E}}$.

According to the result of Problem 1

$$\frac{w_2}{w_1} = \exp\left(\frac{\tilde{E}}{\tilde{R}}\left(\frac{\tilde{T}_2 - \tilde{T}_1}{\tilde{T}_1\tilde{T}_2}\right)\right) = \exp\left(\frac{\tilde{E}}{\tilde{R}}\frac{\tilde{R}\tilde{T}_1^2}{\tilde{E}}\frac{1}{\tilde{T}_1\left(\tilde{T}_1 + \dfrac{\tilde{R}\tilde{T}_1^2}{\tilde{E}}\right)}\right) = \exp\left(\frac{\tilde{T}_1}{\left(\tilde{T}_1 + \dfrac{\tilde{R}\tilde{T}_1^2}{\tilde{E}}\right)}\right)$$

$$= \exp\left(\frac{1}{\left(1 + \dfrac{\tilde{R}\tilde{T}_1}{\tilde{E}}\right)}\right) = \exp\left(\frac{1}{(1 + Ar)}\right) \approx \exp(1 - Ar) \approx e; \quad Ar \ll 1 \tag{P2.1}$$

For small values of Arrhenius number $Ar \ll 1$, reaction rate increases approximately by the factor of e.

3 Energy generation function is (in the range of realistic combustion temperatures) convex; energy dissipation function for the Semenov model is linear.

These functions are plotted schematically in Figure P3.1. This plot gives an alternative way of illustration of the Semenov model, compared to Chapter 2 where the combination of the two functions, the single EGD function, is considered.

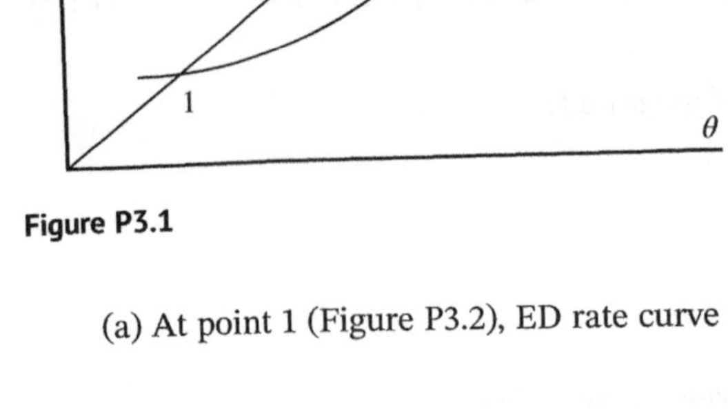

Figure P3.1

(a) At point 1 (Figure P3.2), ED rate curve tangent is larger than the EG rate curve tangent.

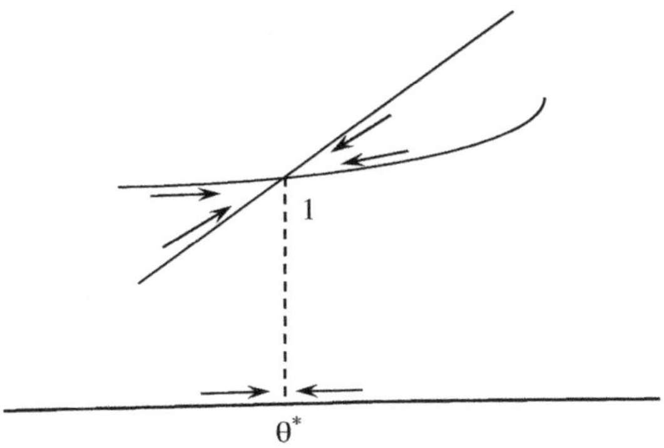

Figure P3.2

Small deviation of the system (i.e. temperature θ) from point 1 ($\theta = \theta^*$) leads to (see Eq. (2.4))

$$\exp\left(\frac{\theta}{1 + Ar \cdot \theta}\right) > \frac{1}{Se}\theta \Rightarrow \frac{d\theta}{d\tau} > 0 \tag{P3.1}$$

(deviation to the left) or

$$\exp\left(\frac{\theta}{1 + Ar \cdot \theta}\right) < \frac{1}{Se}\theta \Rightarrow \frac{d\theta}{d\tau} < 0 \tag{P3.2}$$

(deviation to the right).

In both cases, the perturbed system approaches point 1 (from the left or right, respectively) with the absolute value of the temperature change rate approaching zero (possibly asymptotically).

Thus, under small deviation (perturbation) from point 1, the system evolves back to this point, and in the limit ($\frac{d\theta}{d\tau} = 0$) remains at this point forever, as point 1 is the equilibrium point of the system.

Therefore, under small deviation from point 1, the system not only remains forever in the vicinity of this point but actually evolves asymptotically back to this point. In the exact mathematical terms, this means that point 1 is not just stable but possesses a stronger property, that is asymptotic stability.

(b) At point 2 (Figure P3.3), EG rate curve tangent is larger than the ED rate curve tangent. Similar analysis of the system behaviour upon small deviation from point 2 leads to

$$\exp\left(\frac{\theta}{1 + Ar \cdot \theta}\right) < \frac{1}{Se}\theta \Rightarrow \frac{d\theta}{d\tau} < 0 \tag{P3.3}$$

deviation to the left) or

$$\exp\left(\frac{\theta}{1 + Ar \cdot \theta}\right) > \frac{1}{Se}\theta \Rightarrow \frac{d\theta}{d\tau} > 0 \tag{P3.4}$$

(deviation to the right).

In both these cases, the system evolves away from point 2 (to the left or right, respectively). The point 2 is, therefore, unstable.

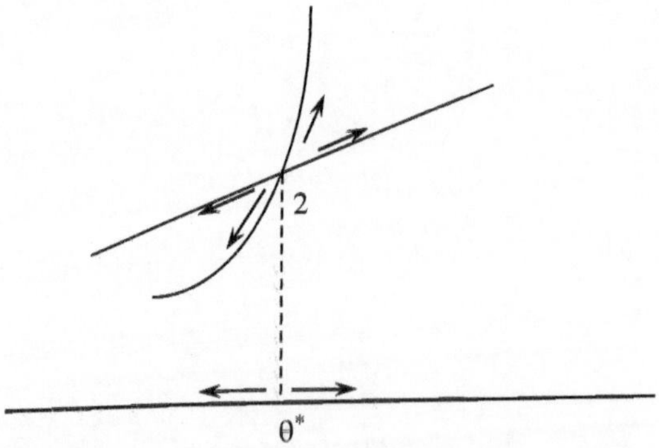

Figure P3.3

4 Units for some of the properties given here are different from those used in the text. However, they are fully consistent and can be equally used to solve the problem. No transformation to other units is required.

As a conservative estimation, assume that ambient temperature stays permanently at 35°C. Since Semenov approximation is being used, the temperature distribution is assumed to be uniform within the tank and heat loss rate uniform over the whole tank surface.

Critical condition for thermal explosion is given by the critical value of the Semenov parameter (Eq. (2.19))

$$Se_{cr} = \frac{1}{e} \tag{P4.1}$$

The Semenov parameter is (Eq. (2.2))

$$Se = \frac{\tilde{Q}\tilde{k}(T_0)\tilde{E}\tilde{V}}{\tilde{h}\tilde{S}\tilde{R}\tilde{T}_0^2} \tag{P4.2}$$

Therefore, the critical tank geometry is given by

$$\frac{\tilde{V}^{cr}}{\tilde{S}^{cr}} = \frac{1}{e}\frac{\tilde{h}\tilde{T}_0}{\tilde{Q}}\frac{1}{\tilde{k}(T_0)}Ar \tag{P4.3}$$

where $Ar = \frac{\tilde{R}\tilde{T}_0}{\tilde{E}}$ is the Arrhenius number, $\tilde{k}(T_0) = \tilde{A}\exp\left(-\frac{1}{Ar}\right)$ (Eq. (1.9)).

Therefore

$$\frac{\tilde{V}^{cr}}{\tilde{S}^{cr}} = \frac{1}{e}\frac{\tilde{h}\tilde{T}_0}{\tilde{Q}}\frac{Ar}{\tilde{A}}\exp\left(\frac{1}{Ar}\right) \tag{P4.4}$$

Further

$$\frac{\tilde{V}^{cr}}{\tilde{S}^{cr}} = \frac{\pi\left(\tilde{R}^{cr}\right)^2\tilde{H}^{cr}}{\left[2\pi\left(\tilde{R}^{cr}\right)^2 + 2\pi\tilde{R}^{cr}\tilde{H}^{cr}\right]} = \frac{1}{2}\frac{\left(\tilde{R}^{cr}\right)^2 5\tilde{R}^{cr}}{\left[\left(\tilde{R}^{cr}\right)^2 + \tilde{R}^{cr}5\tilde{R}^{cr}\right]} = \frac{5}{2}\frac{\left(\tilde{R}^{cr}\right)^3}{6\left(\tilde{R}^{cr}\right)^2} = \frac{5}{12}\tilde{R}^{cr}$$

(P4.5)

From Eq. (P4.4), the critical radius of the tank is

$$\tilde{R}^{cr} = \frac{12}{5e}\frac{\tilde{h}\tilde{T}_0}{\tilde{Q}}\frac{Ar}{\tilde{A}}\exp\left(\frac{1}{Ar}\right) \tag{P4.6}$$

and consequently

$$\tilde{H}^{cr} = 5\tilde{R}^{cr}$$
$$\tilde{V}^{cr} = 5\pi\left(\tilde{R}^{cr}\right)^3 \tag{P4.7}$$

Substituting numerical values

$$Ar = \frac{\tilde{R}\tilde{T}_0}{\tilde{E}} = \frac{8.31 \times 308}{5 \cdot 10^4}\frac{\text{KJ}}{\text{Kmol} \cdot \text{K}}\text{K}\frac{\text{Kmol}}{\text{KJ}} \approx 0.05$$
$$\exp\left(\frac{1}{Ar}\right) \approx 4.86 \cdot 10^8 \tag{P4.8}$$

From (P4.6) and (P4.7),

$$\tilde{R}^{cr} \approx \frac{12}{5e}\frac{10^{-2} \times 308}{2 \cdot 10^4}\frac{5 \cdot 10^{-2}}{0.55 \cdot 10^3}4.86 \cdot 10^8\frac{\text{KW}}{\text{m}^2\text{K}}\text{K}\frac{\text{kg}}{\text{KJ}}\frac{\text{m}^3\text{s}}{\text{kg}} \approx 6\,\text{m}$$
$$\tilde{H}^{cr} \approx 30\,\text{m}$$
$$\tilde{V}^{cr} \approx 3.40 \cdot 10^3\text{m}^3 \tag{P4.9}$$

The maximum safe volume of the tank is $3.40 \cdot 10^3\text{m}^3$ with a radius of $\approx$6 m and a height of $\approx$30 m.

5 Denote, for brevity

$$\left[\frac{1}{2}(A+B)ch(\omega\xi) + \frac{1}{2}(A-B)\cos(\omega\varsigma)\right] \stackrel{\text{def}}{=} [...] \tag{P5.1}$$

Direct differentiation of the function (3.24)

$$\theta(\xi,\varsigma) = -2\,ln\left[\frac{1}{2}(A+B)ch(\omega\xi) + \frac{1}{2}(A-B)\cos(\omega\varsigma)\right] \tag{P5.2}$$

gives

$$\frac{d\theta}{d\xi} = -\omega(A+B)\frac{sh(\omega\xi)}{[...]} \tag{P5.3}$$

$$\frac{d^2\theta}{d\xi^2} = -\omega(A+B)\frac{\left[\omega \cdot ch(\omega\xi)[...] - \frac{1}{2}\omega(A+B)sh^2(\omega\xi)\right]}{[...]^2} \tag{P5.4}$$

$$\frac{d\theta}{d\zeta} = \omega(A-B)\frac{\sin(\omega\zeta)}{[...]} \tag{P5.5}$$

$$\frac{d^2\theta}{d\zeta^2} = \omega(A-B)\frac{\left[\omega \cdot \cos(\omega\zeta)[...] + \frac{1}{2}\omega(A-B)\sin^2(\omega\zeta)\right]}{[...]^2} \tag{P5.6}$$

Note also that

$$\exp(\theta) = \frac{1}{[...]^2} \tag{P5.7}$$

Upon substitution of the results (P5.4), (P5.6) and (P5.7) into Eq. (3.22), the identity that needs to be proven takes the form

$$\begin{aligned} &-(A+B)\left[ch(\omega\xi)[...] - \frac{1}{2}(A+B)sh^2(\omega\xi)\right] \\ &\quad + (A-B)\left[\cos(\omega\zeta)[...] + \frac{1}{2}(A-B)\sin^2(\omega\zeta)\right] = -\frac{\delta}{\omega^2} \end{aligned} \tag{P5.8}$$

or taking into account restriction (3.25)

$$\begin{aligned} &-(A+B)\left[ch(\omega\xi)[...] - \frac{1}{2}(A+B)sh^2(\omega\xi)\right] \\ &+ (A-B)\left[\cos(\omega\zeta)[...] + \frac{1}{2}(A-B)\sin^2(\omega\zeta)\right] = -2AB \end{aligned} \tag{P5.9}$$

Upon bracket expansion in Eq. (P5.9), using the definition (P5.1), the LHS is

$$\begin{aligned} &-(A+B)\left\{\frac{1}{2}(A+B)ch^2(\omega\xi) + \frac{1}{2}(A-B)ch(\omega\xi)\cos(\omega\varsigma)\right. \\ &\left.-\frac{1}{2}(A+B)sh^2(\omega\xi)\right\} + (A-B)\left\{\frac{1}{2}(A+B)ch(\omega\xi)\cos(\omega\zeta) + \frac{1}{2}(A-B)\cos^2(\omega\zeta)\right. \\ &\left.+\frac{1}{2}(A-B)\sin^2(\omega\varsigma)\right\} \end{aligned} \tag{P5.10}$$

or, using basic trigonometric and hyperbolic function identities,

$$\begin{aligned} &-(A+B)\left\{\frac{1}{2}(A+B) + \frac{1}{2}(A-B)ch(\omega\xi)\cos(\omega\varsigma)\right\} \\ &+ (A-B)\left\{\frac{1}{2}(A-B) + \frac{1}{2}(A+B)ch(\omega\xi)\cos(\omega\zeta)\right\} \end{aligned} \tag{P5.11}$$

From here, the LHS of Eq. (P5.9) is

$$\frac{1}{2}\left[-(A+B)^2 + (A-B)^2\right] = \frac{1}{2}(-4AB) = -2AB \tag{P5.12}$$

that is, equal, indeed, to its RHS.
Identity (P5.9), and (P5.8) along with it, has been proven.

6 The major assumptions are those of the monomolecular reaction of the order m and the temperature independence of the pre-exponential factor.

Identically to Eq. (4.1) for the Semenov model, the second term on the RHS of the Frank-Kamenetskii model Eq. (2.8) is modified into

$$\tilde{\rho}\tilde{c}_p\frac{\partial \tilde{T}}{\partial \tilde{t}} = \tilde{\lambda}\tilde{\nabla}^2\tilde{T} + \tilde{Q}\tilde{A}\exp\left(-\frac{\tilde{E}}{\tilde{R}\tilde{T}}\right)\tilde{C}^m \tag{P6.1}$$

Using Eqs. (1.7) and (1.8) (Chapter 1), this term can be written as

$$\tilde{Q}\tilde{A}\exp\left(-\frac{\tilde{E}}{\tilde{R}\tilde{T}}\right)\tilde{C}^m = \tilde{Q}\tilde{k}_0\tilde{C}_0^m\exp\left(\frac{\theta}{1+Ar\theta}\right)C^m \quad \text{(P6.2)}$$

Equation for the Frank-Kamenetskii model takes the form

$$\tilde{\rho}\tilde{c}_p\frac{\partial\tilde{T}}{\partial\tilde{t}} = \tilde{\lambda}\tilde{\nabla}^2\tilde{T} + \tilde{Q}\tilde{k}_0\tilde{C}_0^m\exp\left(\frac{\theta}{1+Ar\theta}\right)C^m \quad \text{(P6.3)}$$

Introducing, in the way analogous to (4.2), the following variables (note that the dimension of reaction rate constant is dependent on reaction order, see Notation section)

$$\theta = \frac{(\tilde{T}-\tilde{T}_0)}{\left(\frac{\tilde{R}\tilde{T}_0^2}{\tilde{E}}\right)};\tau = \frac{\tilde{Q}\tilde{k}_0\tilde{C}_0^m\tilde{E}}{\tilde{\rho}\tilde{c}_p\tilde{R}\tilde{T}_0^2}\tilde{t}$$

$$\delta = \frac{\tilde{Q}\tilde{k}_0\tilde{C}_0^m\tilde{E}\tilde{r}^2}{\tilde{\lambda}\tilde{R}\tilde{T}_0^2} \quad \text{(P6.4)}$$

Eq. (P6.3) is transformed to the following form:

$$\frac{\partial\theta}{\partial\tau} = \frac{1}{\delta}\nabla^2\theta + \exp\left(\frac{\theta}{1+Ar\theta}\right)C^m \quad \text{(P6.5)}$$

or in terms of kinetic function (Eqs. (4.4) and (4.5))

$$\frac{\partial\theta}{\partial\tau} = \frac{1}{\delta}\nabla^2\theta + \exp\left(\frac{\theta}{1+Ar\theta}\right)\varphi(\eta) \quad \text{(P6.6)}$$

Kinetic equation remains identical to the one derived in Section 4.1, i.e. Eq. (4.9)

$$\frac{d\eta}{d\tau} = \frac{1}{Td}\exp\left(\frac{\theta}{1+Ar\theta}\right)\varphi(\eta) \quad \text{(P6.7)}$$

with the Todes number given by Eq. (4.8)

$$Td = \frac{\tilde{Q}\tilde{E}\tilde{C}_0}{\tilde{\rho}\tilde{c}_p\tilde{R}\tilde{T}_0^2} \quad \text{(P6.8)}$$

Initial condition may be set to be identical to (4.10)

$$\theta = 0; \eta = 0 \text{ at } \tau = 0 \quad \text{(P6.9)}$$

7 Dimensional form of the energy equation is

$$\tilde{\rho}\tilde{c}_p\frac{\partial\tilde{T}}{\partial\tilde{t}} = \tilde{\lambda}\frac{\partial^2\tilde{T}}{\partial\tilde{x}^2} + \tilde{Q}\tilde{w} \quad \text{(P7.1)}$$

Time scale may be taken in the same form as in Problem 6 (Eq. (P6.4)), that is for reaction of the second order

$$\tau = \frac{\tilde{Q}\tilde{k}_0\tilde{C}_0^2\tilde{E}}{\tilde{\rho}\tilde{c}_p\tilde{R}\tilde{T}_0^2}\tilde{t} \quad \text{(P7.2)}$$

Upon substitution into Eq. (P7.1), taking into account definition (4.56) and assumption $Le \equiv 1$ (Chapter 4), energy equation takes the form

$$\tilde{k}_0 \tilde{C}_0^2 \frac{\partial \theta}{\partial \tau} = \tilde{k}_0 \tilde{C}_0^2 \frac{\partial^2 \theta}{\partial \xi^2} + \tilde{w} \tag{P7.3}$$

Substituting reaction rate in the form (1.8) (Chapter 1)

$$\frac{\partial \theta}{\partial \tau} = \frac{\partial^2 \theta}{\partial \xi^2} + \frac{1}{\tilde{k}_0 \tilde{C}_0^2} \tilde{k}_0 \exp\left(\frac{\theta}{1 + Ar\theta}\right) \tilde{C}_1 \tilde{C}_2 \tag{P7.4}$$

$$\frac{\partial \theta}{\partial \tau} = \frac{\partial^2 \theta}{\partial \xi^2} + C_1 C_2 \exp\left(\frac{\theta}{1 + Ar\theta}\right) \tag{P7.5}$$

Taking into account Eqs. (4.58) and (4.59),

$$\frac{\partial \theta}{\partial \tau} = \frac{\partial^2 \theta}{\partial \xi^2} + (1 - \eta_1)\eta_1 \exp\left(\frac{\theta}{1 + Ar\theta}\right) \tag{P7.6}$$

which is identical (for the kinetic function (4.61)) to the first of Eqs. (4.60).
Kinetic equation is derived starting from the diffusion equation for the first reactant

$$\frac{\partial \tilde{C}_1}{\partial \tilde{t}} = \tilde{D} \frac{\partial^2 \tilde{C}_1}{\partial \tilde{x}^2} - \tilde{w} \tag{P7.7}$$

This equation is obtained by adding diffusion term to Eq. (4.6).

Using the time scale (P7.2), spatial scale (4.56) and reaction rate term in the form (1.8), the latter equation is transformed to

$$\frac{\partial C_1}{\partial \tau} = \frac{\partial^2 C_1}{\partial \xi^2} - \vartheta \exp\left(\frac{\theta}{1 + Ar\theta}\right) C_1 C_2 \tag{P7.8}$$

where parameter ϑ is given by Eq. (4.61).

Further, due to relations (4.58) and (4.59)

$$\frac{\partial \eta_1}{\partial \tau} = \frac{\partial^2 \eta_1}{\partial \xi^2} + \vartheta(1 - \eta_1)\eta_1 \exp\left(\frac{\theta}{1 + Ar\theta}\right) \tag{P7.9}$$

This equation is identical (for the kinetic function (4.61)) to the second of Eqs. (4.60).

8 Application of Fourier transform F, over variable ξ, to Eq. (4.63), leads to the following ODE:

$$\frac{d\Phi}{d\tau}(\lambda, \tau) = -\lambda^2 \Phi(\lambda, \tau) \tag{P8.1}$$

for the transform function

$$F[\eta_1] = \Phi(\lambda, \tau) = \int_{-\infty}^{\infty} \eta_1(\xi, \tau) \exp(-i\lambda\xi) d\xi \tag{P8.2}$$

with the initial condition

$$\Phi(\lambda, 0) = \Phi_0(\lambda)$$

$$\Phi_0(\lambda) = \int_{-\infty}^{\infty} \eta_1(\xi, 0) \exp(-i\lambda\xi) d\xi \tag{P8.3}$$

Solution of the Problem (P8.1) and (P8.3) is clearly

$$\Phi(\lambda,\tau) = \exp(-\lambda^2\tau)\Phi_0(\lambda) \qquad \text{(P8.4)}$$

Inverse transform of exponential function may be found directly or using Fourier transform tables

$$F^{-1}\left[\exp(-\lambda^2\tau)\right] = \frac{1}{2\sqrt{\pi\tau}}\exp\left(-\frac{\xi^2}{4\tau}\right) \qquad \text{(P8.5)}$$

Inverse transform of the product (P8.4) is convolution of inverse transforms of the factors, therefore

$$\eta_1(\xi,\tau) = \frac{1}{2\sqrt{\pi\tau}}\int_{-\infty}^{\infty}\exp\left(-\frac{s^2}{4\tau}\right)\eta_1(\xi-s,0)ds \qquad \text{(P8.6)}$$

Taking into account precise form of the initial profile $\eta_1(\xi, 0)$ (initial condition (4.62)),

$$\eta_1(\xi,\tau) = \frac{1}{2\sqrt{\pi\tau}}\int_{-\infty}^{\xi}\exp\left(-\frac{s^2}{4\tau}\right)ds \qquad \text{(P8.7)}$$

Changing integration variable

$$\eta_1(\xi,\tau) = \frac{1}{\sqrt{\pi}}\int_{-\infty}^{\frac{\xi}{2\sqrt{\tau}}}\exp(-\varsigma^2)d\varsigma = 1-\frac{1}{2}erfc\left(\frac{\xi}{2\sqrt{\tau}}\right) \qquad \text{(P8.8)}$$

which is identical to Eq. (4.64).

9 The only modification required is taking into account the removal of reactants by the exit stream.

The net rate of reactant removal with the stream velocity $\tilde{u}$, via the exit opening with the area $\tilde{S}$, is $-\tilde{u}\tilde{S}(\tilde{C}-\tilde{C}_0)$ $[N]/[T]$. The corresponding volumetric density of removal is $-\frac{\tilde{u}\tilde{S}}{\tilde{V}}(\tilde{C}-\tilde{C}_0) = -\frac{\tilde{\mho}}{\tilde{V}}(\tilde{C}-\tilde{C}_0)$ $[N]/[L]^3/[T]$, where $\tilde{\mho} = \tilde{u}\tilde{S}$ is the volumetric flow rate through the reactor. In this form, the rate of removal term has dimension identical to that of the LHS of Eq. (4.6).

Adding volumetric density of removal term to the RHS of the latter equation leads to the following equation:

$$\frac{d\tilde{C}}{d\tilde{t}} = -\tilde{w}-\frac{\tilde{\mho}}{\tilde{V}}(\tilde{C}-\tilde{C}_0) \qquad \text{(P9.1)}$$

The reaction rate for the first-order reaction (neglecting temperature dependence of the pre-exponential term) is given by Eq. (1.8)

$$\tilde{w} = \tilde{k}_0\exp\left(\frac{\theta}{1+Ar\theta}\right)\tilde{C}^m \qquad \text{(P9.2)}$$

Then, Eq. (P9.1) is re-written, in terms of non-dimensional concentration $C = \frac{\tilde{C}}{\tilde{C}_0}$ as

$$\frac{dC}{d\tilde{t}} = -\tilde{k}_0\tilde{C}_0^m \exp\left(\frac{\theta}{1+Ar\theta}\right)C^m - \frac{\tilde{\mho}\tilde{C}_0}{\tilde{V}}(C-1) \qquad \text{(P9.3)}$$

Further, in terms of the progress variable $\eta = 1 - C$

$$-\frac{d\eta}{d\tilde{t}} = -\tilde{k}_0\tilde{C}_0^m \exp\left(\frac{\theta}{1+Ar\theta}\right)(1-\eta)^m + \frac{\tilde{\mho}\tilde{C}_0}{\tilde{V}}\eta \qquad \text{(P9.4)}$$

and in terms of the kinetic function $\varphi(\eta) = (1-\eta)^m$

$$\frac{d\eta}{d\tilde{t}} = \tilde{k}_0\tilde{C}_0^m \exp\left(\frac{\theta}{1+Ar\theta}\right)\varphi(\eta) - \frac{\tilde{\mho}\tilde{C}_0}{\tilde{V}}\eta \qquad \text{(P9.5)}$$

Non-dimensionalization is achieved by introducing the same time scale as in Chapter 4, i.e. given by Eq. (4.2).
Eq. (P9.5) takes the form

$$\frac{d\eta}{d\tilde{t}} = \frac{1}{Td}\exp\left(\frac{\theta}{1+Ar\theta}\right)\varphi(\eta) - \frac{1}{Da}\eta \qquad \text{(P9.6)}$$

identical to the second of Eqs. (5.1) with the Todes and Damköhler number defined as

$$Td = \frac{\tilde{Q}\tilde{E}\tilde{C}_0}{\tilde{\rho}\tilde{c}_p\tilde{R}\tilde{T}_0^2}; Da = \frac{\tilde{V}\tilde{Q}\tilde{k}_0\tilde{E}\tilde{C}_0^m}{\tilde{\mho}\tilde{\rho}\tilde{c}_p\tilde{R}\tilde{T}_0^2} \qquad \text{(P9.7)}$$

These definitions are identical to the definitions (5.2) and (5.3).

10 Abrupt drop of ambient temperature at $\tau = \tau_{ctrl}$ to a certain value $0 < \tilde{T}_{ctrl} < \tilde{T}_0$ results in the modification of the energy dissipation term in the dimensional Semenov Eq. (2.1) into

$$-\tilde{h}\frac{\tilde{S}}{\tilde{V}}(\tilde{T} - \tilde{T}_{ctrl}) \qquad \text{(P10.1)}$$

This term can be written in the equivalent form

$$-\tilde{h}\frac{\tilde{S}}{\tilde{V}}(\tilde{T} - \tilde{T}_0 + \tilde{T}_0 - \tilde{T}_{ctrl}) = -\tilde{h}\frac{\tilde{S}}{\tilde{V}}((\tilde{T} - \tilde{T}_0) + (\tilde{T}_0 - \tilde{T}_{ctrl})) \qquad \text{(P10.2)}$$

It follows from here that in the course of further non-dimensionalizations, performed exactly in the same way as in Chapter 2, the energy dissipation term turns into

$$-\frac{1}{Se}(\theta + \theta_{ctrl}) \qquad \text{(P10.3)}$$

where the non-dimensional temperature θ_{ctrl} is

$$\theta_{ctrl} = \frac{(\tilde{T}_0 - \tilde{T}_{ctrl})}{\left(\frac{\tilde{R}\tilde{T}_0^2}{\tilde{E}}\right)} \qquad \text{(P10.4)}$$

Therefore, instantaneous control is described by Problem (8.1) for $0 \le \tau < \tau_{ctrl}$ and by Problem (8.2) and (8.3) for $\tau > \tau_{ctrl}$.

11 Using Eqs. (9.1)

$$\tilde{\rho}(\tilde{t}) = \frac{\tilde{M}_0}{\tilde{V}} + \frac{1}{\tilde{V}}\int_0^{\tilde{t}} \tilde{I}(s)ds \tag{P11.1}$$

or

$$\tilde{\rho}(\tilde{t}) = \tilde{\rho}_0\left[1 + \frac{1}{\tilde{\rho}_0\tilde{V}}\int_0^{\tilde{t}} \tilde{I}(s)ds\right] \tag{P11.2}$$

where $\tilde{M}_0$ and $\tilde{\rho}_0$ are initial mixture mass and density, respectively. Therefore, indeed, non-dimensional density may be defined as

$$\rho(\tilde{t}) = \frac{\tilde{\rho}(\tilde{t})}{\tilde{\rho}_0} = 1 + \psi(\tilde{t})$$

$$\psi(\tilde{t}) = \frac{1}{\tilde{\rho}_0\tilde{V}}\int_0^{\tilde{t}} \tilde{I}(s)ds \tag{P11.3}$$

i.e. Eqs. (9.2) and (9.3) hold.

Time scale is introduced by Eq. (9.4). Equation for the Semenov model is written by modifying Eq. (2.1) of Chapter 2 into

$$\tilde{c}\frac{d(\tilde{\rho}\tilde{T})}{d\tilde{t}} = \tilde{Q}\tilde{A}\,exp\left(-\frac{\tilde{E}}{\tilde{R}\tilde{T}}\right) - \tilde{h}\frac{\tilde{S}}{\tilde{V}}(\tilde{T} - \tilde{T}_0) \tag{P11.4}$$

LHS of Eq. (P11.4), with the help of Eq. (P11.3) and time scale (9.4) may be written as

$$\tilde{c}\frac{d(\tilde{\rho}\tilde{T})}{d\tilde{t}} = \tilde{c}\frac{d(\tilde{\rho}(\tilde{T} - \tilde{T}_0))}{d\tilde{t}} = \tilde{c}\tilde{\rho}_0\frac{\tilde{R}\tilde{T}_0^2}{\tilde{E}}\frac{d(\rho\theta)}{d\tilde{t}} = \tilde{c}\tilde{\rho}_0\frac{\tilde{R}\tilde{T}_0^2}{\tilde{E}}\frac{\tilde{Q}\tilde{k}_0\tilde{E}}{\tilde{\rho}_0\tilde{c}\tilde{R}\tilde{T}_0^2}\frac{d(\rho\theta)}{d\tau}$$
$$= \tilde{k}_0\tilde{Q}\frac{d[(1 + \psi(\tau))\theta]}{d\tau} \tag{P11.5}$$

RHS, upon dividing by the factor $\tilde{k}_0\tilde{Q}$, is written in the form

$$\frac{\tilde{A}}{\tilde{k}_0}\exp\left(-\frac{\tilde{E}}{\tilde{R}\tilde{T}}\right) - \frac{\tilde{h}\tilde{S}}{\tilde{k}_0\tilde{Q}\tilde{V}}(\tilde{T} - \tilde{T}_0) = \frac{\tilde{A}}{\exp\left(-\frac{1}{Ar}\right)\tilde{A}}\exp\left(-\frac{1}{Ar}\right)\exp\left(\frac{\theta}{1 + Ar\theta}\right)$$
$$- \frac{\tilde{h}\tilde{S}\tilde{R}\tilde{T}_0^2}{\tilde{k}_0\tilde{Q}\tilde{V}\tilde{E}}\theta = \exp\left(\frac{\theta}{1 + Ar\theta}\right) - \frac{1}{Se}\theta$$
$$\tag{P11.6}$$

Here, Eqs. (1.7) and (1.9) of Chapter 1 have been used, and Semenov number is given by Eq. (2.2) of Chapter 2.

Combining LHS (P11.5), divided by the factor $\tilde{k}_0\tilde{Q}$, and RHS (P11.6) provides equation identical to Eq. (9.5)

$$\frac{d}{d\tau}[(1 + \psi(\tau))\theta] = \exp\left(\frac{\theta}{1 + Ar\theta}\right) - \frac{1}{Se}\theta \tag{P11.7}$$

Kinetic equation is a modification of Eq. (4.9) of Chapter 4. It is easy to see that in the case of variable density, the factor in front of exponential term in that equation may be written as

$$\frac{\tilde{\rho}(\tilde{t})\tilde{c}\tilde{R}\tilde{T}_0^2}{\tilde{Q}\tilde{E}\tilde{C}_0} = \frac{\rho(\tilde{t}(\tau))\tilde{\rho}_0\tilde{c}\tilde{R}\tilde{T}_0^2}{\tilde{Q}\tilde{E}\tilde{C}_0} = \frac{1+\psi(\tau)}{Td} \tag{P11.8}$$

where the Todes number is now defined explicitly using initial density

$$Td = \frac{\tilde{Q}\tilde{E}\tilde{C}_0}{\tilde{\rho}_0\tilde{c}\tilde{R}\tilde{T}_0^2} \tag{P11.9}$$

Kinetic equation takes the form

$$\frac{d\eta}{d\tau} = \frac{(1+\psi(\tau))}{Td}\exp\left(\frac{\theta}{1+Ar\theta}\right)(1-\eta) \tag{P11.10}$$

identical to Eq. (9.6).

12 Effect of variable density is taken into account in the same way as in Problem 11. That is, the LHS and the first term on the RHS of the Semenov equation are identical to those in Eq. (P11.7).

An additional term that arises in the Semenov equation must account for energy dissipation due to internal mixing between portions of the mixture with different temperatures. In dimensional form, this volume-averaged energy dissipation rate is

$$\frac{\tilde{I}\tilde{c}(\tilde{T}_{in}-\tilde{T}_0)}{\tilde{V}} \tag{P12.1}$$

As far as non-dimensional form of the Semenov equation is concerned, development identical to that in Problem 11 (Eq. (P11.5)) shows that the term (P12.1) must be divided by the factor $\tilde{k}_0\tilde{Q}$. The final form of the equation is therefore

$$\frac{d}{d\tau}[(1+\psi(\tau))\theta] = \exp\left(\frac{\theta}{1+Ar\theta}\right) - \frac{1}{Se}\theta - \wp(\tau) \tag{P12.2}$$

i.e. identical to Eq. (9.10), with the additional heat dissipation term given by Eq. (9.11)

$$\wp(\tau) = \frac{\tilde{I}(\tilde{t}(\tau))\tilde{c}(\tilde{T}_0-\tilde{T}_{in})}{\tilde{Q}\tilde{k}_0\tilde{V}} \tag{P12.3}$$

Index

Thermal Explosion: Theory and Application, First Edition. Vasily B. Novozhilov.

d

e

Printed and bound by CPI Group (UK) Ltd, Croydon, CR0 4YY
17/09/2026
14964267-0006